THE FRONTIERS COLLECTION

THE FRONTIERS COLLECTION

Series Editors:
A.C. Elitzur L. Mersini-Houghton M. Schlosshauer M.P. Silverman R. Vaas H.D. Zeh
J. Tuszynski

The books in this collection are devoted to challenging and open problems at the forefront of modern science, including related philosophical debates. In contrast to typical research monographs, however, they strive to present their topics in a manner accessible also to scientifically literate non-specialists wishing to gain insight into the deeper implications and fascinating questions involved. Taken as a whole, the series reflects the need for a fundamental and interdisciplinary approach to modern science. Furthermore, it is intended to encourage active scientists in all areas to ponder over important and perhaps controversial issues beyond their own speciality. Extending from quantum physics and relativity to entropy, consciousness and complex systems – the Frontiers Collection will inspire readers to push back the frontiers of their own knowledge.

For a full list of published titles, please see back of book or springer.com/series 5342

René Brun • Federico Carminati
Giuliana Galli Carminati

Editors

From the Web to the Grid and Beyond

Computing Paradigms Driven by High Energy Physics

Editors
René Brun
Federico Carminati
CERN
Geneva
Switzerland
Rene.Brun@cern.ch
Federico.Carminati@cern.ch

Giuliana Galli Carminati
Hôpitaux Universitaire de Genève
Unité de la Psychiatrie du
Développement Mental
ch. du Petit Bel Air 2
1225 Chêne-Bourg
Switzerland
Giuliana.GalliCarminati@hcuge.ch

Series Editors:
Avshalom C. Elitzur
Bar-Ilan University, Unit of Interdisciplinary Studies, 52900 Ramat-Gan, Israel
email: avshalom.elitzur@weizmann.ac.il

Laura Mersini-Houghton
Dept. Physics, University of North Carolina, Chapel Hill, NC 27599-3255, USA
email: mersini@physics.unc.edu

Maximilian A. Schlosshauer
Institute for Quantum Optics and Quantum Information, Austrian Academy of Sciences,
Boltzmanngasse 3, A-1090 Vienna, Austria
email: schlosshauer@nbi.dk

Mark P. Silverman
Trinity College, Dept. Physics, Hartford CT 06106, USA
email: mark.silverman@trincoll.edu

Jack A. Tuszynski
University of Alberta, Dept. Physics, Edmonton AB T6G 1Z2, Canada
email: jtus@phys.ualberta.ca

Rüdiger Vaas
University of Giessen, Center for Philosophy and Foundations of Science, 35394 Giessen,
Germany
email: ruediger.vaas@t-online.deH.

Dieter Zeh
Gaiberger Straße 38, 69151 Waldhilsbach, Germany
email: zeh@uni-heidelberg.de

ISSN 1612-3018
ISBN 978-3-642-23156-8   e-ISBN 978-3-642-23157-5
DOI 10.1007/978-3-642-23157-5
Springer Heidelberg Dordrecht London New York

Library of Congress Control Number: 2011941933

© Springer-Verlag Berlin Heidelberg 2012
This work is subject to copyright. All rights are reserved, whether the whole or part of the material is concerned, specifically the rights of translation, reprinting, reuse of illustrations, recitation, broadcasting, reproduction on microfilm or in any other way, and storage in data banks. Duplication of this publication or parts thereof is permitted only under the provisions of the German Copyright Law of September 9, 1965, in its current version, and permission for use must always be obtained from Springer. Violations are liable to prosecution under the German Copyright Law.
The use of general descriptive names, registered names, trademarks, etc. in this publication does not imply, even in the absence of a specific statement, that such names are exempt from the relevant protective laws and regulations and therefore free for general use.

Printed on acid-free paper

Springer is part of Springer Science+Business Media (www.springer.com)

Preface

Modern High Energy Physics (HEP), as a science studying the results of accelerator-driven particle collisions, was born after the Second World War, at the same time as computers. HEP research has been constantly limited by technology, both in the accelerator and detector domains as well as that of computing. At the same time High Energy physicists have greatly contributed to the development of Information Technology.

During all these years, the Conseil Européen pour la Recherche Nucléaire[1] (CERN [1], located in Geneva, Switzerland) has been a privileged place for the evolution of HEP computing. Several applications conceived for HEP have found applications well beyond it, the World Wide Web (see Chap. 2) being the most notable example. During all these years HEP computing has faced the challenge of software development within distributed communities, and of exploiting geographically distributed computing resources. HEP computing has been very successful in fulfilling its mandate, and the quantity and quality of scientific results produced by High Energy Physics are a direct consequence of this. In this activity, HEP computing had to face problems and situations that anticipated those that Information Technology at large would meet years later. In many respects, HEP computing has been today where computing was going to be tomorrow.

This book describes the evolution of HEP computing, and in particular those aspects that have been most innovative. These aspects are described by contributions from different authors. The reader is presumed to be familiar with the field of computing, from software engineering to programming languages. No specific physics knowledge is required. The subject has been treated in the past mostly in conference proceedings or in specialised books on physics computing, but, to our knowledge, no other book has given a complete account of the evolution of HEP computing.

---

[1]The Conseil Européen pour la Recherche Nucléaire, or European Council for Nuclear Research is the name of a provisional body founded in 1952 with the mandate of establishing a world-class fundamental physics research organisation in Europe. At that time, pure physics research concentrated on understanding the inside of the atom, hence the word "nuclear".

High Energy Physics in a Nutshell

The discovery in 1896 of the natural transmutation of the elements via radioactive decay by the French scientist Henri Becquerel, while working on phosphorescent materials, coincided with the discovery of "rays" composed by sub-nuclear particles of varying energy. These rays were soon used to probe the structure of matter, and using them the New Zealand scientist Ernest Rutherford discovered in 1909 that matter is composed by atoms with a very dense and small positively charged nucleus and negatively charged electrons floating in virtual emptiness. In 1912 the Austrian-American physicist Victor Hess during a hot-air balloon flight discovered the existence of highly energetic rays coming from outer space and hitting the earth after having traversed the atmosphere. While the origin and nature of cosmic rays still poses some of the most formidable puzzles in physics, this discovery was instrumental in many ways. The history of the universe and the nature of matter were since then intimately linked; the study of our origin and of the infinitely big now relies on advances in our understanding of the microscopic nature of matter, and vice versa.

> Cosmic rays and the discovery of antimatter.

In 1936 the American physicist Carl David Anderson, studying cosmic rays, discovered antimatter, earlier postulated by the British physicist Paul Dirac. Cosmic rays also provided physicists with probes of energies much higher than those coming from nuclear transmutation. This was an important step in the investigation of the nature of matter. In his 1924 doctoral thesis, the French physicist Louis-Victor-Pierre-Raymond, 7th Duc de Broglie, introduced the successful hypothesis that with each particle is associated a wave whose length varies as the inverse of the particle energy. The "resolution power" of an instrument, that is its ability to distinguish small details, varies with the inverse of the wavelength of the radiation employed. Highly energetic particles are excellent probes into the sub-nuclear world, as the associated waves have very short length.

While cosmic rays provided very short length waves, their energy, as well as their arrival angle, was subject to a wide statistical distribution, making systematic studies very difficult. Physicists then decided to produce high energy rays in the laboratory by accelerating charged sub-nuclear particles and nuclei with electromagnetic fields. In the 1930s the British physicists John D. Cockcroft and Ernest Walton, the American engineer Robert Van de Graaf and a Soviet team in Kharkov managed to accelerate nuclei via the creation of an intense electric field. Vacuum tubes were used to accelerate electrons, and this led to the discovery of the photoelectric effect in 1887 by the German physicist Heinrich Rudolf Hertz. The explanation of this effect by Albert Einstein in 1905 led to the formulation of a corpuscular theory of light and was one of the founding steps towards modern Quantum Mechanics.

The birth of High Energy Physics

The use of very intense electric fields was however impractical due to the difficulty to generate and control them. Particles could be accelerated by guiding them repeatedly through lower intensity electric fields, but this required very long "linear" machines to achieve high energies. A major breakthrough was achieved in 1929 by Ernest O. Lawrence at the University of California, Berkeley. He managed to accelerate electrons with a relatively low voltage by passing them several times through a potential difference. Lawrence's machine used a magnetic field to keep the electrons on a circular path, and an alternating voltage gradient to accelerate them. Lawrence's first machine, called a cyclotron, was made out of brass, sealed with wax and had a diameter of ten centimetres; it is said to have cost $25 in all. It was the first particle accelerator and it opened the way to the creation of sub-nuclear and nuclear rays of ever-increasing energy (hence the name High Energy Physics). Almost all high energy particle accelerators today derive from Lawrence's machine. Due to fundamental physics laws, the diameter of the accelerator has to increase with energy, so modern machines have diameters of several kilometres and are hosted in underground tunnels. Several particle accelerators have been built since the end of the Second World War. Major laboratories exists in the United States, Japan, Europe and Russia. The largest of them all is at the European Conseil Européen pour la Recherche Nucléaire (CERN) in Geneva, Switzerland.

The role of CERN

Built on the Franco-Swiss border, CERN was founded in 1954 and it is currently funded by 20 European nations. CERN is now operating the Large Hadron Collider (LHC) which is colliding proton beams at energies up to 7 Tera (10^{12}) electronVolts (to be increased to 14 Tera electronVolts after 2013), and lead nuclei beams at energies up to 2.76 Tera electronVolts per nucleon (to be increased to 5.5 Tera electronVolts after 2013). The LHC is the largest scientific machine ever built. CERN has financed and built most of the accelerator complex, with important contributions from the United States, Russia and Japan. Four experiments have started to analyse the results of particle collisions around the 28 km LHC ring. Experiments are designed, financed and built by large international collaborations of hundreds of physics institutes, university departments and laboratories – including CERN's own Experimental Physics Division – comprising thousands of physicists from four continents.

A modern High Energy experiment is typically built at the "intersection point" of a particle accelerator. It is here that counter-rotating beams of particles cross each other generating particle collisions. The energy of each collision transforms itself into mass, according to Einstein's famous formula $E = mc^2$, generating a host of particles not observed on Earth under normal conditions. The "detectors" are very large scientific instruments that surround the interaction points and detect the particles. Several million collisions occur each second, and, after discarding the uninteresting ones, the surviving ones are registered for further processing at data rates that can reach few Gigabytes per second.

It is perhaps important to note that particle collisions create a very high temperature and energy density, similar to what is believed to have existed at the beginning of our universe immediately after the Big Bang. In this sense, a particle accelerator is also a "time machine" which reproduces the condition of the universe soon after its birth, even if only for a very tiny amount of matter. Again, the understanding of our origins is linked with the microscopic composition and behaviour of matter.

Computing in HEP

Since its beginnings, High Energy Physics has relied heavily upon computers to extract meaningful physics results from observations. This is due to several reasons, which have to do with the statistical or probabilistic nature of the underlying theory and with the large amount of collected data that has to be treated electronically.

Very large computing needs and complexity.

This has taken on very large proportions with the latest generation of experiments, which have produced several hundred TeraBytes of data. The new experiments which came on-line in 2009 at CERN are producing a few PetaBytes (10^{15}) of data per year. These data need to be pre-processed several times before being properly analysed. The design and understanding of the experimental apparatus require the computer simulation of its response, and this is a very demanding task in terms of computing resources, as it generates an amount of data comparable with that of the real experiments. However the complexity of computing for High Energy Physics is not only due to the sheer size of the resources needed, but also to the way in which the software is developed and maintained. Laboratories that have an accelerator, such as CERN in Geneva, Fermi National Accelerator Laboratory in Chicago, the Brookhaven National Laboratory, or Stanford Linear Accelerator Centre host high energy experiments. As we have mentioned, an experiment is built by a large international collaboration comprising thousands of scientists coming from hundreds of institutes in a few tens of countries. These researchers work rather independently on the same code, meeting only rarely and with little hierarchical structure between them. Requirements change very frequently and the problems to be solved often push the boundaries of scientific knowledge, both in physics, and also in computer science. It can be said that scientific research in fundamental physics has its limit not in the creativity of the researchers, but in the current state-of-the-art of engineering and computer science.

It has happened several times in the past that physicists have been at the origin of important developments in the field of computing. Notable examples are the invention of the Web at CERN and, more recently, the CERN-led development and deployment of the largest Grid (see Chap. 3 for the description of the concept) in operation in the world for the storage and processing of the data coming from the LHC experiments. And while the Grid is still developing its full potential, the attention of the HEP computing world is already moving to computing Clouds and virtualisation (see Chap. 6) at the boundary of knowledge in distributed computing. These results have been given considerable attention by the media. However there is relatively little published material on the general history and development of computing in High Energy Physics, which presents many other interesting and innovative aspects. For instance, programme code has always been shared between physicists, even beyond the boundaries of a single experiment, in a very similar way to what has now become known as "open source" software development. The management of the development of a large piece of software driven by very dynamic requirements has given rise to techniques very similar to those that have been independently developed by "Agile Software Engineering"technologies.

Most scientific communication between fundamental physicists is centred on their scientific results, with little attention paid to the discussion of their activity as Information Technology professionals (see Chap. 5).

> IT innovation originated in HEP went often unnoticed outside the field.

The result is that several computing techniques and concepts, developed and applied in the context of High Energy Physics, have had broader success after being "reinvented" and developed independently, maybe years later. The main reason for writing this book is to describe the world of computing in High Energy Physics, not only through its results, but by describing its challenges and the way in which HEP is tackling them. The aim is to provide a needed "theoretical" justification for the techniques and "traditions" of HEP computing with the intention of making them better known, and possibly promoting dialogue and interdisciplinary collaboration with other disciplines that use computing either as a research field per se, or as an important tool in their activity. This seems to be a very appropriate moment to do it, because a generation of experiments has now come on-line after almost 20 years of development, and therefore we can now describe a mature and accomplished situation in computing for HEP. The next generation of experiments will almost certainly introduce radical changes to this field, or at least we hope so.

A large part of this book talks about the development of HEP computing at CERN and therefore the reader may question the legitimacy of the claim that it deals with the history of HEP computing worldwide. This is indeed a delicate subject, and CERN has been at times accused of "software imperialism" and of trying to impose its software choices on the worldwide HEP community, while hindering or undermining the development of independent software projects elsewhere in institutions collaborating with its experimental programme.

The special role of CERN in the development of HEP computing.

These criticisms may not be totally unfounded, but we also believe that the software development process in HEP is essentially a Darwinian process, where the ideas that eventually prevail have some degree of fitness to the environment. This does not make them necessarily the best in all possible respects, but it definitely tells us something about their adaptability to different experimental conditions, size of collaborations, and computer hardware and operating system diversity.

HEP is a very distributed and collaborative activity, and so is its computing. It is therefore an undeniable fact that only a fraction of HEP code has been written at CERN, but it is also true that the fundamental software infrastructure adopted by all HEP experiments in the 1980s and through the 1990s was the CERN Program Library, and from the late 1990s the ROOT (see Chap. 1 for a description of the ROOT genesis and evolution) package, developed at CERN by two of the authors of this book and their team. While the whole HEP community has largely contributed to both projects, their design, development, maintenance and distribution has been centred at CERN.

From the point of view of the subject, this book is largely complementary to existing literature. However the information contained in this book is original in the way it is treated and presented. Our objective is not to describe algorithms and computing techniques, nor do we concentrate only on a few specific aspects such as programming language or development methods. Our aim is to give a comprehensive description of the evolution of computing in HEP, with examples. The closest analogy will be with the material presented at the Computing in High Energy Physics (CHEP) conference series, but presented in a set of chapters providing a methodological and historical account and covering the major aspects and innovations of HEP computing.

A book for education and research.

This book is intended both for education and for research. For students of physics and other scientific disciplines it provides a description of the environment and the modalities of the development of HEP software. It often takes young researchers a relatively long time to become active in HEP computing, due to the relative uniqueness of its environment and the fact that it is not documented or even less taught at university. For physics researchers this book will be a useful documentation of the history of one of their most important tools, i.e. computing. For researchers in computing and its history, the book will be an important and original source of information on the evolution of HEP computing in the last 20 years from a very privileged point of view.

The Structure of the Book

As mentioned above, this book is written by different authors, each one of them having his or her own style and point of view. As we said, this is yet an untold story, which spans more than 20 years of intense development, and some of the main actors often held widely diverging point of views on the course to take. The subject matter itself is very diverse, ranging from computer hardware to programming languages, software engineering, large databases, Web, Grid and Clouds (see Chaps. 3 and 4). It is not surprising that this is reflected in the material of this book. While we believe that this is one of the main assets of the book, we are also aware that it may introduce some disuniformity and lack of coherence. To help in alleviating this danger, we have tried to give some formal coherence to the appearance and structure of the different chapters. Each chapter starts with a short statement, followed by an introductory paragraph. It ends with a conclusion section on the lessons learned and, where appropriate, a look at the future. Grey filled boxes indicate important concepts developed in the text. We have tried, where possible, to cross-reference the various chapters, in order to allow the reader to easily recognise related concepts in material with which he or she may not be familiar.

It is important to mention at this point that the opinions expressed in this book are those of the authors and may not reflect the official position of their institutes.

Chapter 1 is a broad overview of the evolution of the programmes, the technology and the languages of HEP computing over the last 20 years. The sheer evolution of the computing hardware in these years has seen a 10,000 increase in processing speed and a 100,000 increase in live (RAM) memory. This is probably the largest technological evolution of any field of technology. The chapter describes how HEP computing coped with this breathtaking evolution to provide the needed resources for the scientific endeavours of HEP.

Chapter 2 recalls the story of the invention of the World Wide Web inside the HEP community. This is one of the best-known, but also one of the most striking examples of "spin-off" from fundamental research. It was a paradigmatic event that deserves to be studied in detail. Few inventions have changed the face of our world so quickly and so deeply as the Web has done, and yet it was invented with the immediate purpose of helping to share images and information related to HEP. In an era where there is so much debate on the relations between applied and fundamental science and on their respective roles and impact on society, this extreme example of serendipity in fundamental science is certainly worth considering in detail.

Chapter 3 describes the genesis and evolution of the worldwide LHC Computing Grid, the largest Grid in operation in the world to date. Extending the Web paradigm of sharing distributed information to include the sharing of distributed computational resources, the Grid aims at homogenising the computing infrastructure, making it ubiquitous and pervasive, and freeing it from the space and time availability constraints of any single physical instance of computing and storage. The consequences of this process are yet to be fully understood and appreciated, but their impact on society at large can be potentially larger than that of the World Wide Web. While the idea was born in the U.S. academic world, the first large scale realisation of the Grid paradigm was realised for the needs of the LHC experiments via a CERN-led international effort.

Chapter 4 describes the Grid in operation. While the middle-ware technology is extremely complex and still in a state of "frantic" evolution, the deployment of the Grid in over 50 countries and 300 institutes around the world has opened a whole new set of opportunities and challenges for international collaborations. Scientists and system administrators from widely different cultures and time zones are learning to work together in a decentralised and yet tightly coupled system, where the global efficiency depends on the efficiency of every single component. Researchers in economically and digitally challenged countries suddenly find themselves with the opportunity to access the largest computing infrastructure in the world and to perform frontier research with it. With the access to data and computing resources becoming completely democratic and ubiquitous, a new set of rules has to be invented for this new virtual community to work together.

Chapter 5 describes the software development process in HEP. Software Engineering has been invented, or rather postulated in the 1950s to apply "solid engineering methods to software development". Although a staggering amount of software has been developed, some of which very successfully, the varying degree of success of software projects is still surprisingly wide. Software Engineering has offered some very interesting insights in the process of software development, but has failed to develop "solid engineering methods" that can transform software development into a predictable process. HEP has developed software for many years, often very successfully, without ever adopting a specific software engineering method. Nevertheless it had its own method, although this was never formalised or described. This chapter describes the software development methodology which is proper to HEP and discusses its relations with the recent advances in software engineering.

Chapter 6 describes the impact of virtualisation on Grid computing. The Grid computing paradigm proposes a conceptual virtualisation of the physical location and other features of a set of computational resources. It is therefore quite natural to push virtualisation one step further and virtualise the hardware within computing centres, in order to increase the uniformity of the Grid components and to help application developers and end users to cope with the different physical platforms. Moreover this makes reconfiguration and upgrade of the underlying systems more transparent for the end users, increasing the overall robustness and resilience of the Grid infrastructure. Finally, virtualisation techniques offer opportunities to increase security of the applications running on the Grid, which is one of the major concerns for such a widely distributed system. The newly born Cloud concept adds another level of abstraction, but also of opportunity, to this virtual infrastructure.

Chapter 7 deals with the exploitation of parallel computing. HEP computing features a kind of parallelism which has been defined as "embarrassing", as the workload comes in units of "events" (the result of particle collisions) which are independent from each other. The result of the processing of each event can be calculated independently and statistically combined with the other results in any order. As a consequence of this, parallelisation has been used since a long time in HEP to reduce the time-to-solution and to allow the interactive optimisation of the analysis algorithms. However, parallelisation had stopped at the "boundary" of the code that treats each event. The advent of multi and many-core machines and of powerful Graphics Processing Units (GPUs) and the large memory requirements of modern HEP codes make it less efficient to continue assigning an event to every core, and therefore parallelisation has to happen also within the code that treats each event.

Chapter 8 deals with the legal aspects of intellectual property law for HEP software developers. The question of the intellectual property of the software developed by large HEP collaborations is technically very complicated. The experimental collaborations are not legal entities, and the software is produced by tens, at times hundreds of authors from different institutions who spend some time in the collaboration and then move on either to other experiments or to a different career. Questions such as who is the legal copyright holder or what is the legal status of the software are in fact complex problem of international law that must be understood and correctly addressed in collaborations that can last several years and involve thousands of researchers.

Chapter 9 deals with databases. HEP computing is essentially a data-centric problem. The principal activity is data transformation and data mining. In this sense HEP computing is similar to other forms of computing that deal with large data-sets, which are becoming more and more commonplace as sensors and measurement instruments become inexpensive and ubiquitous. It is therefore natural that databases occupy a central place in HEP computing. The efficiency of data and meta-data access is perhaps the most important feature in determining the success of an experiment's computing infrastructure. As the focus is put on ordered data transformation and data mining, the transactional aspect of relational databases is not particularly critical for HEP computing, and therefore classical or object-oriented databases occupy a "niche" in HEP computing, while the bulk of the data is held in custom designed data stores.

Chapter 10 deals with obtaining fast and reliable access to distributed data. Grid computing proposes a paradigm for transparent and ubiquitous access to distributed high-end computing resources, but globalised data access is still a very hard problem to solve. For truly transparent data access, local and remote operation should be transparent for the applications. Moreover the different features of data access over Wide Area Networks (WAN's) and Local Area Networks (LAN's) should be optimised and, in some sense, "masked" by the protocol. The systems providing data access should also hide the diversity of the data storage technology from applications and virtualise the data repository, thus eliminating the need for an application to worry about which storage element holds a given file or whether it has been moved or not. These are very difficult, and in some sense contradictory requirements to satisfy, and HEP computing has accumulated very substantial experience in dealing with these problems over the years.

Chapter 11 attempts to go beyond the technical description of HEP computing to consider the larger picture of the development of an increasingly interconnected system of machines and people sharing computing resources and information in a virtual universe, ubiquitous and largely independent of actual space and time. This virtual interconnected world resembles more and more the realisation of a human nervous system on a planetary scale. As already said, HEP computing is today where everyday computing will be tomorrow, and it is interesting to draw a parallel between the capacity of the human brain to represent and understand the universe, and the construction of a world-wide network of connections and nodes where information is stored, exchanged and processed as in the neurons of our own brains. This introjection of the universe and projection in the physical (and virtual) reality of our own brain is intriguing and worth considering at the end of our book.

The Authors

René Brun is a physicist working at CERN since 1973. While working on the detector simulation, reconstruction or analysis phases of the NA4, OPAL, NA49 or ALICE experiments, he has created and lead several large software projects like GEANT, a general detector simulation system, the data analysis system PAW, and ROOT a general experiment framework for data storage, analysis and visualisation. All these systems have been or are used today by thousands of people in the scientific community.

Predrag Buncic. Following the studies at Zagreb and Belgrade Universities, Predrag Buncic stated his carrier as a physicist in NA35 experiment at CERN where he worked on a streamer chamber event reconstruction and where he quickly discovered his passion for scientific computing. In 1994 he joined the NA49 experiment to work on challenging problems of data management as well as reconstruction, visualisation and data processing. In 2001 he moved to ALICE experiment at LHC where he initiated AliEn project, a lightweight Grid framework that later served as an inspiration for the first gLite prototype. For this reason he joined EGEE project in 2004 and worked for two years in CERN/IT on Grid middleware architecture. Since 2006 he is working in CERN/PH Department and currently leading Virtualisation R&D project in PH/SFT group.

Federico Carminati is presently Computing Coordinator of the ALICE experiment at LHC. After getting his Master in Physics at the University of Pavia, Italy in 1981 he worked at Los Alamos and Caltech as particle physicist before being hired by CERN in the Data Handling Division. He has been responsible for the CERN

Program Library and the GEANT detector simulation programme, the world-wide standard High Energy Physics code suite in the 1980s and 1990s. From 1994 to 1998 he worked with Nobel Prize winner Prof. Carlo Rubbia at the design of a novel concept of accelerator-driven nuclear power device.

Fabrizio Furano is a Staff member at CERN, working for the Grid Technology group, in the IT department. He has an extensive experience related to software engineering, performance, simulation and distributed systems, acquired in his activity in the telecommunications industry, university and High Energy Physics computing. From a position related to software development and contact center technology, he moved to work with INFN (Italian Institute for Nuclear Research) for his Ph.D. studies in Computer Science and for a post-doctoral grant, during which he was contributing to the Computing Model of the BaBar experiment at the Stanford Linear Accelerator Center. He taught C++ Programming as Assistant Professor at the University of Padua (Italy) and Software Engineering as Adjunct Professor at the University of Ferrara. He moved to CERN in 2007, to work on Data Management technologies for the LHC experiments.

Giuliana Galli Carminati is presently senior psychiatrist responsible for the Unit of Mental Development Psychiatry (UPDM), University Hospitals of Geneva (HUG), Switzerland. After getting her degree in Medicine at the University of Pavia (Italy) in 1979, she obtained specialisations in Laboratory Medicine and in Psychiatry and Psychotherapy, as well as a Master in Group Therapy and a Doctorate in Psychiatry (Geneva University) in 1996. In 1998 she also got a doctorate in physics (Laurea) at the Tor Vergata University (Rome). In 2008 she obtained the title of Privat Docent at the University of Geneva. Her research activities deal with the treatment of Intellectual Disability and autism and the Quality of Life of intellectually disabled patients. She is particularly interested in the relations between matter and mind, and in particular in the application of Quantum Information theory to the modelling of human psyche, subject on which she has authored some papers.

Andy Hanushevsky obtained his B.A. in Geology at the University of Rochester and his M.S. in Computer Science at the Cornell University. After having worked at the Xerox Corporation, the NPD group and the Cornell University as system programmer, he joined the Stanford Linear Accelerator Laboratory in 1996 as Information System specialist. There he worked on extremely scalable data access systems, high performance protocols for distributed systems and parallel and multi-threaded algorithms. Together with F. Furano he developed the Xrootd – Scalla system which has become a standard tool for high performance local and distributed data access for High Energy Physics experiments worldwide. He was awarded a U.S. patent in 1996 for a method to read hierarchical distributed directories without obtaining locks and the 1993 IEEE Certificate of Appreciation for MSS Standards Work.

Patricia Méndez Lorenzo was born is Salamanca (Spain) where she studied Physics at the University of Salamanca. She completed her Diploma thesis and her Ph.D. in Particle Physics in Munich (Germany) at the Ludwig Maximilians University. Since 2002 she is working at CERN. She has been part of the WLCG

Project until January 2010. Inside the WLCG Project she has been the Grid Support responsible of the ALICE experiment and the responsible of the NA4 effort for Particle Physics inside the EGEE-III project. She has also collaborated with a large range of applications ported to the Grid.

Lawrence Pinsky is the current chairperson of the Physics Department at the University of Houston, an experimental particle physicist who recently served as the Computing Coordinator for ALICE-USA as a member of the ALICE Collaboration at CERN and the ALICE-USA WLCG representative. He is also a licensed Attorney at Law in the State of Texas as well as being a licensed U.S. Patent Attorney with both a J.D. degree and an LL.M. in Intellectual Property and Information Law. He also has regularly taught courses in Intellectual Property Law at the University of Houston Law Center.

Fons Rademakers received his Ph.D. in particle physics from the Univ. of Amsterdam in 1991 for his work on event displays and data analysis for the DELPHI experiment at LEP. Since then he has worked at CERN and been involved in designing and developing data analysis programs. In 1991 he joined the PAW team of René Brun where he developed amongst others the column wise-ntuples, the PAW GUI and the PIAF system. In 1995 he started with René Brun the ROOT project and has been involved in all aspects of the system since then. In 2001 Fons joined the ALICE collaboration and has worked as software architect on the initial version of AliRoot. In recent years his special attention has gone to high performance parallel computing using PROOF.

Les Robertson was involved in the development and management of the computing services at CERN from 1974, playing a leading role in the evolution from super-computers through general purpose mainframes to PC-based computing fabrics and finally distributed computing Grids. He was active from the beginning in planning the data handling services for the experiments that use the Large Hadron Collider (LHC), and led the Worldwide LHC Computing Grid Project (WLCG), set up in 2001 to prepare the computing environment for LHC. The project included the development and support of the common tools, libraries and frameworks required by the physics applications, the preparation of the computing facility at CERN, and the development and operation of the LHC Grid, which integrates resources in more than 140 computing centres in 35 countries around the world. Prior to coming to CERN he worked in operating systems and network development in the computer industry in the UK. He has now retired.

Ben Segal got his BSc in Physics and Mathematics in 1958 at Imperial College, London and his Ph.D. in Mechanical and Nuclear Engineering in 1951 at Stanford University (U.S.). He started working at CERN in 1971 on high-speed computer networks as well as an early satellite data transmission system "STELLA" (1978–83). He also coordinated the introduction of the Internet Protocols at CERN (1985–89) and encouraged the introduction of Unix (then Linux) at CERN since the early 1980s. He participated in setting up the Internet Society (ISOC) Geneva Chapter in 1995. He was a mentor to Tim Berners-Lee who invented the World Wide Web at CERN (1989–91). From 1990 was a member of a small team which developed the "SHIFT" system, migrating CERN's computing capacity from central mainframes to distributed Unix clusters. He has worked since 2004 in the developing area of "volunteer computing", with its great potential for public involvement in science and education. Projects include LHC@home, MalariaControl.net and Africa@home. He is also co-founder in 2009 of the CERN-based Citizen Cyberscience Centre.

Jamie Shiers currently leads the Experiment Support group in CERN's IT department. This group plays a leading role in the overall Worldwide LHC Computing Grid (WLCG) project, with a strong focus on service and operations. He has worked on many aspects of LHC computing since the early 1990s, moving to the Grid service area in 2005 when he led two major "Service Challenges" designed to help bring the service up to the level required for LHC data taking and analysis. He is a member of the Management Board of WLCG and has authored numerous articles on Grid and Cloud computing. Dr. Shiers received a Ph.D. in physics from the University of Liverpool in 1981, following a degree in physics at the University of London (Imperial College) in 1978. He has worked in the IT department at CERN for the past 25 years in a wide variety of positions, including operations, application development and support, databases and data management, as well as various project leadership roles. Prior to this he worked as a research physicist at the Max Planck Institute for Physics in Munich, Germany and as a guest physicist at CERN.

Acknowledgements We gratefully acknowledge the help of the many colleagues with whom we have discussed the idea and the content of this book. Trying to mention them all would inevitably mean forgetting some. But a special thanks goes to Michelle Connor for having proofread the manuscript with intelligence and attention.

Geneva, Switzerland   *Federico Carminati*

Reference

1. http://www.cern.ch

Contents

1 **Technologies, Collaborations and Languages: 20 Years of HEP Computing** 1
   R. Brun

2 **Why HEP Invented the Web?** 55
   B. Segal

3 **Computing Services for LHC: From Clusters to Grids** 69
   L. Robertson

4 **The Realities of Grid Computing** 91
   P.M. Lorenzo and J. Shiers

5 **Software Development in HEP** 115
   F. Carminati

6 **A Discussion on Virtualisation in GRID Computing** 155
   P. Buncic and F. Carminati

7 **Evolution of Parallel Computing in High Energy Physics** 177
   F. Rademakers

8 **Aspects of Intellectual Property Law for HEP Software Developers** 201
   L.S. Pinsky

9 **Databases in High Energy Physics: A Critical Review** 225
   J. Shiers

10 **Towards a Globalised Data Access** 267
   F. Furano and A. Hanushevsky

11 **The Planetary Brain** 289
   G.G. Carminati

Glossary 311

Index 337

Name Index 351

Chapter 1
Technologies, Collaborations and Languages: 20 Years of HEP Computing

René Brun

Research in HEP cannot be done without computers. The statistical nature of the data analysis process, the sheer amount of data to be processed and the complexity of the algorithms involved, to be repeated several times over millions of single collision data require large amounts of computing power. However the sheer computing power is only one part of the story. The data treatment required to extract the physics results from the data is very specific to the discipline, as are data formats and algorithms. The consequence is that HEP physicists have to develop most of their code "in house", and they can only very rarely rely on commercial products. This has led HEP to develop very large and complex software systems for data simulation, reconstruction and analysis.

One additional complication comes from that fact that HEP experiments are one-off endeavours, and each one is different from the others, otherwise there would be little point in building and operating it. This hinders code reuse from one experiment to the other and also it imposes additional constraints on the framework. Another complication comes from the fact that HEP is a "computer resource hungry" activity, where science is actually limited by the amount of computing that can be bought by the available budget. So physicists must be able to move their code to any new hardware and Operating System (OS) that offers the best price-performance ratio appearing on the market.

This chapter tells the history of the evolution of the programmes written by HEP physicists for their research over the last 20 years.

R. Brun (✉)
CERN Geneva, Switzerland
e-mail: Rene.Brun@cern.ch

1.1 Introduction

The major components of HEP computing are simulation, reconstruction and analysis. It is important to understand their relations and their respective role in HEP research in order to appreciate the constraints and the requirements that have guided their evolution.

> Simulation, the third way to scientific knowledge after theory and experiment.

Modern HEP detectors are huge engineering endeavours. The ATLAS [28] detector at LHC is about 45 meters long, more than 25 meters high, and weights about 7,000 tons. Millions of detecting elements have to be positioned with micrometer level accuracy (10^{-5} cm). Data is collected by millions of electronic channels that collectively consume electric power of the order of several tens of thousands of Watts. Temperature has to be rigorously controlled and several components require a constant gas flow, superconducting temperatures, near absolute zero ($-272.9°$C) cooling, very high voltage and so on. But, most important of all, the detectors must respond to quite stringent "physics performance" criteria, i.e. they must be able to detect particle properties (position, speed, energy, charge) with the necessary precision to obtain the physics results for which they have been built. To verify that this is actually the case, before a detector is built, it is simulated on a computer, sometimes for many years. Simulation programmes are complex code systems that simulate the collisions between elementary particles at the foreseen experimental conditions, and then "transport" the particles generated during the reaction through the matter of the detector. The finest details of the interaction of nuclear radiation with the material of the detector is reproduced, with its intrinsic statistical fluctuations via so-called Monte-Carlo [76] methods. The response of the sensitive elements to the passage of the particles is reproduced, as well as the planned electronic treatment of the generated electric signals, and finally the coding of these signals in binary form to be stored in computer files. Millions of events are thus "simulated" to assess the design of the detector and to train the reconstruction and analysis programmes. Once the detector is built, simulation is still essential. The simulation is compared to the actual experimental results and "validated", i.e. tuned to provide results similar to those actually found in reality. This allows to simulation to estimate the "corrections" to be applied to the observed rate with which a given configuration is detected. By agreeing with what is seen, the simulation allows us to estimate how much is "lost" due to the detector geometry and efficiency. Simulation is mostly a "CPU bound" activity, where the input data are limited to small configuration files and the output is similar to the output of the detector.

Reconstruction, from electric signals to particles.

The output of a detector is a set of electric signals generated by the passage of particles through matter. These are the so-called "raw data" or simply "raw". The "reconstruction" process aims at processing these signals to determine the features of the particles generating them, i.e. their mass, charge, energy and direction. In some sense this process has some relations with image processing, in the sense that the signals are localised in space and time, and they are a sort of "pixels" that have to be recognised as part of a particle trajectory through space-time, or "track". This process is also called "tracking" or track reconstruction. It is usually a very complex procedure where the information from different detectors is combined via sophisticated statistical methods. The results of this process are the tracks, i.e. a set of parameters identifying a trajectory in space together with their statistical uncertainties and correlations. A very important element in the reconstruction is the calibration and alignment of the detectors. Each sensitive detector element has a specific relation between the intensity of the emitted signal and the physics quantity that is at the origin of the signal (energy, time, velocity, charge). To obtain optimal precision, all detecting elements have to be "calibrated" in order to give the same response to the same input. Moreover the actual position of each detecting element can differ slightly from its ideal position. Both calibration and alignment of the detectors can be determined from the recorded signals, using recursive procedures which take into account the response to the passing particles. The output of reconstruction are the Event Summary Data (ESD) containing the description of the tracks. Usually raw data are reconstructed several times, as the knowledge and understanding of the detector, as well as the reconstruction algorithms, improve. Depending on the complexity of the algorithms deployed, reconstruction can be an "I/O bound" or a "CPU bound" activity, where the input are the raw data and the output the ESD files.

Analysis, the final step before publication.

Once the tracks are identified, together with the degree of statistical uncertainty of the related information, analysis can start. This is the process by which the experimental results are compared with the theory to verify or falsify it. This is also when new phenomena, requiring novel theories, are discovered and described. During the analysis the ESD are read and they are used to produce mono- or multi-dimensional statistical distributions. The identification of the interesting events is made via selections that are called "cuts" in the HEP jargon. Beyond the isolation of the interesting classes of events, the analysis activity aims at determining the frequency with which these events occur. This is a very delicate operation as the detector efficiency is never 100% and in particular is different for the

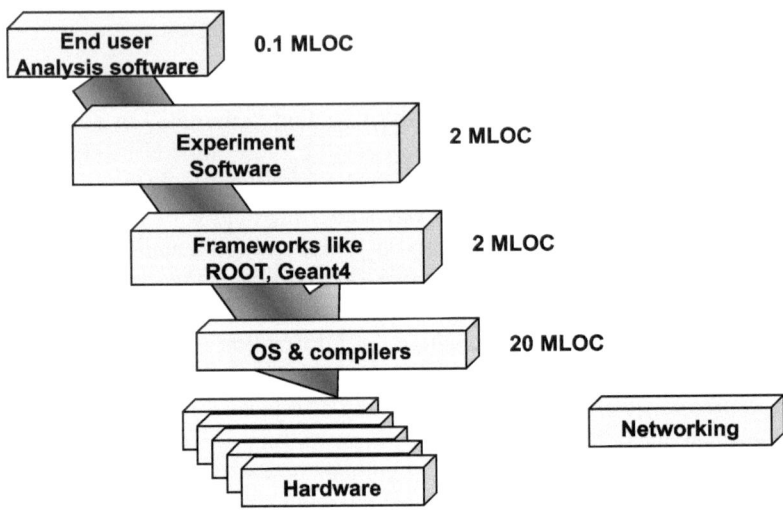

Fig. 1.1 Machines and operating systems

different event classes. While simulation and reconstruction tend to be centrally organised activities, analysis is both a centrally organised activity and a personal one, performed by each physicist willing to test an hypothesis. Analysis is usually an I/O bound activity, where the input are the ESD, or a subset of them, and the output are various kinds of statistical distributions.

All these activities require heavy usage of computing resources (see Fig. 1.1) and the development of programmes and frameworks specific to HEP. In the rest of this chapter we will describe the story of these programmes over the last 20 years.

1.2 Hardware and Operating Systems

At the end of the fifties, the first computer at CERN was a man, Wim Klein. He was hired to help physicists in computing complex mathematical expressions. His operating system and clock cycle were such that he was able to evaluate in a few seconds any trigonometric, square root, exponential or logarithmic function. His memory was able to store several MegaBytes of data where he could store the phone directory of Amsterdam at the end of the Second World War, then later tables of logs, prime numbers or large matrices. He was able to compete with the first CERN computer, a Ferranti Mercury in 1958, or even more elaborate machines in the early sixties. It was only when the first really powerful computer, a Control Data CDC 6600, came to CERN in 1964 that Wim "retired" and gave TV shows or famous lectures for CERN summer students. For about one decade the Control Data CDC 6600 was the main work-horse at CERN. This was a 60 bit machine with 64 kilo-words of RAM and a few 32 MegaBytes disks. It took several months to tune the system to the CERN environment. Because of many instabilities in the native operating system CDC Scope, a special CERN version called CERNScope was developed. The system could run only one job at a time. The job scheduler was an operator taking the next job from one of the input queues, looking at the requests for mounting tapes from a rack to one of the many tape units and dumping the result of the job to the line printer.

The pioneering years: punched cards and line printers. New opportunities and technical limitations.

Users were writing their programmes on special FORTRAN-oriented [81] (FORmula TRANslation) sheets of 80 columns and up to 40 lines per sheet. Once the user was satisfied by his creation, he converted his hand written programme into a computer readable form using one of the many card punchers available in the card puncher room next to the computer centre, not far away from the CERN main cafeteria. The card puncher room was an interesting area, a focal point where people helped each other and discussed the latest news. Once the programme had been converted to punched cards (typically a few hundred cards), it was put in a special card box and the box put in one of the job rack input queues (short, medium, long, very long). The order and priority to process the jobs from the different queues was in general a strong function of the quality of the user-operator relationship.

Starting from 1971, users could feed their jobs to the system themselves by using one of the new Remote Input-Output Stations (RIOS) available on the site. It was possible to see the status of the job (or jobs) by typing a command like "inquire job brun 124" on a teletype next to the card reader. Once the job had been run, the log file and results (typically a few numbers or a few histograms) were printed on the printer next to the card reader. It was a quite tedious operation to get a programme running. Most of the time people were fixing compilation errors or the like. A task that would take today less than one minute on any system, was taking hours.

Sometimes it was hard to find an error when the programme crashed (the infamous "segmentation fault" error). The ultimate solution was to follow the execution of the programme on the main operator console and take note of the stack trace (a few octal numbers) and try to figure out the position of the crash in the executable module. In 1973 the revolutionary FOCUS system was introduced. This was a CDC 3100 machine used as a front-end to the CDC 6600. FOCUS controlled the card readers and printers on several CERN sites. But the big plus was the possibility to use a teletype instead of the card puncher to create programmes on a disk file, edit this file and submit it directly to the CDC 6600. This was a decisive improvement (at least for me). The teletype was replaced one year later by a fantastic Tektronix 4010 display giving, in addition, the possibility to display results in graphics form.

Because disk space was extremely expensive and unreliable, the FOCUS session file(s) were dumped to a tape at log-out time. To log-in to the system, one had to type a magic formula like "focus login brun". This triggered a red lamp to flash in the computer room to attract the attention of an operator to mount your previous session from tape to disk. And of course in a symmetric way your current session on disk was dumped to tape at log-out time. It was possible to mount an old session from yesterday, last week or even last month in case of a catastrophic event.

> The arrival of the CDC 7600, CERN's first supercomputer. A change of era: interactivity and CRT terminals.

A new machine, a CDC 7600, was added to the computer centre in 1973 as well as two other front-end machines, a CDC 6400 and a CDC 6500. These were running an interactive subsystem called INTERCOM where small programmes could be developed and tested before submitting them to the CDC 7600 batch queues. With the advent of INTERCOM and the availability of more and more functional graphics displays like the Tektronix 4002, more and more buildings had serial line connections to the CDC system at typical speeds of 4,800 baud (bits per second) and later 9,600 baud.

In 1977 a new IBM 370/168 computer was introduced. This was a 32 bit machine but with more RAM (a few MegaBytes). The machine was running the IBM MVS batch operating system, with the Wylbur interactive front-end subsystem developed at the Stanford Linear Accelerator Centre (SLAC). Users could edit their programmes, submit them to the batch queues and inspect results. It was also possible to share a common documentation system and exchange mails between CERN users. Because Wylbur had been designed as a light-weight system, the machine could support up to a few hundred people concurrently. It was also possible to run interactive programmes under the native TSO system. However, for economical reasons, only very few users could run concurrently.

In 1976 smaller Norsk-Data machines (super-minis from a Norwegian company) appeared as a solution for data taking and analysis in the large experiments at the SPS (CERN Super Proton Synchrotron particle accelerator). These machines were used as alternatives to the conventional PDP 11 machines for data acquisition. In particular, the SPS accelerator control was operated by a farm of Nord10 machines. A Nord50 machine with more memory (about 256 kiloBytes of RAM) was purchased by the SPS NA4 experiment where we ported the full CERNLIB, and where the first versions of the GEANT simulation package, GEANT 1 and GEANT 2, were developed.

> The golden age of the mainframes. IBM-VM/CMS offers true interactive access to the mainframe.

In 1982, the old IBM 370 was up-graded to a more powerful IBM 3081 running the VM/CMS (Virtual Machine) operating system. VM/CMS was offering

1   Technologies, Collaborations and Languages: 20 Years of HEP Computing   7

a page-oriented alphanumeric user interface that was appropriate for the VT100 (or compatible) terminals installed gradually in all offices at CERN. IBM 370/168 emulators (168E) were built by SLAC and CERN and used quite extensively for the data acquisition of the big UA1 experiment, and later came the 3081E, built in Israel in view of the upcoming Large Electron Positron (LEP) experiments (in particular OPAL [41]). These emulators were running with a reduced instruction set and were pretty inexpensive to build. However, porting general software on these machines proved to be more difficult than originally thought. After a few years, it became more economical to switch to the new generation of Motorola 68000 processors. These processors also came with a VME board, making them extremely attractive for data acquisition (DAQ) systems.

> CERN enters the UNIX world thanks to the Cray-XMP supercomputer. We discover that HEP code cannot take advantage of vectorisation.

In 1987 the CDC machines were replaced by a more powerful Cray-XMP machine. The Cray had been used successfully in other scientific organisations to run large codes that had been vectorised. However, as we will see later, most attempts to vectorise our programmes failed, but the Cray was a popular machine to run pure scalar code. With the growing amount of data collected on tapes, several generations of robotic systems were introduced in the basement of the computer centre.

In the early 1980s, the VAX 780 machines from Digital Equipment Corporation (DEC) became available in many places. The VAX featured the very user-friendly operating system VMS with a very good FORTRAN compiler and debugger. Large VAX systems were introduced at CERN and many other laboratories. It is hard to understand how DEC could fail as a company after having built this golden machine. In 1980 a VAX 780 computer was installed in the "barn" (next to the CERN computer centre) for the OPAL experiment. It was on this machine that the first versions of the GEANT 3 [11] detector simulation programme were developed.

> The 1980s was the era of the workstations. The Apollo machines came to CERN.

In 1982, the first workstations appeared on the market. In 1981, a PERQ workstation (sold in Europe by the English company ICL) was given to the CERN computing division. This machine ran a very user-friendly but exotic operating system implemented in Pascal, a very popular language at the end of the 1970s. However, it was not easy to port our software on this station (due to its poor networking, and bad FORTRAN compiler). In June 1982, I was invited to visit the Apollo company in Chelmsford (MA). In less than one week, I was able to port a large fraction of the GEANT 3 simulation system and associated packages, including an interactive graphics system HTV [12] to display histograms or detector pictures.

Two Apollo DN 400 machines were given to CERN in July 1982. A DN 400 built around the new Motorola 68000 had 512 kiloBytes of RAM, a 32 MegaBytes disk and a nice bitmap screen. The volume of the processor, disk and connections was about one cubic meter and was a room-heating system in itself. It is on this machine that the major components of GEANT 3 were developed and later the PAW [13] system.

Directives at CERN were encouraging the purchase of machines built in Europe. To justify the acquisition of a new model Apollo DN 600 (costing around 100,000 CHF before discount), we used the tactic of purchasing a graphics tablet with a mouse built by a Swiss company Logitech and a "peripheral equipment" to operate the mouse, the DN 600. This went through! The DN 600 had 1 Megabyte of RAM and 128 MegaBytes disk, something sufficient to run our large programmes at that time. The Apollo system also featured a very advanced 12 Megabit/sec token ring networking system (Apollo Domain) where each machine with a 96 bits address had a unique identifier in the world. The very advanced operating system AEGYS was such that it was possible to see and use files from other Apollo Domain machines or to run and debug programmes remotely. A few hundred Apollo machines were purchased during the 1980s. Later on DEC-VMS workstations also became popular.

CERN pioneering activity in network. CERNET was one of the first local area networks ever deployed.

During the 1970s, a very innovative networking system called CERNET had been developed. It connected the various data acquisition systems to the central machines and operated at growing speeds up to 2 Megabit/s. IBM introduced their BITNET wide area networking system in 1978. Electronic mails could be exchanged via BITNET with the main computing centres in the world. During the 1980s, Ethernet TCP/IP became the de facto standard for networking, in contrast with the expectations of the OSI proposed protocols. At the same time Unix was also gaining ground. Many SUN machines were purchased and also several DEC-Ultrix machines, used to operate a central Unix service. At the end of the 1980s, Reduced Instruction Set CPU (RISC) machines, such as Apollo DN 10000 or SUN, became attractive both performance and price-wise.

From mainframes to clusters. The Scalable Heterogeneous Integrated Facility (SHIFT) is deployed.

In 1991, Hewlett-Packard (HP) came onto the scene with their new model HP 720 and HP 750 workstations. These PA-RISC machines became very interesting performance and price-wise with a scalar performance nearly equivalent to the Cray XMP processor. In the fall of 1991, IBM released their first IBM RISC 6000 workstation series. Although superior to the HPs in principle, they turned out to run large codes like the simulation codes based on GEANT 3 much slower than the HP machines. The reason was their small 32 kiloBytes cache while the HP's had a 128 kiloBytes cache. Also in 1991, as the first step in the "SHIFT" project, two SGI PowerSeries machines were purchased by the Indiana OPAL group. SHIFT, the "Scalable Heterogeneous Integrated Facility", (see Chap. 3) was the computing division's networked cluster solution for replacing mainframes with more cost-effective systems. This cluster attracted interest from the other LEP experiments and the SHIFT system, with a large number of SGI, and then later SUN, DEC and HP machines, became very popular for LEP computing.

In the early 1990s, MPP (Massively Parallel Processor) systems were thought the way to go, as many people predicted that the computing power of conventional scalar systems would not exceed the power of a Cray processor. However, this prediction proved to be totally wrong with the advent in 1994 of the Intel x86 architecture and the associated explosion of inexpensive computing that has witnessed the multiplication in one decade of the CPU, RAM and disk by at least two orders of magnitude. A big revolution came with the first Pentium Pro processors. These inexpensive processors quickly invaded most Personal Computers (PC) running Microsoft software or the emerging Linux system. In 1995 the NA49 [40] collaboration pioneered the introduction of Linux at CERN. It took 2 years to convince the Information Technology (IT) Division to provide a better support for this system that was anyhow used (despite the small support) by a rapidly growing user community. The main computer centres in the world moved to Intel (or AMD) based systems. Between 1994 and 2002 the increase in performance continued to follow the famous Moore's law [80] (see Chap. 7).

> Parallel processing slowly imposes itself as the only way to achieve the needed performance. Code re-engineering is a major obstacle.

However in 2001, it was already obvious that the race to increase the processor clock had to stop because it was becoming expensive to cool a large farm of processors. The dual core, then quad core machines operating at a lower frequency appeared in 2003. If we trust the micro-processor vendors, we can expect multi core (up to 16) and later many-core machines with a mixture of fat and light core engines in the coming years. Are we today in the same situation as in 1990? The only way out again seems parallel architectures and we are now forced to re-engineer our programmes designed for sequential processing to take advantage of multi core CPUs or the like. The question of parallelism will be treated in detail in Chap. 7.

With the x86 processors, the number of desktop machines exploded very quickly. Starting in 2000, laptops changed the modus vivendi in HEP and other fields. Conference rooms were populated by attendees each with laptops on their knees communicating with the outside world via a wireless network, such that they could participate in many conferences simultaneously or continue to do their normal work. Presentations that were using transparencies moved quickly to Microsoft PowerPoint with a substantial quality improvement. PowerPoint (or compatible systems like Open Office) became the standard communication medium. While at the beginning most people were running Windows on their laptop, Linux gained ground with dual boot systems or Windows run under VMware or other virtualisation products. With an improved user interface, Apple machines running the popular MacOSX system are now gaining in popularity.

1.3 Languages and Compilers

FORTRAN has been the main language for scientific computing since the origin of computing. The first FORTRAN compiler was a milestone in the history of computing; at that time computers had very small memories (a few kiloBytes), they were slow and had very primitive operating systems. This first compiler was designed and written from scratch in 1954–57 by an IBM team led by John W. Backus who was also system co-designer of the computer that ran the first compiler, the IBM 704.

FORTRAN was the first high-level language that gained large popularity and was extensively used in HEP.

Thanks to this high level language, programmes computing nuclear power reactor parameters now took hours instead of weeks to write, and required much less programming skills. Another great advantage of the new invention was that programmes now became portable. FORTRAN won the battle against Assembly language, the first in a series of battles to come, and was adopted by the scientific and military communities and used extensively in the Space Programme and military projects.

In 1958 FORTRAN II (followed the same year by FORTRAN III) was a significant improvement; it added the capability for separate compilation of programme modules. Assembly language modules could also be "linked loaded" with FORTRAN modules.

In 1961 FORTRAN IV improved things with the implementation of the COMMON and EQUIVALENCE statements, and eliminating some machine-dependent language irregularities. In 1962 an American Standard Association (ASA) committee started developing a standard for the FORTRAN language, a very important step that made it worthwhile for vendors to produce FORTRAN systems for every new computer. The new standard was published in 1966 and it was the language of choice for the implementation of the main scientific libraries, including CERN-LIB. The next standard was published in 1977, (and corresponding FORTRAN version was known under the abbreviation of F77) adding new features such as: DO-loops with a decreasing control variable (index), block if statements IF ... THEN ... ELSE ... ENDIF (before F77 there were only IF GOTO statements), CHARACTER data type (before F77, characters were always stored inside INTEGER variables) and, finally, apostrophes delimited character string constants.

> FORTRAN 77 marked the apogee of the FORTRAN language. An efficient and easy-to-learn standard language for scientists and engineers.

Learning FORTRAN 77 was quite easy. Most physicists learnt this language by extending or copying pieces of code from colleagues. Very efficient implementations of the language were available on all machines. The main strength of the language was the fact that only basic and easy constructs were available and the data types restricted to characters, Boolean, integers and floating point. Of course, the price to pay for this simplicity was "spaghetti style" code and the wheel reinvented many times, in particular when more complex data structures had to be implemented. A macro extension of FORTRAN 77, called Mortran (More than FORTRAN) was developed at SLAC in the early 1980s. With Mortran, it was possible to write more compact code by writing macros with parameters. Mortran was a pre-processor generating standard FORTRAN code. The Mortran language became popular in some circles but remained a niche market.

The fact that the wheel was reinvented many times led to the implementation, with time, of standard mathematical subroutines, memory management and histogramming packages, and graphics systems.

In the early 1980s some attempts to introduce Pascal as a better language appeared in the data acquisition systems, or derivatives of Pascal in the accelerator controls. However, because the Pascal compilers were not generally available and the generated code not as efficient as FORTRAN (at least from some reports), Pascal was quickly abandoned.

> Pascal, ADA, PL1 and LISP are tested, but cannot replace FORTRAN 77 as the main workhorse for physics programming.

In 1979, the DOD (Department of Defence in the U.S.) decided that the next programming language would be ADA and major companies in the U.S.

(e.g. Boeing) switched to ADA. As a result we were very strongly pushed to investigate ADA. In 1984, a young CERN fellow Jean-Luc Dekeyser implemented a large part of the GEANT 3 [11] geometry package in ADA. The advantage of ADA was that it forced serious thinking to design the data and programme structure before making an implementation. However the only compiler available was on VAX machines and the time to compile a short programme of 1,000 lines was several minutes compared to only several seconds with FORTRAN 77. This exercise was nevertheless interesting as an eye-opener on the pros and cons of more sophisticated languages. When the ideas around the PAW project emerged in 1984, Rudy Böck and I visited the Deutsches Elektronen-Synchrotron (DESY) laboratory in Hamburg to discuss with Erik Bassler who had implemented a popular data analysis system on the IBM, written in the PL1 language. PL1 was a nice language that became available also on other systems like VMS. However we never considered this language as a general solution as it was too specific to IBM.

At the end of the 1980s, many voices were pushing for an alternative to FORTRAN. LISP was very popular with computer scientists. The SmallTalk language was gaining ground in the circles pushing for object orientation. I remember a very nice "small SmallTalk talk" by David Myers in 1989. At that time, we were deeply embarked in the implementation of the interactive analysis system PAW with substantial manpower devoted to an interactive language called KUIP [15]. I was personally strongly influenced by two CERN fellows working for short periods with our team. Andy Hansom from Indiana University was a LISP expert and in 1990 trying to implement a PAW front-end based on LISP. None of his attempts were successful and looked more complex to use than our non-standard KUIP system. Our second expert (Andy Kowalski from Tel Aviv) had some expertise with C++ and his recommendation was to use a native language like C++ to control PAW instead of a home grown system. He was with our team only for a few weeks in 1992, but we retained his message when we started development of the ROOT [74] system three years later.

FORTRAN 90: too little too late, or too ambitious? A lost occasion for HEP or a narrowly escaped blind alley?

The FORTRAN 90 standard was published many years after FORTRAN 77 was out. Many new features were added to the language, in particular:

- Free format source code form (column independent)
- Modern control structures (CASE and DO WHILE)
- Records/structures – called "Derived Data Types"
- Powerful array notation (array sections, array operators, etc.)
- Dynamic memory allocation
- Operator overloading
- Keyword argument passing

- Control of numeric precision and range
- Modules – packages containing variable and code

In the early 1990s, FORTRAN 90 seemed to be the obvious evolution to the well-established FORTRAN 77. The leader of the computing division said "I do not know what will be the features of the next language, but I know that it will be called FORTRAN". However, this did not happen and this will be the subject of another section. We spent many months evaluating FORTRAN 90, in particular with the hope that the native support for derived data types could eliminate a large part of our ZEBRA [16] system. Mike Metcalf was our FORTRAN guru. He was a member of the FORTRAN committee, had written a book about FORTRAN 77 [77] and written several prototypes showing good examples of how FORTRAN 90 could improve and simplify our software. I spent a lot of time in 1991 and 1992 working with Mike encouraging him to provide examples of code that could replace our traditional bank-oriented software used extensively in the implementation of our major packages.

> Dynamic memory management with fixed memory stores. A programming tour-de-force in FORTRAN 77 that was not possible to migrate to native FORTRAN 90.

It is perhaps important here to make a short digression to explain what has been known as the *bank-system* in HEP programming, because it occupies a central place in the evolution of programming language choices in HEP. Each event is inherently different in size from the others. Events are read in memory, processed and then deleted to make space for the next one. This calls for the possibility to have dynamic, variable-size memory structures that can be easily created and deleted thousands or millions of times. However not only the size has to be dynamic, but also the layout of the data structure. One event has tracks, which end at a decay vertex, generating other tracks variable in number. FORTRAN 77 has fixed-size arrays, which are very impractical for this kind of structure, particularly if high performances are required and memory is scarce. The solution was to implement a dynamic memory management system within fixed arrays in COMMON blocks. Portions of memory allotted inside the COMMON blocks were called "banks".

The discussion of the pros and cons of this system, that in its latest incarnation was called ZEBRA, would take us very far. It may be sufficient to remark that the system was complex and the code using it tended to be very obscure. The system was dynamic but only within the limits of the memory available; when memory was full, the programme died mercilessly with a long and infamous "ZFATAL dump". Memory management was done with "pseudo-pointer" stored in integers, which made the transition to 64 bits problematic. On the bright side the system did work, and it was very efficient and rich in functionality, particularly I/O, as explained below.

At this point one could ask why we did not use the C language, which indeed offered the possibility to create and delete dynamic structures in memory. Again this would require a long discussion exposing the internals of ZEBRA. It may be enough to say that in the late 1990s we did move to C, in fact to C++, and we realised that `new ... delete` was only a small part of the story. Even today, with fully mature C++ frameworks in production, some of the functionality and nice features of ZEBRA have not been completely recovered. This will be explained in more detail in Sect. 1.6.

One major stumbling block in the move to FORTRAN 90 was the question of Input/Output. With ZEBRA, we had a simple way to describe data structures (banks) built out of basic types (typically integers and floats). Because FORTRAN 90 supported derived data types, it was theoretically possible to implement the most complex data structures that we used to model with ZEBRA. In particular ZEBRA was able to write and read these data structures from machine independent files.

> The need for introspection to deal with derived data types doomed the efforts to move to FORTRAN 90. It was going to be as hard (if not more) with C++, but we did not know this at the time.

Using FORTRAN 90, it appeared pretty hard to make a general implementation equivalent to ZEBRA without parsing the data type description in the FORTRAN 90 modules. In fact, we encountered the same problem later with C++, but we naively ignored at that time how much work it was to implement a data dictionary or reflection system describing at run time the type of objects. Mike Metcalf was aware of the problem and we reported this to a special session of the FORTRAN committee at CERN in 1992. As most members in the committee had no experience with this problem and thought that this was a database problem and not a language problem, the enhancements that we were expecting in the language did not happen.

1992 was an interesting year. In September, at the CHEP (Computing in High Energy Physics) [75] conference in Annecy (France), the Web was the big attraction. Tim Berners-Lee and his small team had developed the first World Wide Web servers and clients using the Objective C language on a NexT workstation. The language seemed to be very attractive, but unfortunately without much attraction from the industry. Other new languages also looked appealing. An example was Eiffel. Some people in the recently created RD41 Moose [6] project had already positively evaluated this object-oriented language with a better support for parametrised types than C++. However most people in the Moose project were known to be more game players than people to trust to propose a solid solution, and they were pretty slow to come up with valid proposals.

As a result the situation was quite confusing. At that time, I was running the Application Software group in the CERN Computing Division. Following CHEP in Annecy, I organised a small workshop in the week just before Christmas 1992 to discuss possible orientations of our software, with the hope that we could make a concrete proposal to the recently born large LHC collaborations. At the time of this small workshop, the group was supporting some large projects:

- A general detector simulation system GEANT 3, implemented in FORTRAN 77.
- An interactive data analysis system PAW, implemented in FORTRAN 77, but with a large subset in C for the graphics and user interface systems.
- A bookkeeping software HEPDB [84] and a file catalogue system FATMEN [83].
- General FORTRAN packages in CERNLIB.

I realised that the group was strongly divided into three groups: those proposing to go on with FORTRAN 90, those proposing an evolutionary approach, but with an evaluation of object-oriented languages like C++, and finally those who supported the use of commercial software for things like databases and graphics.

No conclusion was reached by this workshop. During the following year, the situation was even more confusing. The Moose project had been launched in the context of the Electronics and Computing for Physics (ECP) division that later became famous when it was stopped in 1995 for its numerous failing projects. A subtle equilibrium had to be maintained between the ECP Division and the Computing and Networks (CN) division. In fact the two division leaders were totally unable to follow what was happening in the various corners of software and they were known to make strategic alliances rather than propose and encourage innovative and trusted solutions.

> The transition to C++ is launched. It would take an unforeseen effort and a long time before converging to a satisfactory and largely consensual situation.

In 1994 several Research and Development (RD) projects were launched at CERN, in particular the RD44 project to implement in C++ a new version of GEANT known since as GEANT 4 [31], and the RD45 project aiming to investigate object-oriented database solutions for the coming experiments. The consensus was reached to use C++ as the mainstream language for the two projects. This consensus was reached following talks by several gurus (in particular Paul Kunz from SLAC) showing the Stanford Computer library with plenty of books about C++ and only one about FORTRAN. It was a funny period where experienced software experts (but using FORTRAN) left the field open to plenty of computer scientists with no experience. One strong argument in favour of C++ was that most physicists were going to industry after their thesis and it was a better investment to be familiar with C++ than with FORTRAN. This highly debatable argument was largely accepted and not many people questioned the move to C++ at that time. Most physicists were resigned to the idea that they would have to learn C++ at some point in the near future, but most were unaware that the process to move from the FORTRAN world to C++ would take more time than initially anticipated.

When Java appeared in 1995, some of the people in the defunct Moose project who were in favour of Eiffel instead of C++ took this opportunity to push for Java. As the two official projects were in the implementation phase, the work continued with C++, but most physicists thought that C++ was just an interim solution and that our scientific world would move to Java (perceived as a more modern language and easier to learn) in a few years once the Java compilers had matured and the performance approached the FORTRAN performance. The Java fans made several declarations in 1997 that Java with its just-in-time compiler would become faster than C++ and could become faster than FORTRAN.

> 1995, the rise and fall of the Java dream as a language for HEP to replace FORTRAN, and one more rift in the community.

As we will see later in the section about packages, a few systems were implemented in Java. However, we must recognise today that none of them succeeded despite the claims that Java was easier to learn, making implementations faster. The only successful system in our field implemented in Java is the MonALISA system [32] that takes advantage of the Web oriented features of the language.

High Energy Physics software is now implemented in C++, and because of inertia and the time to develop, it is likely that C++ will remain the main language in the coming two decades at least. In fact we are seeing several levels of use of C++: FORTRAN-like "spaghetti code", simple C++ with well designed classes with no abuse of inheritance, and more sophisticated C++ code using templates extensively. Most physicists will admit that the complexity of our codes today is higher than it used to be in the old FORTRAN era.

Moving from the FORTRAN world to the C++ world took about ten years. Probably the LHC software could have been implemented in FORTRAN with less frustrations for the majority. However, I am convinced that this decision, taken about 14 years ago, was a good one and not many people question this point today.

If C++ has become the main stream language, it is not the only language. Perl has been used by several projects as a scripting language. Most systems requiring a scripting language tend to use Python these days. We will discuss the debate about scripting languages in another section.

1.4 CERN Program Library

The need to collect standard utility routines, provide some documentation and distribute the source code and compiled libraries was already obvious in the early days of computing. During the 1960s a large collection of algorithms, typically one single FORTRAN subroutine for each algorithm, was accumulated and organised into chapters. The CERNLIB office was created and employed a few mathematicians for about 20 years. The library manager was responsible for accepting new proposals, often helping providers to write portable code. The office was a focal point answering not only questions about the library, but also any question about programming, FORTRAN or the use of computers. Due to the limitations in the first versions of FORTRAN, many routines were written in assembler for the CDC 7600 and later implemented in IBM assembler too.

The library structure was quite simple for the most basic mathematical functions like computing integrals, linear algebra, least square fits, Bessel functions or random number generators. Several chapters were specific to the computer environment, like access to tapes, job queues, printing or specialised graphics systems. CERNLIB was organised in 22 chapters, identified by one of the letters of the alphabet, and about 1,100 entries where each entry was documented by a short write-up explaining the basic user interface and a long write-up explaining the algorithm in more detail.

For many years the documentation was produced on a type-writer. Then in the early 1970s the first computerised documentation system appeared: BARB, with input using only upper case letters, followed rapidly by BARBascii with the full character set. When the IBM machine was introduced, the documentation was moved to an IBM specific format DOC. Write-ups were automatically generated from the documentation database and accessible directly from an on-line help system called FIND. The FIND system, developed for the IBM, was extended to XFIND to access documentation files stored on other machines. XFIND was complemented by XWHO to find people, offices, phone numbers, email addresses, etc. In 1990, XFIND developed by Bernd Pollerman was very successful. Other projects were flourishing at CERN to organise or access documents, e.g. in the CERN central library to access pre-prints or published papers in journals.

> The various CERN documentation systems and the birth of the World Wide Web. Radical innovation versus gradual improvement.

Because Tim Berners Lee (see Chap. 2) had suggested a more general approach, he had moved to the computing division in the same section as Bernd with the idea that a super XFIND system could be produced. While Bernd was proposing a centralised system with a database keeping trace of all registered documents, Tim, who was familiar in his previous job with remote procedure calls to communicate between computers, proposed a more decentralised and scalable approach. However, Tim's approach required a standardisation process so that a documentation server could process the requests from clients. He was aware of all the standardisation processes going on in the U.S. (e.g. the Gopher project) and he was extremely active in this area, participating in several conferences and workshops. The fact that Tim was spending a lot of time in this attempt to standardise communications was highly criticised. XFIND extensions were seen to have higher priority and to be more useful to the community. As a result, Tim's efforts were not encouraged by CERN in general. He wrote a proposal in a memorandum explaining his ideas about a World-Wide-Web. Tim and Bernd were moved to the Application Software group in 1991. Tim had already developed a document server on his NexT workstation and a nice client using Display Postscript. Chapter 2 is dedicated to the development of the World Wide Web.

Looking a posteriori, Tim was right to target the development of a nice client on the NexT first. However this machine was perceived as being exotic and the work not very useful for the vast majority of users who were using either very cheap alphanumeric terminals (Falcos) or workstations running X11 graphics. A young technical student Nicola Pellow implemented an alphanumeric client to access the very few documents available on Tim's server machine. The quality of this first "web client" was not competitive with the full screen alphanumeric interface familiar to most people using the IBM machines. This generated even more criticism. A few of us tried to convince Tim to implement an X11 interface (maybe using the MOTIF graphical widget library). Because Tim did not have enough experience with X11 and MOTIF, he thought that this would be far too much work. Instead he concentrated his work on improvements on the server side and some improvements to the line mode interface. When the first X11/MOTIF based client MOSAIC [2] was announced in early 1993, the Web became a tremendous success. This client had been written by Marc Andreessen from the National Centre for Supercomputing Applications (NCSA) and it was free for internal use by any organisation. Meanwhile the Web had been the main attraction at the CHEP conference in September 1992 in Annecy and what follows is history.

1.5 Code Management Systems

Code Management systems have always been among the most debatable subjects. They have always been necessary, but none of them has been fully satisfactory. Simple projects require simple solutions, while large collaborations have to organise millions of lines of code and describe their dependencies.

In the early 1970s, Julius Zoll's PATCHY was the first system developed to support the development and the maintenance of the large programmes written for the Bubble Chamber experiments. A project was organised in "Patches" and each Patch in "Decks". A Deck was typically a FORTRAN subroutine. A "Sequence" directive was used in the same way that we use an "include" statement today. Conditional code was possible with "IF" "ELSEIF" directives. The input file myproject.car in readable format or myproject.pam (Patchy Master file) in binary and more compact formats could be processed by utilities like Ypatchy to generate myproject.F as input for the compiler and linker producing the archive library myproject.a. Once the version of a project was released, individual users could apply a patch containing bug fixes or new code. The patch was applied to the original released version and the corresponding object files produced would be loaded before the old versions in the archive file. This system proved to be very efficient and robust when working in batch mode. The system was less user-friendly for developers. They preferred to edit a complete deck in their favourite editor rather than to apply a patch.

Code Management Systems: one size does not fit all.

On CDC machines, the UPDATE system was somehow simpler to use than PATCHY but restricted to CDC architecture. CERNLIB was maintained for several years with UPDATE (the librarian at the time did not like PATCHY).

The CMZ [17] system developed in the late 1980s was a PATCHY compatible system and offered a simpler and better interface when working with editors. The system was developed by a commercial company CodeME but was freely available to the HEP community. With CMZ (like with PATCHY) one could select any old version or generate the delta files between versions.

The first requirement when porting code to a new system was to port PATCHY first. PATCHY itself depended on a very small subset of CERNLIB distributed with PATCHY such that the installation of PATCHY was not dependent on CERNLIB itself.

With the move from FORTRAN to C++, PATCHY or CMZ were no longer required, as some of their functionality was available with the CPP (C pre processor) and associated Makefiles. The CVS [67] system became the standard tool to manage large projects in a distributed environment. A central software repository is accessible via the network. Users on different sites can check-in new code or update existing code in a simple and reliable way. It is easy to check-out a specific version of the code or see the differences between different versions.

Large packages like GEANT 4, ROOT were developed with CVS and Makefiles only. The SubVersioN system (SVN) [48] is a more modern version of CVS. It offers many advantages compared to CVS. In particular, it is easy to move or rename files without losing the history. The development of the ROOT system moved from CVS to SVN in a transparent way during 2007. This is also the case of the most popular Open Source projects in the world.

However some of the big collaborations felt the need to develop special tools like SCRAM [85] (used by CMS [29]) or CMT [58] used by ATLAS and LHCb [38] at CERN or FERMI [61] at SLAC. SCRAM and CMT can use CVS or SVN for their software repository; they were developed essentially to manage the huge number of dependencies in their software. These systems are seen as essential by the software managers, but disliked by users. The FERMI collaboration at SLAC decided to discontinue the use of CMT in 2009.

1.6 Data Structures and I/O Systems

As FORTRAN did not support derived data types, nor dynamic memory allocation, special packages initially called "memory managers", then "data structure managers" were developed starting in 1973. The first of these packages, HYDRA [86], was developed for the analysis of Bubble Chamber pictures. The package was quite powerful, but complex. A static FORTRAN floating point array with a fixed dimension was created by the user, say Q(MAXDIM). Two other integer arrays IQ(MAXDIM) and LQ(MAXDIM) were equivalenced with the array Q. To create a dynamic structure (a bank) containing 2 integers, 10 floating points and 5 pointers, one had to do:

```
call mqlink(q,lstruct,12,5,3)
```

lstruct was returned by the routine and pointed inside the big array (Q/IQ) to the dynamic structure. To access the first two integers, one had to do iq(lstruct+1) and iq(lstruct+2). To access the first floating point q(lstruct+3). The bank lstruct could have structural pointers to other banks owned by lstruct or reference links to banks not owned by lstruct. In the example above, a bank was created with five links in total, of which three are structural links. To access the first structural pointer, one used to do lq(lstruct-1) and to access the first reference link lq(lstruct-4). Of course, the code written in this way was quite difficult to read. Several pre-processors to HYDRA were developed such that the code written was easier to read and to maintain. These pre-processors were coupled with the bank documentation system and took advantage of the FORTRAN PARAMETER type. For example, it was frequent to find code like Q(ltrack+kPx) to address the element at the constant address kPx in the struct/bank ltrack. In C the above example would be declared as

```
struct {
integer i1,i2;
float px,py,pz,...;
Track *t1, *t2, *t3;
Vertex *v1, *v2;
} Track;
```

and used like

```
Track t; // or Track *t = new Track();
t.i1= ..;
t.i2=..;
t.px = ..;
t.t1 = new Track();
```

HYDRA also had an I/O sub-system to write and read a complete data structure to a file in a machine independent format. We will discuss this important feature later.

In 1974, ZBOOK [18], an alternative to HYDRA, was developed in the CERN Data Division. ZBOOK was much simpler to use than HYDRA. While in HYDRA only one single storage area was possible, with the store description in common blocks, in ZBOOK it was possible to have as many stores as necessary and the book-keeping of each store was part of the store itself. This feature was particularly interesting when multiple dynamic stores had to be created and managed separately by an application. It was, for instance, frequent to have a dynamic store for histograms, another one for the main structures of one physics event and another one to describe the detector geometry. ZBOOK also had its own I/O system including the possibility to support direct and random access to structures in a file. The ZBOOK system was gradually adopted by a growing number of experiments at the SPS and was used by the first versions of the detector simulation system GEANT 1, 2 and 3. More and more features were added over the years.

In 1980, a similar system called BOS (Bank Organisation System) had been developed by Volker Blöbel at DESY. BOS was used by the PETRA [42] experiments at DESY and some interest for BOS appeared in the LEP ALEPH [25] experiment in 1982.

In 1983 (as we will see in Sect. 1.7) the GEANT 3 system had been used very successfully to simulate the OPAL and L3 [33] detectors at LEP. Because ALEPH also intended to use GEANT 3, the director for computing, Ian Butterworth, decided in 1983 that only the ZBOOK system would be supported at CERN and that the development of HYDRA was frozen. Being the main developer of ZBOOK, I was strongly congratulated by my boss and other group members. To explain the situation, it is important to say that the Data Division (DD) and the Nuclear Physics (NP) Division where HYDRA was developed had two rival software groups for many years. Ian Butterworth's decision was seen as a victory of DD against NP. His decision could have been the inverse, given the fact that the HYDRA system had more features than ZBOOK and more senior people supporting it. I considered this decision as a profound injustice for Julius Zoll, the main developer of HYDRA (and many other systems too like PATCHY).

The very same day that the decision in favour of ZBOOK was taken, and following the congratulations of my colleagues, I was kindly invited to a meeting by some people (Francis Bruyant, Martin Pohl and Harvey Newman) from the L3 collaboration. They all accepted Butterworth's decision, but were still hoping that some features from HYDRA could be incorporated inside ZBOOK. We all met with Julius Zoll late in the afternoon and came to an agreement to develop in common a new package that we named ZEBRA (combination of ZBOOK and HYDRA). We agreed on most specifications of ZEBRA and the day after we proudly announced our alliance to the great surprise of most of my supporters. Julius and I were immediately called to Ian Butterworth's office to give more explanations. Julius had made a quick estimate that merging the two systems should not take more than one month. The message from Ian could not have been more clear: "If in 3 months, the ZEBRA system was not functioning, we were both fired from CERN". As it is well known that in the software area, one can always claim that something is ready on time, we were proud to show a running ZEBRA system less than 2 months later.

In addition, GEANT 3 had been fully converted from ZBOOK to ZEBRA. This seemed like a convincing argument to OPAL, L3 and DELPHI [30] collaborations who decided to base their framework on ZEBRA immediately. However, the ALEPH collaboration (many people there did not like HYDRA at all) decided to use the BOS system instead, even when using the GEANT 3 package.

The BOS system was further developed by Volker and an extension called YBOS used for many years by the CDF [50] collaboration at Fermi National Accelerator Laboratory.

> The ZEBRA memory management and I/O system set very high standards of functionality and efficiency, still competitive with the modern C++ frameworks.

ZEBRA combined the features of ZBOOK and ZEBRA. The concept of multiple stores in ZBOOK was implemented in ZEBRA by having multiple dynamic divisions within the main store. We also imported from ZBOOK the possibility to expand or shrink dynamically the number of data words in a bank or the number of links. Users could define a link area in their private FORTRAN common blocks. These links were automatically updated whenever the bank was moved to a different position in the store or the link reset to zero when the bank was deleted. A garbage collection mechanism could be triggered automatically or forced by a call. Another interesting feature of the memory management system was to wipe all the banks in one division. For example it was common to wipe the "event division" when finishing the processing of an event. It was very hard to create memory leaks with ZEBRA. We are today badly missing in our C++ systems the concept of multiple heaps that made ZEBRA particularly efficient.

The memory manager was an important part of ZEBRA, but even more important was the input/output system writing and reading complex data structures to files (either in sequential or direct and random access formats). The ZEBRA sequential format has been extensively used to write large collections of events. The direct access package was the basis of the HBOOK [14] and PAW storage systems and is still in use today in many corners of the world. The debug and documentation system were interesting tools to graphically draw a complex data structure in memory.

However, in 1993, it was clear that ZEBRA's days were counted. We had to use the native constructs provided by all modern languages and at the same time preserve the nice capabilities of our data structure persistency tools. Following our failure to take advantage of FORTRAN 90 for ZEBRA, the ZOO proposal to develop our persistency system in the object-oriented world was presented in the spring 1994, but was judged not the direction to go at a time when many people were expecting object-oriented databases to dominate the world a few years later.

> The lost hope for Objectivity to become a one-stop-solution for all HEP needs in terms of data storage and memory management.

You can read more about the experience with Object DataBase Management Systems (ODBMS) in Chap. 9. The years 1994 to 1998 were extremely confusing. The vast majority of users did not understand the implications of using an ODBMS. It was only in 1998, when the first signs of problems appeared with the chosen implementation based on the commercial product Objectivity [68], that more and more people started to question the direction taken by the RD45 project which proposed to use ODBMS systems for HEP data. We will come back to this point later when describing the history of the ROOT system. It is clear today that the choice of an ODBMS like Objectivity would have been a total disaster for storing HEP events:

- Objectivity was directly mapping objects in memory to disk. We know today how important the support for transient data members in the I/O system is. Without this support, one would require a transient version of the class and a persistent version, making the class management rapidly non-scalable when the number of classes increased and when these classes evolved with time.
- Class schema evolution was not supported without rebuilding the full database with the latest versions of all classes. In large experiments with thousands or tens of thousands of classes and databases in the PetaBytes range, this would have been impossible.
- Storing data in a central database being accessed at the same time by thousands of programmes was impossible performance-wise. This effect was in fact rapidly seen by the BaBar [55] collaboration at Stanford. A considerable effort was spent by this collaboration to get something running. However, it is unlikely that BaBar would have survived with Objectivity and the increasing database size. Most HEP data are write-once and read-many-times. There is no need to assume that somebody might be modifying your data set while you are reading it.
- With the development of the Grid, a central database is now a central point of failure. If something goes wrong (for example a deadlock) all programmes are aborted or in a deadlock too. Network bandwidth and latency, despite the tremendous developments in this area, are not yet able to support a generalised scheme where readers on the wide area network can efficiently access a central database. Where file transfers are now optimised to copy large chunks of data (GigaBytes per transfer), we are still working on implementing the necessary tools to access data sets with a smaller granularity (GigaBytes or kiloBytes).
- Many users prefer to develop their analysis software on their laptop, even when the laptop is not connected to a network. For code development, only small read-only data sets are required. This was one of the motivations pushing people to use the ROOT system.

1.7 Detector Simulation

For many years, the design of a new detector was based on the experience acquired with the previous generation of detectors and a new design was tested on special test beams in the accelerators. However in the mid 1970s, the complexity of the new detectors was such that it was increasingly hard to predict the behaviour when the final system was installed. Detector simulation tools had been available for many years, but mainly to understand and tune some simple calorimeters. The EGS (Electron Gamma Showers) [60] programme had been designed at SLAC and was the de-facto standard for electromagnetic calorimetry. Some ad hoc programmes were available for hadronic calorimetry like the Alan Grant system or TATINA from Toni Baroncelli. Transporting neutrons was the speciality of MCNP [66] from Oak-Ridge in Tennessee. However, these programmes were either very specialised or missing a very important component, a geometry package. When using the EGS package for instance, one had to implement two routines answering the questions, *Where Am I?* and *Where Am I Going?* Given a particle position x, y, z and its direction, the user had the difficult task of implementing these two functions: not too complex when transporting particles only in a few hundred detectors, but becoming much more difficult and error prone when simulating detectors with many thousands, or millions of components. None of the existing programmes produced graphical results and no general simulation framework existed.

> Geometrical modelling is an essential component of a HEP simulation programme.

In 1976 a combined effort (by the NA3, NA4 and Omega experiments) to implement a very simple simulation framework was successful. The product was named GENEVE (for GENeration of EVEnts), but several other products had already a similar name. Jean-Claude Lassalle suggested the name GEANT (Generation of Events ANd Tracks) and soon we were releasing the GEANT version 2 using the newly developed ZBOOK system. GEANT 2 had been adopted by a growing number of experiments. Working in the deep inelastic muon scattering experiment NA4 led by Carlo Rubbia, I was in charge of the simulation and reconstruction software and spending a substantial fraction of my time in describing more and more complex geometries. I had also simulated new devices, in particular studying the resolution of a di-muon detector with thousands of components that turned out to be a large subset of the future UA1 detector where the Z^0 particle was discovered. I was supposed to follow Carlo Rubbia to UA1 in 1980, but the new NA4 spokesman was opposed to my transfer to UA1. In the fall of 1980, I was invited to join the brand new LEP experiment OPAL and in particular to implement a GEANT 2 based simulation for this new detector. For several months we had prototyped several ideas to automatise the description of large geometries and all of my colleagues reached the conclusion that this was an impossible exercise.

> The origin of GEANT 3, the worldwide standard for detector simulation for more than 20 years.

It turned out that in OPAL I met Andy MacPherson who was working for Carleton University in Ottawa. Andy was extremely interested in investigating a more general solution to describe OPAL in an easy and scalable way. After one night of discussion and a bottle of Beaujolais, we had basically designed the main features of what later became the GEANT 3 geometry system. I designed the data structures and the Application Programming Interface (API) and Andy implemented the mathematical algorithms computing distances to boundaries. Less than one month later in early 1981, we had succeeded in modelling a large part of OPAL and this was recognised by the collaboration as a big success. Of course it took much more time to consolidate the GEANT 3 geometry system, but by 1983 most shapes were coded and meanwhile the L3 collaboration had successfully encoded their detector geometry with GEANT 3. Following this success with the geometry system, we started importing many of the algorithms from the EGS 4 (Electron Gamma Shower) system, thanks to Glen Patrick from OPAL, and later to Michel Maire and Laszlo Urban from L3. We implemented an interface with the hadronic shower package TATINA from Toni Baroncelli and in 1985 Federico Carminati made an interface with the GHEISHA [21] hadronic shower package developed at PETRA by Harm Fesefeldt who was also joining L3.

In 1983 we started the development of an interactive version of GEANT 3 with a graphics driver showing pictures of the detectors and showers. The user interface was based on a command line interface ZCEDEX [10] and a graphics system based on the Apollo graphics package. Both systems were developed by Pietro Zanarini. When we started the development of the PAW system in 1985, ZCEDEX was upgraded to KUIP and the graphics system converted to the HIGZ [19] package, another PAW component.

In the coming years and until 1993, GEANT 3 saw many more developments. With Francis Bruyant we made several unsuccessful attempts to vectorise the geometry and tracking system. The overhead in creating vectors and the bookkeeping was surpassing the gains coming from vectorisation (both on Cray and IBM systems). I already mentioned the exercise by Jean Luc Dekeyser to implement the geometry package in ADA. In 1989, we implemented an interface with the hadronic package of the FLUKA [3, 20] system. But the major work was in making the system more robust, implementing better physics algorithms for energy loss, multiple scattering, and tracking in non homogeneous magnetic fields. The development stopped in 1993. However GEANT 3 has been used for the design of all LHC experiments and is still used today by many experiments, in particular with the Virtual Monte-Carlo (VMC) system developed in 1997 by the LHC ALICE [26] collaboration.

From GEANT 3 to GEANT 4, a still unfinished transition.

In 1994, it was decided to start from scratch the development of a new system called GEANT 4, fully implemented in C++. It took many years for the GEANT 4 team to develop a system competing with the FORTRAN version of GEANT 3. Like most projects started in 1994, the team had to face the discovery of a new and immature language, move gradually to the Standard Template Library (STL) and upgrade year after year to the latest compiler generations. While GEANT 3 including the ZEBRA system was around 400,000 lines of code, GEANT 4 grew to more than one million lines of C++ and more than 1,000 classes. The current implementation is still facing severe performance problems, being about three times slower than GEANT 3. Compared to GEANT 3 several physics packages were improved with the addition of new models, in particular in the area of hadronic showers. GEANT 4 has not yet reached the necessary maturity to become a prediction tool without having to tune many parameters, but the situation is improving gradually. There are still many areas to be improved in the system. The graphics system to generate nice detector pictures is still batch oriented. The interactive control is quite primitive offering a regression of the KUIP system as a command line interface or a primitive Python interface. There is no Graphical User Interface. The possibility to save and retrieve the geometry data structures was implemented only recently and it is up to the user to implement input/output of classes generated by the system. The GEANT 4 and ROOT teams being in the same group, there is some hope that a better integration between the two systems will happen in the not too distant future. This will offer GEANT 4 better graphics capabilities, a better user interface and an object persistency solution.

The FLUKA system is the HEP standard for radio-protection and dose calculations.

The FLUKA system, originally developed in the 1980s by a collaboration between the Italian INFN and the CERN radio-protection group, was greatly improved in the middle 1990s to include state of the art physics for everything concerning radiation and shielding calculations. The main developers Alfredo Ferrari and Paola Sala have received strong support from INFN in Italy and after years of fighting with the GEANT 4 team, the two projects seem to evolve progressively with more cooperation. FLUKA is still implemented in FORTRAN and because FORTRAN is now a small niche market, it is becoming more and more difficult to port the system to the latest machines and compilers. The FLUKA system has been used extensively by the LHC design team, and by most LHC experiments to evaluate the level of radiation in the detectors. FLUKA is also used by the ALICE collaboration and the new collaborations at the FAIR [63] machine at the centre for research on heavy ions (GSI) at Darmstadt.

The VMC system defines an abstract interface such that the same geometry description can be used with several navigators. The current implementations support interfaces with GEANT 3, GEANT 4 and FLUKA.[1] It is developed by the ROOT and ALICE teams. Besides ALICE, a growing number of experiments, e.g. MINOS [53] at FNAL, OPERA at Gran Sasso National Laboratory [65], and HADES [52], CBM [62] and Panda [54] experiments at GSI use the VMC system. Andrei and Mihaela Gheata from ALICE have developed an implementation of the GEANT 4 navigator system using a ROOT based detector geometry. This package G4ROOT [44] is now part of the standard VMC distribution. The ROOT geometry modeller has the big advantage that it can be used not only with several transport engines, but also in the programme reconstruction classes, event display systems or on-line software.

1.8 Graphics

For many years graphics systems were all batch and plotter oriented. A graphics metafile was generated and sent to a specialised minicomputer driving a pen plotter. The GD3 [79] library developed by Mike Howie and included in CERNLIB was particularly successful and could be smoothly ported to many generations of plotters between 1965 and 1985. The documentation of the GD3 system was produced by GD3 itself (an impressive work). Most packages developed in the 1970s and early 1980s were using the GD3 library as a picture generator. This was in particular the case of the HPLOT [7] package that was specialised in drawing histograms generated by the HBOOK [14] histogramming system. Carlo Vandoni and Rolf Hagedorn were developing since 1971 an interesting system called SIGMA [8] running only on CDC systems. SIGMA had an array based user interface and powerful graphics functions. One part of the SIGMA software was adapted to the PAW system later in 1986 by Carlo Vandoni.

When the first Tektronix terminals became available in 1974, the GD3 library was interfaced to the very basic Tektronix control sequences to display simple vector graphics; then in 1982, with the arrival of Apollo workstations, GD3 was interfaced to the Apollo graphics software. An interactive version of HBOOK and HPLOT called HTV was developed (the main predecessor of the PAW and ROOT

---

[1] It has to be noted that the interface with FLUKA breaches the licensing conditions of the FLUKA programme, and therefore it can be used only after having obtained a special permission from the authors. This permission has been granted only to the ALICE experiment at LHC.

system). HTV was running on Apollos, VAX/VMS and also the IBM under the TSO operating system. On the Apollo, HTV was upgraded to HTVGUI to exploit the bitmap capabilities of the machine. HTVGUI attracted the attention of Rudy Böck and Jean-Pierre Revol from the UA1 collaboration. Rudy and Jean-Pierre came with many suggestions for improvements of HTVGUI and this led to the proposal of the Physics Analysis Workstation (PAW) system in the fall of 1984.

> The beginning of computer graphics in HEP coincides with the advent of the graphics terminal and the personal workstation.

In 1981, some exotic and expensive devices were also available. In particular a Megatek display controlled by a VAX 750 machine was a big success in the UA1 collaboration. It was on this system that the first Z^0 and W particles were visualised in 1983. Jurgen Bettels and David Myers were the main engineers developing the high level interfaces to operate the Megatek driven from the VMS machine. Jean-Pierre Vialle, a physicist from the UA1 collaboration, implemented the experiment specific software running under HYDRA to generate the graphics primitives on the Megatek. The early 1980s were quite confusing years in the area of graphics. A standardisation effort in the U.S. produced the CORE [23, 78] system defining the basic principles behind graphics 2-D and 3-D architectures. Andy Van Dam from Brown University had been very active in this standardisation process and his book quickly became the bible for anybody dealing with 3-D graphics systems. At Brown University, Andy bought about 100 Apollo workstations to teach graphics principles to his students. Each student had his own machine on the university's Apollo Domain network.

At the same time we were also buying many Apollo machines at CERN and Andy was invited for several weeks to drive us in the right directions. However at the same time a European sponsored initiative produced the GKS [24] (Graphical Kernel System), one of the main topics at the newly born EuroGraphics conferences organised to counter-balance the powerful SIGGRAPH [71] association in the U.S.. Carlo Vandoni was an active member of EuroGraphics. Together with Jurgen Bettels and David Myers, our graphics experts, they pushed CERN to write software consistent with the GKS standard and they decided to buy the GTSGRAL software system from a company in Darmstadt.

> The HIGZ system, a high-level graphics visualisation package.

GKS had been proposed as a follow-on (I would rather say as a competitor) to the CORE system. While the CORE system had been designed mainly at the time of vector graphics, GKS was attempting to cover more modern hardware such as bitmap workstations. However a better proposal emerged. Proposed by the Massachusetts Institute of Technology, the X11 system appeared in 1984 as a better choice than GKS. When the PAW project was proposed in 1984, however, our mandate was to use GKS and nothing else. CERN had to support a European product, full stop. However, we managed to circumvent this political problem by implementing the HIGZ package. HIGZ provided a standard API for all functions required by PAW or GEANT 3 and we had interfaces to the CORE system (thanks to Harald Johnstadt from FNAL), GKS, X11 and later the PHIGS [82] system too.

The GTSGRAL package was not particularly good, containing many bugs across successive releases, and the performance was not the one expected. In addition a commercial license was necessary, such that very rapidly the X11 HIGZ interface became popular and the GKS interface was left to the IBM and VMS systems only. Except for the UA1 case with the Megatek, 3-D graphics was marginal at CERN, because expensive. A 3-D version of GKS became available, but its performance was far from expectations. Jurgen Bettels and David Myers proposed the PIONS [5] system (Partial Implementation of Our New System). This package was used as a successor of the original Megatek software. It featured a client-server approach. Despite the big effort involved in this development, nobody could ever understand what were the attractive features of PIONS. Probably Jurgen realised this and decided to leave CERN to join the DEC company. In the early 1990s, a new standard for 3-D graphics emerged, PHIGS. We implemented an interface HIGZ/PHIGS, but this interface remained anecdotal too.

> Graphics standards have been for many years a moving target due to the rapid evolution of the hardware, both CPU, storage and terminals.

For many years, Graphics meant passive vector graphics only. Although the hardware and standards like GKS were supporting the concept of segments holding graphics data structures that could be erased separately without clearing the entire screen, this possibility was rarely used in our environment. It was sometimes used by some early primitive graphical user interfaces to display dynamic menus, but this type of interface was never very popular. With HTVGUI, we had prototyped a more advanced style of interface exploiting bitmap capabilities, but because we had to be GKS-compatible, this work was never moved to the PAW system, despite the fact that this type of interface became popular with the first Macintosh machines. Because we had to provide solutions for the majority, and the majority was using VT100 compatible terminals, the committee launching PAW in 1984 required a graphical user interface compatible with VT100 capabilities that was really outdated at this time.

For many years the main interface provided for systems like PAW and GEANT was a command line interface only. It was only in 1991 that we implemented PAW++ that offered a MOTIF [69] based graphical user interface. MOTIF was the newly proposed standard by the Open Software Foundation (OSF) group. OSF was assumed by many to become the next standard operating system. In fact the only implementation came from DEC on ALPHA machines. MOTIF was built on top of X11 and provided advanced widgets to design nice user interfaces. MOTIF was quite complex and came with a large memory foot-print. The system is still in use today on X11 based systems, but more modern systems have now appeared on the market. On Windows systems, Microsoft Foundation Classes (MFC) provide all the widgets to design nice interfaces. The latest implementations on Windows support modern fashionable interfaces with 3-D effects, vanishing windows, non rectangular and semi transparent windows, etc. Apple is known, since the Macintosh success in 1984, for the most elegant and advanced user interfaces in their successive operating systems. However, not many applications in our field provide direct interfaces to these Microsoft or Apple widgets.

With the advent of Linux or Linux-like systems, new Graphic User Interface (GUI) systems have emerged. The most popular is QT [49] from a Norwegian company TrollTech. QT is the basis of KDE [64] the most popular windowing system under Linux. Some applications in HEP have direct interfaces to QT. Several versions of QT have appeared on the market, causing some confusion. For example the Linux RedHat [70] distributions until recently featured old versions of KDE built with an ancient version 3.1 of the QT system. A newer version QT4 has been released but is incompatible with the previous version. QT has pioneered the so-called "signal-slots" mechanism that replaces the traditional callback mechanism in MOTIF or the messaging system in MFC. Objects can emit signals (for example when the mouse leaves or enters a window, or when the mouse is clicked on top of an object). Users can subscribe to any signal and implement a "slot" function that will be called automatically by the system when the signal is emitted. Compared to the traditional callback mechanism, this has the advantage of separating the code in the class emitting the signal from the receiver code and making implementations of user interface systems more modular. In January 2008, the TrollTech company was bought by Nokia. It is likely that the company will target the new developments of QT for mobile phones, making its future rather uncertain in our field. This proves once more that our field can hardly rely on commercial systems over ten or twenty years.

1.9 Interactive Data Analysis

In HEP or Nuclear Physics, a data analysis process loops over a long list of events, each one being the description of the results of a collision. The typical analysis scenario consists of creating a set of histograms and conditionally fill these histograms in the event loop. The physicist applies cuts to some variables to find possible correlations and repeats this process until an interesting subset (possibly just a few events) has been identified.

> Interactive Data Analysis, a new dimension for physics data mining made possible by PAW.

Traditionally this analysis process was performed with short jobs and the main tool used for this work was a histogramming package. The most popular package used for this task was HBOOK for many years and the system is still widely used today within the experiments that did not move to C++. I was hired as a CERN staff member in July 1973 into a group of the "DD" (CERN Data Division) developing special hardware processors that were supposed to be used for the trigger and fast reconstruction of the R602 experiment at the Intersecting Storage Rings (ISR) accelerator.

The ISR was the first particle collider in the world when it entered operation in 1971, producing a huge (for the time) amount of data. Many people thought that the only way to select the good events and analyse them quickly would be special hardware processors. Hence my job was to collaborate on this development. A career in the electronics domain seemed to be more favourable than computing, that was then considered more as a formula computation system than an event processing system. However, for many reasons, the delivery of some hardware components did not happen in time, and anyhow the principle was flawed. As a result, I moved from hardware electronics to software within a few months.

During my thesis work, I had to analyse data coming from a very simple telescope of two detectors: one germanium detector measuring the energy loss of particles produced by the collision of 600 MeV protons on various targets, and another NaI (Sodium Iodide) detector measuring the total energy of the stopping particle. The analysis was pretty simple and consisted of producing correlation plots of the energy loss versus the total energy to identify particles like protons, deuterons and alpha particles. I had developed a simple tool called BIDIM to display the correlation plots on a line printer. So I was already an expert in histogramming systems. Given the fact that the job for which I had been hired was not a full time occupation and in agreement with my supervisors, I started the development of the HBOOK system. In 1974, together with Paolo Palazzi who was just back from a year of sabbatical leave at SLAC, we considerably extended the package (he wrote the documentation and I wrote the code).

> The birth of HBOOK, the HEP standard histogramming package for more than 30 years.

The system started to be widely used very soon for the data analysis of the ISR experiments and later for the SPS experiments. For the bubble chamber experiments, a system called SUMX, developed originally at Berkeley, was the main analysis engine. Data were written in a special but simple format. SUMX was a kind of pre-processor where users specified via input data cards the quantities to be histogrammed and the set of cuts to be applied. This was indeed a good idea that did not imply writing complex code for a simple analysis. However, the system was quite inflexible for more sophisticated data analysis. Users quickly showed their preference for a histogramming library with which they could create and fill the histograms according to their special needs.

The people developing the HYDRA system realised this trend too, but the HYDRA histogramming package was not as simple to use as HBOOK and HBOOK won the contest. Together with Howard Watkins, we extended HBOOK with the graphics system HPLOT based on the CERN GD3 system. A huge quantity of HPLOT output was produced on the rather expensive Calcomp pen plotters. A 35 mm microfilm output system was introduced in 1978 together with a few microfilm viewing machines in the computer centre. This system became obsolete when more graphics terminals became available.

With more and more people using the HBOOK and HPLOT systems, an increasing number of requests appeared for a more interactive system. We already had a system called TV to visualise the GD3 metafiles. It was possible via simple commands like "v 7" to show the picture number 7 in the GD3 metafile, but it was not possible to change or improve the output format. In 1980, I started the development of HTV (HBOOK/HPLOT +TV). HTV featured a simple command line interface of the style "Menu/Command param [optional param]". With HTV it was possible to view a histogram inside a HBOOK file with any of the HPLOT supported formats and generate a GD3 metafile. This was indeed a big progress, but of course users were requesting more and more. In particular, it was quite tedious to repeat the same commands to set the viewing attributes. The simple command line interface was extended with the participation of Pietro Zanarini and became the ZCEDEX system to support the execution of macros.

ZCEDEX was rapidly used by HTV and also by GEANT 3. In fact, GEANT 3 was developed from day one as an interactive system, making it hard to believe that 15 years later and despite the advances in technology, the GEANT 4 system was developed as a pure batch system with no tools to help the debugging process. As indicated earlier, the HTV system was extended to HTVGUI on the Apollos with a graphical user interface showing the menus and commands as pull down menus. In 1984, it became obvious it had to be further extended. The ZCEDEX system had grown from a trivial menu/command interface to a more complex package supporting optional arguments and a help system, but it was not possible to execute macros with conditional code based on some global or local variables.

> The quest for more interaction with the computer graphics: HTV and ZCEDEX. A new paradigm slowly comes into existence.

It was not possible to fill histograms based on some cuts. Everything had to be done in batch compiled programmes a priori. The previous SUMX users were requesting SUMX like facilities to create and fill histograms based on a similar cut logic. Erik Baessler from DESY had already developed the GEP system. GEP featured a simple n-tuple[2] analysis facility. The data to be analysed had to be stored into vectors (n-tuples) and it was possible to create histograms, by looping on these n-tuples, and apply cuts at the same time. For many years Rudy Böck had been very active in developing and coordinating the development of software for bubble chambers. Since 1980, he was an active member of the UA1 collaboration. He was very interested by the developments around HTV and HTVGUI and was pushing for a possible common development with the GEP system. We visited DESY together in the autumn of 1984. We attended presentations of GEP, and the way it was used at the DESY network. GEP was implemented in PL1 on the IBM machines. Erik was a strong supporter of PL1 and did not want to discuss any other language than PL1 for any future collaborative project. So Rudy and I came back from DESY, convinced that it was impossible to embark in a common development based on the PL1 language only available on IBM and somehow on VAX. We had many more machines to support and also the graphics system used by GEP, although with a lot of features was not the right direction.

Rudy set up a small committee including Carlo Vandoni, Luc Pape, Jean-Pierre Revol and myself to design a new system based on our previous experience with HTV. After a few meetings, the recommendation of the committee was to develop the PAW system, originally based on HTV and ZCEDEX, but with a menu interface appropriate for the VT100 terminals and the graphics system based on GKS, but with no support for n-tuple analysis. The committee was supposed to watch the development of the new system. However, as I was not happy with the conclusions of the design phase, I managed to escape the committee control process and PAW was developed quickly in substantially different directions, but it took us at least two years to recover from the limitations in the original design.

Pietro Zanarini upgraded the ZCEDEX system (meanwhile renamed KUIP: Kernel User Interface Package) to support variables and conditional macro execution. KUIP was continuously extended until 1993 and gradually rewritten in C by an exceptional programmer Alfred Nathaniel.

---

[2] An n-tpule or ntuple is an ordered sequence of n elements. Each sequence represents a given physical object, most frequently an event or a track. n-tuples can be considered as lines of a matrix whose length is determined by the number of "objects" represented. This representation is very practical for High Energy Physics data. The most "natural" way to store an n-tuple is "row-wise", i.e. rows are stored one after the other on disk. This is not however the most "efficient" way to store the data, particularly taking into account the fact that, most of the time, only a few of the "columns" are used in a given analysis. Better efficiency can be reached by storing the data "column-wise", i.e. the data belonging to the same "column" are stored sequentially on disk. This technique provides substantial gains in performance: however it substantially complicates data handling for the application.

The low level graphics system HIGZ offered a standard layer to many machine and system dependent libraries like GKS and X11. HIGZ was considerably extended over the years, in particular the X11 interface that rapidly became the de facto standard.

> PAW becomes the standard analysis tool for High Energy and Nuclear Physics for more than 20 years.

We extended the HBOOK system to support row-wise n-tuples in HBOOK files and we rapidly implemented utilities to loop over n-tuples with pre-defined cuts. Row-Wise N-Tuples (RWN) were somehow similar to a table in a relational database and our n-tuple query system looked like a small subset of SQL. Very soon, we realised that RWN had a big performance penalty when processing only one or a few columns from the n-tuple set. In 1989 we implemented a new format CWN (Column-Wise-N-tuples) where each column had its own buffers. The FORTRAN interpreter COMIS [4] (developed by Vladimir Berezhnoi from Protvino in Russia) was an interesting complement to KUIP macros when more and more complex analysis tasks had to be executed. A typical analysis command looked like "N-Tuple/Draw 10 select.f" where 10 was the n-tuple identifier and select.F a FORTRAN file containing the analysis function to be called for each event. The COMIS interpreter was typically 5 to 20 times slower than the native FORTRAN compiler. To speed up the process, we implemented a direct interface to compile the select.f file on the fly with the native compiler. The above command had to be changed simply to "N-Tuple/Draw 10 select.f77".

After a difficult start, PAW rapidly became a success and was the main tool for data analysis for nearly 20 years.

In 1991, we implemented the PIAF [9] system (Parallel Interactive Analysis Facility). PIAF was an extension to PAW allowing the analysis of large n-tuple files in parallel on a small farm of nine HP workstations connected with a high speed HiPPI network. The hardware was a gift from HP that considered PIAF as a very interesting development at the time. This subject will be developed in detail in Chap. 7. Most developments around PAW stopped in 1994 with an attempt to replace the KUIP system by a TK/TCl interface instead, but this interface had plenty of limitations and was never used in production.

1.10 Development of the ROOT System

1994 was an interesting year, somehow the end of the old software world, but also the beginning of the "dark years". By the end of the year the CERN Detector Research and Development Committee (DRDC) had approved the two major projects: RD44 (GEANT 4) and RD45 (object-oriented databases). After having managed the two major projects GEANT 3 and PAW for so many years, it was clear to me that I had to look into different directions. However, it was not easy to move to a different field. My interest remained in the development of general purpose software systems, but what could the next one be? For a few months, I considered the possibility of joining the ATLAS experiment, but the leaders of the two Divisions involved in my move thought that my presence in ATLAS could be seen as an obstacle to their two new born projects. The Leader of the Computing Division, David Williams, invited me very politely to take a sabbatical leave to Berkeley or SLAC. It was clear to me that he was making a lot of effort to remove me from the software arena, at least for several months. In October 1994, he invited me to join the NA49 experiment, in desperate need of software manpower. Looking at my face, David certainly realised that I was not happy with his proposal and effectively I declined his offer. However after one week of further thinking, I realised that it was, maybe, an opportunity. All NA49 members had offices on CERN's remote Prevessin site. Moving to this site was a good way of disappearing from the main CERN site and possibly getting an opportunity to start something new. I sent a mail to David, telling him that I was finally accepting his offer. This was also an opportunity to clean my office. I took my HP workstation in my car with a few other documents and after a brief discussion with Andres Sandoval (NA49 software coordinator), I installed my system in a nice quiet office with a view of a nice pine tree frequented by squirrels. This was indeed a good environment to think and to work hard. Like everybody else, I had followed a few courses about object-oriented programming, but C++ was new to me. I thought that the best way to learn the language was to re-implement in C++ a large subset of the HBOOK package.

The long road to C++ and object-oriented. Revolution or evolution?

I spent one month in a first implementation that looked more like FORTRAN with semi colons than good C++. Following some remarks from some C++ aficionados, I jumped to the other extreme with an attempt to use a templated version of my classes to more elegantly support the different histogram types. In 1994, the best C++ compiler was the HP implementation. Igor Stepanov had implemented a template system while working with HP before his move to Silicon Graphics where he extended his system that eventually became the Standard Template Library accepted by the C++ committee. The first compilers digesting STL appeared only much later (in 2002 at best) and support for templates in general was not appropriate. After one month of hard work with my templated implementation, I concluded that it was quite inefficient and anyhow not portable to other compilers with no template support at all. In January 1995, I changed the templated version for a more conventional C++ implementation with a base class and derivatives for each histogram type.

I was quite happy with this new version that was competing performance wise pretty well with the old HBOOK FORTRAN version. I decided to make more developments around this initial exercise to get my hands familiar with C++. ROOT was born. I found this name appropriate looking at the nice pine tree in front of me, and as some people suggested that this could be an acronym for "Rene object-oriented Technology", I had nothing against the principle. However, in my mind ROOT meant a system with solid roots, on top of which you can grow more goodies. At the end of January 1995, Fons Rademakers, with whom I had been working on PAW and PIAF since 1990, joined me in NA49. Fons's contract with CERN had expired, but because we had excellent working relations with HP, the company decided to sponsor Fons at CERN to work with me. This gift was initially for one year, but in fact Fons succeeded in various different ways to be funded by HP for many years until 2001 when he was finally offered a CERN position. Thanks to Fons's invaluable contribution, we spent several months designing, coding and recoding a first version of the ROOT system.

In November 1995, I presented the ROOT system in the Computing Division auditorium, completely crowded for this event. Fons and I had been ignored for about one year. Suddenly, several people realised that we were back and our new product generated considerable interest. In the first few months of 1996, the Computing Division decided to launch another project called LHC++ [72], with the clear objective to counter the development and the adoption of ROOT, and this was for us the beginning of a lot of trouble. In July 1996, the CERN director for computing, Lorenzo Foa, sent a letter to the four LHC experiment spokespersons stating that the ROOT project was only a private initiative that was accepted in the context of NA49, but that no support would be given to this product in the context of the LHC. In fact, this letter was an excellent thing for us.

> ROOT is officially forbidden at CERN for LHC, becoming the underdog of LHC computing, both a curse and a challenge.

All the other official projects were run under the umbrella of the LHC Computing Board (LCB) chaired by Mirco Mazzucato. The "official" projects had to report on a regular basis to the LCB. The LCB members had been active in their previous careers either as software coordinators or with other managerial duties. However, they had no experience with the new style software and could only rely on the experts' opinions that were all in the official projects. As a result, over the course of a few years, we witnessed an ineffective refereeing mechanism with reports indicating that the projects were making good progress, but nobody in this committee made a serious risk analysis until, in the late 1990s, it was clear that all projects were late, with products impossible to use and many more problems. In particular when the LHC++ project was stopped in 2002, it would have been interesting to see a report objectively analysing the reasons for its failure.

Meanwhile we were developing ROOT in many directions. Our previous experience in the development of large systems was a great help. In particular we were well aware of the limitations and main user criticisms against our previous projects. One of the key areas where we invested a lot of thinking was the Object persistency system. The ZEBRA I/O system had a major drawback making the implementation complex. Because of its FORTRAN heritage, direct access files had to be created with fixed block sizes (typically 4 kiloBytes). When writing out data structures, we had designed a complex "padding" system when writing arrays across block boundaries, in particular when streaming double precision floating point numbers. With C and C++ I/O, we could remove this complexity and simply write variable length blocks, leaving the task to the OS to flush out memory buffers to disk at the most appropriate time. The layout of a ROOT file was much simpler and could be described in a one page document. Of course, streaming C++ classes turned out to be far more complex than writing ZEBRA banks with embedded I/O descriptors. To achieve this, one must have a description of the C++ class, including the type and the offsets of each data member within an object. In a first phase in 1996, we decided to generate the necessary code to stream objects by using the information coming from our C++ interpreter CINT [43].

We implemented a tool called rootcint to do this job and this technique survived a few years. The technique was robust and simple to explain and use. We did not develop the CINT system ourselves. Via our HP contacts, we were informed that Masaharu Goto from HP Japan had already developed a C++ interpreter. We invited Masa to CERN for a first visit in April 1996 and we immediately took the decision to replace our very primitive C++ interpreter by Masa's implementation. We have continued our cooperation with Masa since. While implementing the ROOT I/O system, we were also following closely the RD45 project and in particular looking at the way people were using the Objectivity product, the flagship of the project.

Objectivity was writing each object to a file with an offset (16 bits) into one page, itself addressed with a 16 bits quantity inside a segment. Each object had a unique 64 bits identifier (the OID) that was a member of a base class OOobject to be included into all user persistent classes. This trivial and naive mechanism implied that the objects on disk were a direct copy of the objects in memory. Objectivity did not support an automatic class schema evolution mechanism. A change in a class schema implied a complete copy of the database referencing these classes. In addition the database system was a central system with a locking mechanism that was extremely expensive when supporting concurrent writes and reads. Sometimes one may have a doubt concerning the quality of an algorithm. One of the nice things with software is that one can always propose a better solution.

Object-oriented Database Systems are not suited for HEP data. ROOT and its lightweight I/O system has emerged as a clear winner.

However, in the case of Objectivity, it was clear from day one that the principle was wrong. In the ROOT team we were totally convinced that the system would fail and could not understand that a more "objective" analysis was not done. After having expressed our own criticisms (in particular at a dedicated LCB meeting in Padova in 1996), we decided to keep quiet and let people come sooner or later to the same conclusion.

The BaBar collaboration at SLAC had been the first large collaboration deciding to move to C++ in 1995. Following this courageous decision, they had to face many problems and in particular the implementation of Objectivity in their data management system. They received good help from the Objectivity company located a few miles away, but nevertheless more and more problems were accumulating. This fact was quite well known in 1998 when the CDF [50, 51] and D0 experiments decided to investigate their software approach for the so-called RUN II that was supposed to start in 2001 at FNAL. These two experiments created two projects: one aiming to find a solution for object persistency, and the second one for interactive data analysis. They made calls for proposals and analysed three solutions for each. During the CHEP conference in Chicago in September 1998, they announced that ROOT had been selected as the solution for the two projects. This announcement was immediately followed by a similar announcement from the four experiments at the Relativistic Heavy Ion Collider (RHIC) [57] at the Brookhaven National Laboratory [56].

These announcements, of course, generated a big turmoil at CERN. ROOT was becoming an embarrassing system that was now difficult to ignore. The following year saw the first signs of cracks in the official Objectivity-based line in BaBar. Marcel Kunze had developed a highly controversial ROOT-based analysis system called PAF (Pico Analysis Framework) that was orders of magnitude faster than the official tools reading data directly from Objectivity. At the same time, the Objectivity supporters at CERN and SLAC were engaged in a difficult negotiation with the company to get additional features badly missing in the system. With the 64 bits OID, it was not possible to create large databases. A request to increase the OID to 128 bits was never implemented (in reality this would have been quite difficult to implement in an efficient way). Meanwhile, BaBar moved from Objectivity to a ROOT solution and Andy Hanushevsky, the most experienced Objectivity user started an implementation of the xrootd [73] server. This will be described in detail in Chap. 10.

In the fall of 1999, the CERN Director for Computing, Hans Hoffmann realised that there was a potential problem in the medium term and he set up a general software review that lasted several months. We made a few presentations of ROOT to the CMS and ATLAS collaborations. ATLAS launched its own software review committee in January 2000. During this review I stressed the importance of tools to generate object dictionaries for object persistency, interpreters and graphical user interfaces. Not many people understood all the implications at the time.

> ROOT success comes from outside. The major experiments running in the U.S. decide to adopt ROOT as a software framework.

Most people were still thinking in terms of data dictionaries and a project was launched by an ATLAS team from LAPP in Annecy to create a new language ADL (ATLAS Description Language). This project went on for several months and was then stopped when the collaboration realised that the delivery of a satisfactory system would not be in time. Instead the collaboration decided to use the GAUDI [37] framework developed by the LHCb collaboration. However GAUDI in 2001 was a very primitive system, essentially an empty shell. ATLAS decided to use GAUDI as a model for their ATHENA [27] framework and to embark on the design of an extremely complex system called STOREGATE that was defining a protocol to put/get objects in a store. The system is still alive today, a big nuisance to most users and one of the reasons why ATLAS users are systematically looking for ATHENA-free solutions for data analysis.

In 2002, the LHC Computing Grid project [34] (LCG, see Chap. 3) was launched with a considerable amount of resources. About 80 new people were injected for LCG Phase 1. A few Research Technical Assessment Group (RTAG) meetings were organised and by June 2002, the so-called Blueprint [35] RTAG was published. This document was the result of a compromise between the pro-GAUDI and the previous proponents of the Objectivity system. ROOT was not excluded from the picture, but considered mainly as one of many possible alternatives to Objectivity. In LCG Phase 1 (2002–2005), we had the so-called "user-provider" relationship between ROOT and the other projects.

Several projects were launched in 2002 under the coordination of Torre Wenaus from Brookhaven. The SEAL [47] project was supposed to deliver a C++ reflection system. The PI [46] project, was the Physicist Interface without specifying exactly what that meant. The POOL [36] project was a neutral software layer on top of Objectivity, or ROOT or Oracle. In addition POOL had to deliver a file catalogue system.

At the end of LCG Phase 1, the Objectivity system had been abandoned and ROOT was the only serious alternative for object persistency and data analysis. In LCG Phase 2, ROOT became fully integrated in the LCG planning and substantial manpower resources were finally allocated to the project. The PI project had never produced anything tangible. The main product of the SEAL project, REFLEX, was supposed to be integrated in ROOT as a replacement of the CINT dictionary. This integration is turning out to be much more difficult than foreseen initially, and the project has been put on hold.

The above may give the impression that substantial time and resources have been lost in parallel developments. However this strategy, even if not the most "elegant", may well have been the best possible. For the ROOT team, it was great to be a challenger for many years, without the burden of official reports or internal reviews.

We greatly benefited from the numerous comments and criticisms that we took into account in our successive releases.

The number of people using the ROOT system increased drastically after 1998 and the system spread to other branches of science and into the financial world too. In 2010, we had on average 12,000 downloads per month of our binary releases.

ROOT has been in constant development and there are always requests for new features or improvements. Many developments were made in the I/O system. We moved from the primitive generated streamers to a system totally driven from our C++ reflection system in memory. We implemented support for automatic schema evolution and are still improving it. This is an essential feature when processing data sets generated by multiple versions of user classes. New data members can be added, removed, their type modified, moved to a base class or the inheritance scheme changed.

> The storage model is the heart of ROOT and one of the main reasons for its success. Schema evolution is an essential element of it, for a project that has to last several years.

The main ROOT storage model is a Tree. With Trees, the database containers are automatically built following the user object model. A Tree is the natural extension of the concept of Column-Wise n-tuple into the object-oriented world. Rows are no longer vectors of entities but classes, whose data members take the role of the entities in the n-tuple. The top level object is split into as many branches as data members. If a member is a collection, for instance an STL vector of objects, new sub-branches are created corresponding to the members in the STL vector class. This model has at least two advantages compared to conventional object streaming models as in Java: it is possible to read selectively only the branches used by the analysis algorithm and, because the leaves of the branches contain more homogeneous data types, a gain (typically 25%) is obtained during the data compression. The Tree split mode is now the standard model for all experiments.

In 2002, the QT graphical user interface system had been the recommended GUI toolkit. However, today, only a few applications use QT directly. ROOT has an interface to QT, but the vast majority prefer to remain QT free and use now the native GUI toolkit in ROOT that works on all platforms.

1.11 Frameworks and Programme Management

The development of software systems has always required discipline and conventions: a common code management and versioning system, a common data structure management system, a common data persistency system, a common user interface (command line or graphical), common mathematical libraries, etc. With the growing size of the experiments, common conventions and systems to manage the storage and access in memory of more and more complex data structures and algorithms has become a key requirement to avoid anarchy. For many years, this has been the task of dedicated systems such as HYDRA, ZBOOK, ZEBRA or BOS. Access to one data structure or data set was achieved by following pointers in a hierarchical data structure or by requesting this structure by name to a central service. With the advent of object-oriented programming, and parallel developments by independent teams developing the sub-detector systems (more on this in Chap. 5), the access to data structures has become more complicated. It became essential to use a publish-and-retrieve mechanism common to all sub-components, like a white board. This system must be such that accessing data structures across sub-components should not imply the inclusion of the corresponding class declarations, otherwise a simple change in one class layout would imply a complete recompilation of all sub-components. Different solutions have been implemented by the experiments using a mixture of access by name in class collections assuming some base class. These have also the advantage that it becomes possible to browse complex data structures via a graphical user interface. In a similar way, most modern frameworks provide a way to describe a hierarchy of algorithms and to browse them too. The class libraries including these algorithms can be dynamically loaded at run time.

> Access to complicated data structures is one of the major challenges in modern HEP computing. Object-oriented programming offers tools towards a solution, but probably there is still a lot of work ahead.

The GAUDI system developed by LHCb, and also used sometimes by ATLAS, is an example of a framework with tools to manage data structures and algorithms in memory. ROOT itself provides a TFolder class to organise objects and a TTask class to organise a hierarchy of algorithms. Much work still remains to be done in this area such that the infrastructure can also be used in parallel environments, in particular multi-threaded systems that are required to take advantage of multi-core CPU systems. The design of complex data structures is a non trivial task. One of the advantages of object-oriented programming has been to group in a class both data and functions. However it is a common requirement that objects generated by one programme have to be used by a different programme without having to link with all the libraries referenced by the code in this class. Most experiments came to the conclusion that basic objects shared by many components and a chain of different programmes should be only data objects with a minimum dependency on the code. The services to operate on these data objects can be loaded on demand via plug-in managers. Invoking these services is typically achieved via abstract interfaces that allow multiple and competing implementations of these services. The frameworks of the current large experiments in HEP have grown to an extent where only very few people understand the global structure, and installing a new release of these frameworks is a major task and may represent several days of frustrating work. One of the reasons is the excess of dependencies on external libraries and tools. This situation is really bad as it prevents porting the experiment software on new platforms and compilers in a short amount of time. Some large experiments are only able to run on a specific version of Linux with access to a common file system (AFS) and with one version of the compiler. As a result, there is a growing demand to provide a small subset of these frameworks for the vast majority of users who simply want to process data sets for physics analysis. Ideally users would like to analyse their data by using the ROOT framework only.

1.12 Trends and Challenges

If we look at the past few decades, we see a big latency in the processes of collecting requirements, making a design, implementation and the effective use of any system. It takes nearly 10 years to develop large systems like ROOT and GEANT 4, then it takes a few more years for users to become familiar with the new products and yet a few more years to see de facto standards emerging. For example the LHC experiments are now discovering the benefits of the "split-mode" style I/O with the ROOT Trees designed in 1995 and promoted in 1997.

This indeed poses a problem because while development cycles are so long, evolution in the Information Technology world happens at an amazing pace, and seems to be accelerating. A product started today will be ready in several years, and so one has to use a "crystal ball" to guess which will be the technology that, while offering a solid development platform today, will still be viable when the product is deployed. This exercise has proved very difficult, as can be seen by a few examples.

1.12.1 The Crystal Ball in 1987

1.12.1.1 Programming Language

FORTRAN 90X seemed the obvious way to go. It was unthinkable that HEP could move to another language than FORTRAN. We were all hoping that the FORTRAN committee would come with enough language extensions or new constructs that one could consider a simple replacement or upgrade of our data structure management systems like ZEBRA.

> 1987: FORTRAN, OSI Networking protocols and Massively Parallel Processing (MPP) computing.

1.12.1.2 Networking

OSI protocols to replace TCP/IP. TPC/IP was seen as a dead end and a non-scalable system. Networking gurus in the Computing Division were pushing towards the OSI protocols.

1.12.1.3 Processors

Vector or MPP machines. Massively Parallel Processors were considered as the only possible alternative in the medium term for our conventional mainframes or clusters of workstations. A tightly connected network of RISC processors seemed to be a likely solution.

1.12.1.4 Package Evolution

PAW, GEANT 3, BOS, ZEBRA: Adapt them to F90X. The obvious move for all these large packages was a gradual adaptation to the new FORTRAN standard.

1.12.1.5 Methodology Trend

The Entity Relationship Model (ER) was considered to be the state of the art. The ADAMO [22] system developed by Paolo Palazzi for the LEP experiment ALEPH had been adopted by a few experiments, in particular at DESY. ADAMO implemented a table system, using either BOS or ZEBRA underneath, and had many similarities with most front-end systems used with relational databases. The system had the advantage of providing good documentation for the tables and their relations.

But it was hard to anticipate that the Web would come less than 4 years later, or foresee the 1993/1994 revolution for languages and projects and the rapid growth in CPU power starting in 1994 (Pentium family).

1.12.2 Situation in 1997 and the New Crystal Ball

1.12.2.1 Programming Language

All the LHC projects were in the process of moving to C++. Some projects proposed to use Java when Just In Time compilers would be able to beat the static compilers. Some groups (ATLAS in particular) thought that it would be possible to use Java, C++ and FORTRAN90 in the same application. Java was very clearly gaining ground and many would see it as the future language.

> 1994: C++, Internet, OODB and the WEB. The Pentium CPU provides all the needed performance without the need of massive parallelism.

1.12.2.2 Databases

Starting in 1994 and following the decisions to move towards object-oriented programming languages, a huge effort to investigate the use of object-oriented databases (see Chap. 9) started in all experiments. In 1997 the hype peak for this project had been reached. The ODBMS Objectivity system was in use by the BaBar experiment at SLAC and also considered by most LHC experiments. A centrally managed event store seemed very attractive to many, despite the fact that wide-area networking was still very primitive (2 MegaByte/s) and not anticipated to grow as fast as it happened a few years later. However, the first signs of problems were appearing with BaBar: performance issues due to the fact that many programmes only reading data had to suffer from the locking mechanism assuming a general read and write situation, and reliability problems coming from programmes crashing during execution and leaving locks that had to be cleared manually. It is interesting to note that nobody at the time had in mind a distributed architecture like the ones available today with the Grid systems.

1.12.2.3 Commercial Tools

These were evaluated for data analysis and visualisation. The LHC++ project had been launched in 1996, mainly as a concurrent to the new born ROOT system started in 1995. A letter from the director of computing Lorenzo Foa to the LHC experiments was a clear message against ROOT.

The vast majority of users were very sceptical about the new official line. Most users were still programming in FORTRAN 77 and using PAW as a data analysis engine. The management in the experiments and the various committees (in particular the infamous LHC Computing Board) had not yet understood that the first thing to do to move users massively to an object-oriented environment was to provide a data analysis and visualisation system that could also deal with objects. The only large scale object-oriented package, GEANT 4, was not yet able (and still not in 2010!) to write objects or geometries on files.

It is hard to remember today that in 1997 a typical hardware configuration had about 256 MegaBytes of RAM. A typical big programme used less than 32 MegaBytes, and comprised less than 500,000 lines of code, statically linked or using less than 10 shared libraries. If users had been aware that 10 years later, programmes would reach the 2 GigaBytes range and several hundred shared libraries, it is likely that they would have been even more reticent to move to the new environment.

1 Technologies, Collaborations and Languages: 20 Years of HEP Computing

Fortunately a few people did not believe in the official line based on commercial systems including Objectivity. Despite the fact that the ROOT project was discouraged by the management, more and more people were considering it as a possible alternative in case all the other projects failed. The fact that Fermilab chose ROOT as a data storage and analysis solution in 1998 was a shock for many.

1.12.3 Situation in 2011

1.12.3.1 Move to C++

It took far more time than expected to move people to C++ and the new frameworks. The main reason, already mentioned above, is that there is no point in producing objects if there is no programme able to analyse these objects. The development of a large framework like ROOT has taken a lot of time and effort. A robust, efficient and distributed storage system supporting an automatic class schema evolution facility has been a big enterprise with evolving requirements once users discovered the potential of all the provided solutions. The implementation of a set of mathematical and statistical classes was a sine qua non condition before one could consider phasing out the FORTRAN-based CERNLIB routines. ROOT is the de facto standard for data storage and interactive analysis. However a big effort still remains to be done to make the package more robust and easier to use.

> 2011: Grid and Cloud computing; Parallel computing is back as clock speed is plateauing. Mega-frameworks provide new challenges in software development and management.

1.12.3.2 The Grid

The Grid projects started in 2000 have been major enterprises, consuming a lot of manpower with many failing systems. More pragmatic solutions have emerged with most of the time experiment-specific solutions to run bunches of batch jobs in parallel. See Chap. 3 for more details.

1.12.3.3 The Experiment Software

Experiment frameworks are becoming monsters that are more and more difficult to install. Running these frameworks on 64 bit machines and using less than 2 GigaBytes of RAM is a real challenge. Programmes take forever to start because of too much code linked (shared libraries with too many dependencies) and it takes time to restructure large systems to take advantage of plug-in managers. A normal user has to learn too many things before being able to do something useful. A general tendency is to develop a light version of the framework such that some interesting classes and utilities can still be used for data analysis and easily ported to new systems.

1.12.4 The Future Challenges

1.12.4.1 The Framework Evolution

The impressive changes in HEP software over the past 10 years (see Fig. 1.2) have required a big effort of adaptation from thousands of users. The trend towards larger and larger applications will continue. This will require more discipline to organise the application in a set of libraries that can be dynamically configurable such that the running code in memory is a small fraction of the total code. This will push for better services to organise collections of objects in memory and more formalism in the organisation of tasks. Data and tasks should be easily browsable to facilitate their understanding. Today we organise the code in shared libraries that can be dynamically loaded via a plug-in manager. It would be good if instead of using pre-compiled shared libraries, an application could reference only source files on the network. For instance a statement like:

```
use http://root.cern.ch/root520
TH1F hist("myhist","test",100,0,1)
```

will give a source path to be looked at when an unknown class (here TH1F) is found. In this case the class containing the source for TH1F will be downloaded, compiled on the local system and its binary file cached for further executions. This simple feature would considerably simplify the installation of large systems.

1.12.4.2 The Quest for Performance

It looks like in the foreseeable future the only way to increase performance is to use the parallelism coming with the many variants of multi-core CPUs. Two levels of parallelism must be considered: coarse and fine grain. Coarse grain parallelism assumes that the problem can be decomposed into large units. For example, in HEP one can run the same code on many events in parallel or even on many tracks within each event. Fine grain assumes that data vectors are produced and short algorithms applied to these vectors. The problem is to produce large enough vectors such that the algorithms can run on independent data units without interference. This is in general quite difficult to achieve because at some point the result of a computation must be inserted into another vector at the level above. This requires a

1 Technologies, Collaborations and Languages: 20 Years of HEP Computing   47

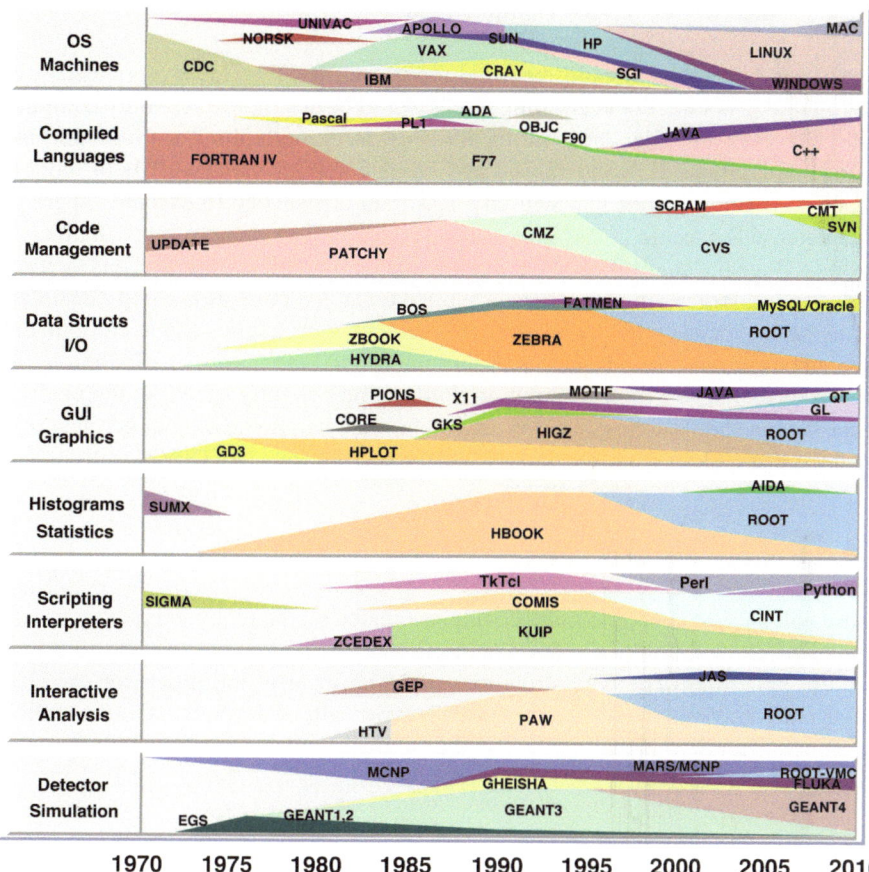

Fig. 1.2 A compilation of the main tools and packages used in HEP offline software since 1970. This compilation includes systems used by at least four experiments. The Y scale indicates the relative importance of the system with time

synchronisation mechanism. Most programmes designed so far have been designed for sequential processing. Changing the algorithms to be thread-safe or thread-aware is going to be a non-trivial task. Organising a programme in a set of processes running in parallel is a bit simpler (like in the PROOF [45] system) (see Chap. 7). The right mixture of multi-threading versus multi-processes will depend on the evolution of parameters like the amount of RAM, size of caches, disk and network speed. In addition to parallel architectures, we will continue to see more and more client-server type applications, e.g. data servers like xrootd or dCache [59] interfaces to relational databases for data set catalogues or calibration objects.

1.12.4.3 Interpreted Versus Compiled Code

Ideally an interpreter should be used only for the thin layer organising the programme structure. For everything else it would be desirable to use only compiled code. However compilers are still far too slow to compile the few thousand lines of code typically used in an analysis session. A compiler may take 10 s or more to compile an analysis code that will take less than one second to execute, where an interpreter will execute in 2 s. Interactive queries should not take more than a few seconds. Once the analysis code has been tested on a small data sample, it is then run on larger and larger samples where compiled code is a must. It is clearly desirable to run the same code with the interpreter in case of short queries and to compile this code when running longer queries. Using interpreted languages like Python in the data analysis phase is not the solution. Python may slightly speed-up the analysis phase for the short queries, but it is not a solution when processing larger data sets. We hope that a clean solution could be brought by product like LLVM [39].

1.12.4.4 Graphical User Interface Systems

These will have to support dynamic widgets created on the fly by interactive scripts. These widgets will be able to browse complex data structures and react to the objects in these structures. The data structures could be local to the processor or active objects on remote machines. More and more work will be devoted to GUIs to follow the general trend with new OS's and the impressive developments with graphics processors.

1.12.4.5 Class Reflection

These systems must be part of the language and the compiler. Once this becomes available in the standards, one will have to adapt systems like ROOT to use the new facility in the interpreters, input and output systems and also the graphical interfaces that depend on a large extent on the possibility to query at run time the functionality of a class.

1.12.4.6 Distributed Data Access

With more and more distributed processing on wide area networks with growing bandwidths but constant latency, sophisticated data cache systems will be developed to optimise not only the file transfer speed, but also, and in particular, the direct access to remote data sets. These techniques that start to appear in systems like xrootd or dCache will become increasingly important.

1.12.4.7 Distributed Code Development

Because applications are becoming more complex, written by many people anywhere on the Internet [1], stricter rules to control the code quality will have to be introduced. Code checkers, rule checkers and profilers are already essential elements for the LHC experiments. The development of test suites ranging from class level validation programmes to more and more complex suites will take a substantial fraction of the scientists developing simulation, reconstruction or data analysis applications. This will be the price to pay to avoid (or at least minimise) running millions of programmes on tens of thousands of processors to discover after a few months that all these programmes had one or more serious bugs making the resulting data useless.

1.12.4.8 Code Transparency

Users in large international collaborations will be willing to make this effort if the framework in use is not perceived as a huge black box, if it is easily portable to the user's laptop in a small amount of time (minutes and not hours), and, last but not least, is not too experiment-specific. Once users are convinced that their framework is solid and scalable, they are the first to make suggestions to improve it; otherwise they develop their own systems that they can manage to solve their own problems.

The last decade witnessed an intense and momentous development of HEP code. New programming languages have been introduced and HEP code has been migrated to the Grid. A large fraction of the software for the next decade is already in place or is being written and designed now. This development has been characterised by a long elapsed time between the code's design and its actual deployment in production. It is in some sense an important achievement in itself that the whole process has stayed "on track" for such a long time and has actually produced the intended results, while most of the actors have changed over time, moving into and out of the field. The stable development of such a large and complex software infrastructure has only been possible thanks to the Open Source development strategy. This has allowed a community of hundreds of developers and users to collaborate on the same products, ensuring both stability and smooth evolution.

The evolution of computer hardware is continuing at an amazing pace. However it is quite likely that we have reaped the low-hanging fruits of this evolution. The times when the next generation of processors would deliver performance improvements on the same code are probably behind us. Clock speeds seem to be plateauing around 3 GHz, and the opportunities for future performance improvements are linked to the exploitation of parallelism, at many different levels. This will certainly require an evolution of the software development paradigm and a redefinition of the role of the framework in HEP software architecture. The introduction of very powerful Graphics Processing Units (GPUs) further complicates the picture, as they offer the opportunity of obtaining very attractive performance gains, at the price of using special or constrained programming languages.

Most probably this will introduce additional requirements on code quality. We are already seeing that very complex and powerful languages, such as C++, require competence and substantial discipline from the code developers. Errors introduced by careless programming, or simply lack of understanding of the language syntax, can be very hard to spot. Variable scoping and virtual memory allocation and deallocation require very careful handling in large frameworks, and even "correct" user code can lead to unintended nasty side effects on programme memory footprint and performance. This calls for increased user training. FORTRAN could be learned in a week and, although writing good FORTRAN code required advanced computing skills, the amount of "damage" that a careless programmer could inflict on the framework was limited at least in the means if not in the results. To help in improving the quality of code and its readability during debugging sessions, "coding conventions" have been introduced by the different HEP experiments. These are indeed very useful to avoid common mistakes, however they are also very hard to enforce on users, who do not always understand their importance. Moreover, automatic tools to check code compliance with the coding conventions are not easy to find on the market, and several "conventions" are very hard to check automatically. The general lesson to be drawn is that computing training has to be substantially rethought, as well as code convention and architecture at a very early stage of any project.

1.13 Conclusions and Lessons Learned

Several years of software development, deployment and maintenance have taught us quite a number of lessons about producing software for High Energy Physics. We think we can summarise the major points as follows. A large software project cannot be created by a committee. Only a very small team sharing the same views can do it. It is essential to have a demonstration prototype as soon as possible to show the project directions. Once the prototype has been exposed to different working conditions, one can expand the development team step by step. The development team must demonstrate its ability to deliver soon and frequently new releases of their product. Development teams in HEP are distributed on all continents. The development team must put in place a support system using a problem tracking tool and must be ready to answer problems as rapidly as possible. The quality of the support is far more important than new features in the system. The development of a test suite for a large system is of vital importance. It may take as much time to develop as the target software itself. It must include not only unit tests like a class or a package but also real examples testing the combination of all features. Code coverage tools are important to check for untested code. Automatic nightly builds and tests are vital in today's large systems. The development team must proceed in a way which generates as much feedback as possible from all categories of users. In particular negative feed-backs are of crucial importance. They may be the signs of misunderstandings, internal politics, or may be just the result of poor documentation. Once a project is flying, it is important to not sleep on its success. The computing field is moving rapidly, new techniques may appear, forcing developers to make new interfaces or optimise the performance. Current software projects in HEP last far more than 20 years. When a project expands in scope and manpower, it is important to preserve the internal communication and the team spirit. Participants must see the project as their project and be highly motivated to improve it. It is probably the most difficult task for the project manager to sustain the creativity and the motivation of the team members.

References

1. Abbate, J.: Inventing the Internet, New edn. MIT Press, Cambridge (2000)
2. Andreessen, M., Bina, E.: NCSA Mosaic: A Global Hypermedia System. Internet Res. (Bingley, U.K.: Emerald Group Publishing Limited) **4**(1):7–17 (1994) ISSN 1066-2243
3. Battistoni, G. et al.: The FLUKA code: Description and benchmarking. In: Proceedings of the Hadronic Shower Simulation Workshop 2006, Fermilab 6–8 September 2006, M.Albrow, R. Raja (eds.) AIP Conference Proceeding 896, 31-49, (2007)
4. Berezhnoi, V., Brun, R., Nikitin, S., Petrovykh, Y., Sikolenko, V.: COMIS, Compilation and Interpretation System CERN Program Library L210 (1988)
5. Bettels, J., Myers, D.R.: The pions graphics system. IEEE Comput. Graph. Appl. **6**(7), 30–38 (1986)
6. Bos, K.: The Moose project. Comput. Phys. Comm. **110**(1-3), 160–163 (1998)
7. Brun, R., Couet, O., Cremel, N.: HPLOT User Guide – Version 5. CERN Program Library, Y251 (1988)
8. Brun, R., Couet, O., Cremel, N.: HPLOT User Guide – Version 5. CERN Program Library, Y251 (1988)
9. Brun, R., et al.: Data mining with PIAF. Computing in High Energy Physics '94, 13–22. Lawrence Berkeley Lab. - LBL-35822, San Francisco (1994)
10. Brun, R., et al.: ZCEDEX User's Guide, CERN DD/EE/80-6
11. Brun, R., et al.: GEANT3 – Detector Description and Simulation Tool. CERN Program Library Long Writeup W5013. http://wwwasdoc.web.cern.ch/wwwasdoc/geant_html3/geantall.html
12. Brun, R., et al.: HTV – Interactive Histogramming and Visualisation. Documentation file not available anymore.
13. Brun, R., et al.: PAW – Physics Analysis Workstation. The Complete CERN Program Library. Version 1.07 (1987)
14. Brun, R., Lienart, D.: HBOOK User Guide – Version 4. CERN Program Library, Y250 (1988)
15. Brun, R., Zanarini, P.: KUIP – Kit for a User Interface Package. CERN Program Library, I202. (1988)
16. Brun, R., Zoll, J.: ZEBRA – Data Structure Management System. CERN Program Library, Q100 (1989)
17. Brun, R., Brun, M., Rademakers, A. CMZ – A source code management system. Comput. Phys. Comm. **57**(1-3), 235–238 (1989)
18. Brun, R., Carena, F. et al.: ZBOOK - User Guide and Reference Manual Program Library Q210. CERN (1984)
19. Böck, R., Brun, R., Couet, O., Nierhaus, R., Cremel, N., Vandoni, C., Zanarini,P.: HIGZ – High level Interface to Graphics and ZEBRA. CERN Program Library, Q120 (1988)
20. Fassò, A., et al.: FLUKA: a multi-particle transport code CERN-2005-10 (2005), INFN/ TC_05/11, SLAC-R-773
21. Fesefeldt, H.: GHEISHA RWTH Aachen Report PITHA 85/02. L3 Collab. (1987)

22. Fisher, S.M., Palazzi, P.: Using a data model from software design to Data Analysis: What have we learned? Comput. Phys. Comm. **57** (1989)
23. Foley, J., van Dam, A.: Fundamentals of interactive computer graphics. Addison Wesley Longman Publishing Co, MA (1982)
24. Hopgood, F.R.A.: Introduction to the graphical kernel system (GKS). Academic Press, London (1983). ISBN 0-12-355570-1
25. http://aleph.web.cern.ch/aleph
26. http://aliceinfo.cern.ch/
27. http://atlas-computing.web.cern.ch/atlas-computing/packages/athenaCore/athenaCore.php
28. http://atlas.ch
29. http://cms.web.cern.ch/cms/
30. http://delphiwww.cern.ch/
31. http://geant4.web.cern.ch/geant4/index.shtml
32. http://http://monalisa.caltech.edu/
33. http://l3.web.cern.ch/l3/
34. http://lcg.web.cern.ch/lcg/
35. http://lcgapp.cern.ch/project/blueprint/
36. http://lcgapp.cern.ch/project/persist/
37. http://lhcb-comp.web.cern.ch/lhcb-comp/Frameworks/Gaudi/http://ieeexplore.ieee.org/iel5/23/34478/01645038.pdf
38. http://lhcb.web.cern.ch/lhcb/
39. http://llvm.org
40. http://na49info.web.cern.ch/na49info/
41. http://opal.web.cern.ch/Opal/
42. http://petra3.desy.de/
43. http://root.cern.ch/drupal/content/cint
44. http://root.cern.ch/drupal/content/g4root
45. http://root.cern.ch/drupal/content/proof
46. http://seal.web.cern.ch/project/pi/
47. http://seal.web.cern.ch/seal/
48. http://subversion.apache.org/
49. http://trolltech.com/products/qt
50. http://www-cdf.fnal.gov/
51. http://www-d0.fnal.gov
52. http://www-hades.gsi.de/
53. http://www-numi.fnal.gov/
54. http://www-panda.gsi.de/
55. http://www-public.slac.stanford.edu/babar/
56. http://www.bnl.gov
57. http://www.bnl.gov/rhic
58. http://www.cmtsite.org
59. http://www.dcache.org/
60. http://www.egs.org/News.html
61. http://www.fnal.gov/
62. http://www.gsi.de/forschung/fair_experiments/CBM/index_e.html
63. http://www.gsi.de/portrait/fair.html
64. http://www.kde.org/
65. http://www.lngs.infn.it/
66. http://www.mncnp.org/News.html
67. http://www.nongnu.org/cvs
68. http://www.objectivity.com/
69. http://www.opengroup.org/motif/
70. http://www.redhat.com/
71. http://www.siggraph.org/

72. http://wwwasd.web.cern.ch/wwwasd/lhc++/indexold.html
73. http://xrootd.slac.stanford.edu/
74. http:/root.cern.ch
75. 10th International Conference On Computing In High-Energy Physics (CHEP 92) 21–25 Sept 1992, Annecy, France
76. Kali's, M., Whitlock, P.: Monte Carlo Methods 2nd edn. Wiley VCH, Weinheim (2008)
77. Metcalf, M., Reid, J.: Fortran 90 Explained Oxford University Press, Oxford (1990)
78. Michener, J., van Dam, A.: Functional overview of the core system with glossary. ACM Comput. Surv. (CSUR) **10**(4) (1978)
79. Miller, R.: GD3 Program Library long writeup J510. CERN Data Handling Division (1976)
80. Moore, Gordon E.: Cramming more components onto integrated circuits Electronics Magazine, p. 4 (1965)
81. See for instance http://www.nag.co.uk/sc22wg5 and links therein.
82. See PHIGS on http://www.iso.org/iso
83. Shiers, J.: FATMEN - Distributed File and Tape Management. CERN Program Library Q123. CERN (1992)
84. Shiers, J.: HEPDB – Database Management Package. CERN Program Library entry Q180 (1993)
85. Wellisch, J.-P., Williams, C., Ashby, S.: SCRAM: Software configuration and management for the LHC Computing Grid project. Computing in High Energy and Nuclear Physics, La Jolla, California (2003)
86. Zoll, J., Böck, R. et al.: HYDRA Topical Manual, book MQ, basic HYDRA. CERN (1981)

Chapter 2
Why HEP Invented the Web?

Ben Segal

We are going to tell part of the story, little-known by most people, of how one of the most profound and revolutionary developments in information technology, the invention of the World Wide Web, occurred at CERN, the High Energy Physics laboratory in Geneva.

In fact one man, Tim Berners-Lee, invented the Web, not "HEP". So our question should really be re-phrased as: "What was the influence of HEP in leading to the Web's invention?" In the discussion that follows, we will make use of some personal recollections, partly my own but also those of Sir Tim Berners-Lee himself ("TB-L" in what follows) in his book "Weaving the Web" [1] (abbreviated below as "WtW").

The main development and prototyping of the Web was made in 1989–91 at CERN, which by that time had become the world's leading HEP laboratory. But the ideas for what became the Web had germinated and matured in TB-L's mind for many years before those amazing three months in the winter of 1990–91 when the code was written and tested for all the elements composing the Web, namely a browser, a server, a naming scheme (URI, now URL), a hypertext markup scheme (HTML) and a transport protocol (HTTP).

2.1 Introduction

The Web's invention, like many other such leaps, was in fact "coincidental". It was certainly not ordered, planned or anticipated in any way by "HEP", by CERN, or by TB-L's programme of work there. But for that particular leap to occur, its author's inspiration alone would not have been enough without a certain number of

---

B. Segal (✉)
CERN, Geneva, Switzerland
e-mail: B.Segal@cern.ch

supporting pre-conditions, and all these existed at that time at CERN. What were these essential items?

1. Managerial tolerance and vision.
2. A pioneering approach to certain technologies important for HEP.
3. A tradition of pragmatism in working style.
4. A movement from "proprietary" to "open" standards in computing and networking.
5. A link to the new "open-source" movement.
6. Presence of the Internet at CERN.

We will look at each of these important areas in more detail, and try to decide the importance of their influence as we go along. Let us also remember that we are discussing events which took place 20–30 years ago, the equivalent of many "generations" of technology development.

2.2 Item 1: Managerial Tolerance and Vision

CERN is basically a research organisation, but for physics and engineering, not for computer science. The *working style* that TB-L found in the CERN community was decidedly positive. Quoting TB-L directly: "I was very lucky, in working at CERN, to be in an environment ... of mutual respect, ... building something very great through collective effort that was well beyond the means of any one person – without a huge bureaucratic regime". (WtW, p.227).

Managerial tolerance and support was strongly shown from the start by one CERN group leader, Mike Sendall, but much less by any higher CERN levels. Sendall was actually TB-L's first and last manager during his CERN career. In fact, the "spare time" that TB-L used for his Web project was below the threshold of bigger bosses and Sendall's role was crucial: without his quiet encouragement it is fair to say that the Web would probably not have been born at CERN. Later on, when more senior CERN management became involved in discussions with the Massachusetts Institute of Technology, the European Union, and other parties concerning the future of the Web and the W3C, Sendall was again a key player. But even he did not go so far as to insist that the Web's long-term future lay at CERN.

Vision was absent managerially, as practically nobody recognised the actual potential of the Web project. The essential source of vision was TB-L's own, and he held to it with consistency and determination. Thus practically no "official" CERN resources could be found to support the work, either at the start or later. Two students (Nicola Pellow and Jean-François Groff) were "poached" early on from unrelated projects, and both worked for some months with TB-L, but without serious managerial commitment. The only "planned" application of the early Web work at CERN was as a gateway allowing access from VMS, Apollo and other machines to the very popular "XFIND" and "XWHO" phone and staff look-up facilities which had been developed and were running on the IBM VM/CMS service. The success of this gateway (1991) helped to publicise the Web's existence; even demonstrations using the primitive line mode browser convinced some people who used it that the Web perhaps had a future.

Nevertheless, the "managerial space" created by Mike Sendall allowed Tim to continue, and chance also aided Sendall's support for the project, by finding TB-L his principal collaborator, Robert Cailliau. Sendall and Cailliau had moved together from the Computing and Networks division (CN) to a new division Electronics and Computing for Physics (ECP) in 1990 and when Sendall found that Cailliau had done some independent work on hypertext, he introduced him to TB-L and encouraged their collaboration, even though collaboration across divisions was most unusual and therefore again informal.

> Managerial tolerance and vision create the space for innovation.

Cailliau proved an excellent complement to TB-L, not only in the technical realm. He organised International Web Conferences from 1994 and helped to handle developing relations with the European Union. Later on, when Paolo Palazzi became Robert Cailliau's manager in ECP Division, he also recognised the potential of the Web, allocated some students to work on it, and housed the whole team including TB-L and Robert. But TB-L's own computing division CN never went so far, except for allowing TB-L to spend some time in the U.S. from 1991 to spread the word about the Web and meet collaborators there.

Outside CERN, another HEP site (SLAC) very rapidly picked up the WWW scent and started some work on it, again in an informal style based on the interest of a few individuals, Tony Johnson, Paul Kunz, Bebo White and Louise Addis. The SLAC work offered access via the Web to physics reprints hosted on a large database called SPIRES, thus attracting many users. Other HEP sites followed slowly, but with no organised momentum.

We conclude that there was just enough managerial tolerance and vision to allow the work to get started and make some progress, but that the future of the WWW in HEP was never assured.

2.3 Item 2: Pioneering Approach to Certain Technologies

The HEP community, run by physicists and engineers, is not afraid of taking a leading role to develop technologies that it considers essential. Examples include vacuum and low-temperature techniques, magnet and accelerating cavity design, tunneling and other civil engineering methods, as well as some areas of computer technology including high speed data links, large storage systems and computer clusters. Nevertheless, any decisions to develop CERN's own solutions were always dependent on the prevailing staff and financial situation. Many periods were characterised by a "Buy, don't develop" attitude, or – in more straitened times – even by attempts to outsource computer operations or software development.

> Pioneering approach allows direct access to innovative technologies before these are available off-the-shelf.

Fortunately, at the start of the 1990s, the atmosphere was reasonably open in the computer field. In fact two major revolutions were ongoing: the transition to Internet protocols both within and outside CERN, and the move from mainframes to distributed computing using networked clusters of Unix workstations – the "SHIFT" project [2]. This loosening of earlier more centralised attitudes to physics computing had some beneficial effects for the Web project, opening both minds and working practices toward a truly worldwide approach.

Nevertheless, in the end, the laboratory's concern to prevent any distraction from its mission to obtain approval for the LHC project made it impossible to keep WWW or even part of the W3C at CERN, even though financial support for this had been offered by the European Union in 1994.

2.4 Item 3: A Tradition of Pragmatism in Working Style

CERN and the HEP community often favour "engineering" solutions in software development, and are sometimes criticised for this. But this tradition helped TB-L, who made several very important pragmatic choices while developing the Web.

First was his extremely powerful but simple scheme for what he called "universal naming", leading to the notation for today's URL. The entire subject of name spaces at that time was arcane and acrimonious, blocking progress towards any practical consensus. Even the Internet Engineering Task Force (IETF), the Internet's normally rational and productive standards body, took a negative and parochial view of TB-L's proposal for a standard "Universal Document Identifier", objecting to its name "universal" and finally accepting only the watered-down "Uniform Resource Locator" or URL. But this was sufficient for the Web's purposes.

Another good example was TB-L's acceptance of broken hypertext links, which radically aided the practical implementation and growth of the Web. This put him into considerable conflict with the main community of computer scientists then working with hypertext systems. He put it concisely in his book (WtW, p. 30): "Letting go of that need for consistency was a crucial design step that would allow the Web to scale. But it simply wasn't the way things were done". This simplification was analogous to the liberating role played by IP in the success of the Internet (using a connection-less, best-effort lower network layer, in contrast to rival connection-oriented systems such as X.25 which accumulated too much state to scale up easily). In fact a similar sort of "religious war" was fought between the hypertext traditionalists and the emerging WWW community as that already ongoing between the X.25 (and OSI) community and the Internet developers. Internet and WWW proponents both won their respective battles against these powerful opponents.

> Pragmatic "engineering" approach as opposed to "theoretical Computer Science" approach to deliver results quickly and allow evolution of the concept.

Another example of TB-L's pragmatism was his design of HTML, based on the pre-existing standard markup language SGML but drastically simplified. "When I designed HTML for the Web, I chose to avoid giving it more power than it absolutely needed – a 'principle of least power', which I have stuck to ever since". (WtW, p. 197).

His choice of programming language for the Web prototype was also significant, namely Objective C, considered by many at that time as too crude and not sufficiently structured. The choice of the NexT platform and its NexTStep environment was also extremely unorthodox but very astute, reducing the development time to a minimum (three months!), but ruffling feathers at CERN. In fact, purchasing the two NexT machines for TB-L's development work was another administrative feat accomplished by Mike Sendall.

2.5 Item 4: Proprietary Versus Open Standards

The 1980s saw enormous changes in attitudes and practice as the transition occurred from closed proprietary software and hardware systems to open and standardised systems. Typical were: the replacement of proprietary computer hardware architectures (IBM, DEC, etc.) by a small number of microprocessor-based systems; the replacement of diverse and incompatible operating systems by Unix; the increasing availability of programming languages like C which allowed portable application and system development; and the replacement of proprietary and incompatible networking media and protocols by Ethernet LAN's and Internet protocols.

> Open standards have been one of the "enabling" elements of the Information Technology revolution we have witnessed in the past twenty years.

TB-L learned and benefited from many of these advances when he took the major role in implementing a system for Remote Procedure Call (RPC) linking many computer systems at CERN, in collaboration with myself and others in the mid 1980s. RPC was a new paradigm which allowed computing procedures required by one system to be executed on other(s) to increase power and flexibility. The RPC project's implementation, largely carried out by TB-L himself, was an extraordinary accomplishment. Developed between 1985 and 1989, it allowed graceful cooperation between many extremely diverse computer systems of all sizes involved in the physics experiments and control systems associated with CERN's new LEP accelerator. Messages passed over a wide variety of physical networks and bus systems, and the system supported a very wide range of operating systems, from microprocessors and minicomputers to the largest Computer Centre mainframes. This was Berners-Lee's practical introduction to the world of distributed computing with its plethora of competing standards, but with a set of tools allowing its harmonisation and mastery. His ability both as a visionary and as a powerful implementor was fully utilised and became apparent to his close colleagues as a result.

2.6 Item 5: Link to the New Open-Source Movement

The Web idea was based on a vision of *sharing* among groups of people all over the world – sharing ideas, blueprints, discussions, and of course code.

Tim had a particular knowledge of and interest in what is today known as the Open Source movement, but which in the 1980s was just getting started. His awareness of the open source community had led him to the decision to find outside programming resources for WWW, once it had become clear to him that CERN would not provide sufficient support for a serious effort to develop a browser that would run on Windows PC's and mainline Unix desktop systems. He placed some prototype code on the Usenet group alt.hypertext in August 1991 – probably the first time that code written at CERN was published in this way. Very soon this stimulated many browser development efforts including Viola, Erwise, Arena, Lynx ... and of course Mosaic (later to evolve into Netscape, Mozilla and Firefox).The level of interest in the Web increased worldwide and exponentially from that moment. In TB-L's own words: "From then on, interested people on the Internet provided the feedback, stimulation, ideas, source-code contributions and moral support that would have been hard to find locally. The people of the Internet built the Web, in true grass-roots fashion". (WtW, p. 51).

> The Open Source community built the Web as we know it, and in return they got a very effective tool to foster the development of their community.

TB-L was one of very few people at CERN who knew about such things as Gnu Public Licences (GPL). He realised that trying to licence the WWW commercially would be a kiss of death (as it proved to be for the rival Gopher system which rapidly disappeared when licence fees were imposed for it). From 1992 onward he tried to get CERN to grant GPL status for the WWW code, but later realised that even this would be too restrictive and finally in April 1993 obtained CERN's permission for release of the code with no conditions attached.

2.7 Item 6: Presence of the Internet at CERN

This section is somewhat more detailed than those above, as the establishment of Internet protocols at CERN is an area in which the present author played a major role. I have thus permitted myself a more personal approach (including the use of the first person in some parts). The text is based on an article I first wrote in 1995 [3].

2.7.1 In the Beginning: The 1970s

In the beginning was – chaos. In the same way that the theory of High Energy Physics interactions was itself in a chaotic state up until the early 1970s, so was the so-called area of "Data Communications" at CERN. The variety of different techniques, media and protocols used was staggering; open warfare existed between many manufacturers' proprietary systems, various home-made systems (including CERN's own "FOCUS" and "CERNET"), and the then rudimentary efforts at defining open or international standards. There were no general purpose Local Area Networks (LANs): each application used its own protocols and hardware. The only really widespread CERN network at that time was "INDEX": a serial twisted pair system with a central Gandalf circuit switch, connecting some hundreds of "dumb" terminals via RS232 to a selection of accessible computer ports for interactive log-in.

2.7.2 The Stage is Set: Early 1980s

To my knowledge, the first time any "Internet Protocol" was used at CERN was during the second phase of the STELLA Satellite Communication Project, from 1981–83, when a satellite channel was used to link remote segments of two early local area networks (namely "CERNET", running between CERN and Pisa, and a Cambridge Ring network running between CERN and Rutherford Laboratory). This was certainly inspired by the ARPA IP model, known to the Italian members of the STELLA collaboration (CNUCE, Pisa) who had ARPA connections; nevertheless the STELLA Internet protocol was independently implemented and a STELLA-specific higher-level protocol was deployed on top of it, not TCP. As the senior technical member of the CERN STELLA team, this development opened my eyes to the meaning and potential of an Internet network protocol.

> CERN's leading role in European Networking slows down its uptake of Internet in the first half of the '80s.

In 1983, for the first time at CERN, a Data Communications (DC) Group was set up in the computing division (then "Data-handling Division" or "DD"). Before that time, work on computer networking in DD had been carried out in several groups: I myself belonged to the Software (SW) Group, which had assigned me and several others to participate in DD's networking projects since 1970. All my work on STELLA had been sponsored in this way, for example. The new DC Group had received a mandate to unify networking practices across the whole of CERN, but after a short time it became clear that this was not going to be done comprehensively. DC Group decided to leave major parts of the field to others while it concentrated on building a CERN-wide backbone network infrastructure. Furthermore, following the political currents of the time, they laid a very formal stress on ISO standard networking, the only major exception being their support for DECnet. PC networking was ignored almost entirely; IBM mainframe networking (except for BITNET/EARN), as well as the developing fields of Unix and workstation-based networking, all remained in SW Group. So did the pioneering work on electronic mail and news, which made CERN a European leader in this field. In fact, from the early 1980s until about 1990, CERN acted as the Swiss backbone for Usenet news and gatewayed all Swiss e-mail between the EUnet uucp network, BITNET, DECnet and the Internet. As these were precisely the areas in which the Internet protocols were to emerge, this led to a situation in which CERN's support for them would be marginal or ambiguous for several years to come, as the powerful DC Group neglected or opposed their progress.

It was from around 1984 that the wind began to change.

2.7.3 TCP/IP Introduced at CERN

In August 1984, I wrote a proposal to the SW Group Leader, Les Robertson, for the establishment of a pilot project to install and evaluate TCP/IP protocols on some key non-Unix machines at CERN including the central IBM-VM/CMS mainframe and a VAX VMS system. The TCP/IP protocols had actually entered CERN a few years earlier, inside a Berkeley Unix system, but not too many people were aware of that event. We were now to decide if TCP/IP could indeed solve the problems

2 Why HEP Invented the Web?

of heterogeneous connectivity between the newer open systems and the established proprietary ones. We also proposed to evaluate Xerox's XNS protocols as a possible alternative. The proposal was approved and the work led to acceptance of TCP/IP as the most promising solution, together with the use of "sockets" (pioneered by the BSD 4.x Unix system) as the recommended API.

In early 1985 I was appointed as the "TCP/IP Coordinator" for CERN, as part of a formal agreement between SW Group (under Les Robertson) and DC Group (under its new leader, Brian Carpenter). Incorporating the latter's policy line, this document specifically restricted the scope of Internet protocols for use only within the CERN site. Under no circumstances were any external connections to be made using TCP/IP: here the ISO/DECnet monopoly still ruled supreme, and would do so until 1989.

> Internet is finally introduced at CERN in the second half of the 80's.

Between 1985 and 1988, the coordinated introduction of TCP/IP within CERN made excellent progress, in spite of the small number of individuals involved. This was because the technologies concerned were basically simple and became steadily easier to buy and install. A major step was taken in November 1985 when the credibility of the Internet protocols as implemented within CERN was sufficient to convince the management of the LEP/SPS controls group that the LEP control system, crucial for the operation of CERN's 27 km accelerator LEP then under construction, should use TCP/IP. This decision, combined with a later decision to use Unix-based systems, turned out to be essential for the success of LEP. The TCP/IP activity in LEP/SPS included a close collaboration with IBM's Yorktown Laboratory to support IP protocols on the IBM token ring network that had been chosen for the LEP control system.

Other main areas of progress were: a steady improvement of the TCP/IP installations on IBM-VM/CMS, from the first University of Wisconsin version (WISCNET) to a later fully-supported IBM version; the rapid spread of TCP/IP on DEC VAX VMS systems, using third-party software in the absence of any DEC product; and the first support of IBM PC networking, starting with MIT's free TCP/IP software and migrating to its commercial descendant from FTP Software. All this was accompanied by a rapid change from RS232 based terminal connections to the use of terminal servers and virtual Ethernet ports using TCP/IP or DEC-based protocols. This permitted either dumb terminals or workstation windows to be used for remote log-in sessions, and hence to the use of X-Windows. In particular, starting from 3270 emulator software received from the University of Wisconsin and developed by myself and others for Apollo and Unix systems, a full-screen remote log-in facility was provided to the VM/CMS service; this software was then further developed and became a standard way for CERN users to access VM/CMS systems world-wide.

> In 1988 CERN installs its first "supercomputer" running Unix and of course the TCP/IP stack.

Nevertheless, as late as September 1987, DD's Division Leader would still write officially to a perplexed user, with a copy to the then Director of Research: "The TCP-IP networking is not a supported service." This again illustrates the ambiguity of the managerial situation, as these words were written at essentially the same time as another major step forward was made in the use of Unix and TCP/IP at CERN: the choice to use them for the new Cray XMP machine instead of Cray's well-established proprietary operating system COS and its associated Cray networking protocols. Suddenly, instead of asking "What use is Unix on a mainframe?" some people began to ask: "Why not use Unix on everything?". It is hard to realise today how provocative such a question appeared at that time.

The Cray represented CERN's first "supercomputer" according to U.S. military and commercial standards and a serious security system was erected around it. As part of this system, in 1987 I purchased the first two Cisco IP routers in Switzerland (and probably in Europe), to act as IP filters between CERN's public Ethernet and a new secure IP segment for the Cray. I had met the founder of "cisco systems", Len Bosack, at a Usenix exhibition in the U.S. in June 1987 and been very impressed with his router and this filtering feature. Cisco was a tiny company with about 20 employees at that time, and doing business with them was very informal. It was hard to foresee the extent to which they would come to dominate the router market, and the growth that the market would undergo.

2.7.4 Birth of the European Internet

In November 1987 I received a visit from Daniel Karrenberg, the system manager of "mcvax", a celebrated machine at the Amsterdam Mathematics Centre that acted as the gateway for all transatlantic traffic between the U.S. and European sides of the world-wide "Usenet", the Unix users' network that carried most of the email and news of that time using a primitive protocol called "uucp". Daniel had hit on the idea of converting the European side ("EUnet") into an IP network, just as major parts of the U.S. side of Usenet were doing at that time. The news and mail would be redirected to run over TCP/IP (using the SMTP protocol), unnoticed by the users, but all the other Internet utilities "telnet", "ftp", etc. would become available as well, once Internet connectivity was established. Even better, Daniel had personal contacts with the right people at the Internet Network Information Center (NIC) who would grant him Internet connect status when he needed it. All he was missing was a device to allow him to run IP over some of the EUnet lines that were using X.25 – did this exist? I reached for my Cisco catalogue and showed him the model number he needed. Within a few months the key EUnet sites in Europe were equipped with Cisco routers, with the PTT's, regulators and other potential inhibitors none the wiser. The European IP network was born without ceremony.

2.7.5 CERN Joins the Internet

In 1988, the DC Group in DD Division (later renamed CS Group in CN Division) finally agreed to take on the support of TCP/IP, and what had been a shoestring operation, run out of SW Group with a few friendly contacts here and there, became a properly staffed and organised activity. John Gamble became the new TCP/IP Coordinator; he had just returned from extended leave at the University of Geneva where he had helped to set up one of the very first campus-wide TCP/IP networks in Europe. A year later, CERN opened its first external connections to the Internet after a "big bang" in January 1989 to change all IP addresses to official ones. (Until then, CERN had used an illegal Class A address, Network 100, chosen by myself).

> In 1990 CERN becomes the largest Internet site in Europe and the endpoint of the U.S.-Europe Internet link.

CERN's external Internet bandwidth flourished, with a growing system of links and routers. Concurrently with the growth of the new European IP network (later to be incorporated as "RIPE" within the previously ISO-dominated organisation "RARE"), many other players in Europe and elsewhere were changing their attitudes. Prominent among these was IBM, who not only began to offer a good quality mainframe TCP/IP LAN connection product of their own but also began to encourage migration of their proprietary BITNET/EARN network towards IP instead of the much more restricted RSCS-based service. They even began a subsidy programme called EASINET to pay line charges for Internet connection of their European Supercomputer sites of which CERN was one. In this way, the principal link (1.5 Megabit/sec) between Europe and the U.S. was located at CERN and funded by IBM for several years during the important formative period of the Internet.

By 1990 CERN had become the largest Internet site in Europe and this fact, as mentioned above, positively influenced the acceptance and spread of Internet techniques both in Europe and elsewhere. The timing was perfect for providing TB-L with a platform from which to launch the World Wide Web.

2.7.6 The Internet and Commerce

Finally, in 1991, the U.S. Congress passed legislation permitting commercial use of the Internet. This had been forbidden previously, using the argument that the Internet infrastructure had been paid for by the U.S. taxpayer and hence should not allow "for-profit" use on behalf of industry or other groups worldwide. Clearly, without such a change of attitude, the Web's level of success would have been totally different, with its use restricted to research and non-profit organisations in the same way that the Internet had been before this date. Again, a coincidence in time and awareness had made it possible for the Worldwide Web to become truly worldwide and truly revolutionary.

2.8 Conclusions and Lessons Learned

If any simple conclusion can be drawn from the above, it is that major research advances such as the emergence of the Web depend more on chance meetings, coincidence and unexpected insights than on planned programmes or top-down analysis, even though these latter items are of course also needed.

Consequently, the best possible thing that a management structure can do if it aims to foster major inventiveness is to adopt a *hands-off attitude* to its creative personnel, together with methods for detecting and nurturing the existence of truly creative people within its ranks, and providing protection for their activities. I sometimes refer to this process as "making space" for research talent. Apart from our own example of CERN in the case of the Web (where *just enough* space was created!), I could cite as past examples certain IBM and AT&T research laboratories which in their heydays fostered research of Nobel Prize quality in several instances. At the present time, the best example is Google Inc. which encourages its employees to spend a certain percentage of their paid time in the pursuit of their own innovative ideas.

2.9 Outlook for the Future

To predict the future of such a dynamic phenomenon as the World Wide Web is a perilous task. Its creation, as we have stressed above, was dependent on chance and coincidence to a major extent. The Web is in fact a perfect example of what N. N. Taleb would identify as a "Black Swan" [4]. As Taleb puts it: "History does not crawl, it jumps". The later development of the Web's own technology has perhaps been less unpredictable, guided as it has been by the World Wide Web Consortium led by TB-L himself, but the larger lines of media and social development which have been triggered by the Web's presence have been very difficult to foresee.

These include Facebook, YouTube, Cloud Computing, and of course the whole panoply of search engines and their evolving algorithms. In conjunction with the last 20 years of prodigious expansion in computing power, storage capacity and network bandwidth, some radically new problem-solving techniques have emerged, which could not have been envisaged previously and which therefore look today like Black Swans. A striking example is "Google Translate". Many decades of intense research and investment in the field of automatic natural language translation using analytic linguistic techniques have yielded far less success than expected. But recently Google has applied a totally unexpected approach to this problem: by using only massive text-matching and database power, they have achieved remarkable success over a wide range of target languages. So we must conclude that in this age of very rapid technology development, predictions of the future are even less likely to succeed than they have in the past.

References

1. Berners-Lee, T., Fischetti, M.: Weaving the Web : The Original Design and Ultimate Destiny of the World Wide Web by its Inventor. Harper San Francisco, San Francisco (1999)
2. Baud, J.P., et al: SHIFT, the Scalable Heterogeneous Integrated Facility for HEP Computing. In: Proceedings of International Conference on Computing in High Energy Physics, Tsukuba, Japan. Universal Academy Press, Tokyo, March 1991
3. Segal, B: A Short History of Internet Protocols at CERN CERN computer newsletter No.2001-001, section "Internet Services", April 2001
4. Taleb, N.N.: The Black Swan : The Impact of the Highly Improbable. Random House Publishing Group, NY (2007)

Chapter 3
Computing Services for LHC: From Clusters to Grids

Les Robertson

This chapter traces the development of the computing service for the Large Hadron Collider (LHC) at CERN data analysis over the 10 years prior to the start-up of the accelerator. It explores the main factors that influenced the choice of technology, a data intensive computational Grid, provides a brief explanation of the fundamentals of Grid computing, and records some sof the technical and organisational challenges that had to be overcome to achieve the capacity, performance, and usability requirements of the LHC experiments.

3.1 Introduction

Since the construction of the first high energy particle accelerator at CERN physicists have been using the highest performance computing facilities available to help extract the physics from the data emerging from the detectors. The first computer, a Ferranti Mercury installed in 1958, could only process 15 thousand instructions per second, but within a few years CERN had installed a supercomputer from Control Data Corporation (CDC) with three orders of magnitude more capacity. A series of supercomputers from CDC, Cray and IBM followed to meet the growing demand for computing capacity driven by increasingly complex detectors designed to exploit each new generation of accelerator.

> Is High Energy Physics a "computing limited" science?

L. Robertson (✉)
CERN, Geneva, Switzerland
e-mail: les@robertson.net

When the Large Electron Positron (LEP) collider was being constructed in the 1980s it looked as if the ability of the experimental collaborations to perform physics analysis would be severely constrained by the limited data processing capacity that could be acquired with the budget available for computing at CERN, and a programme was initiated to exploit the inexpensive micro-processors that were beginning to be used for embedded industrial applications and specialised graphics workstations. This proved successful and soon after LEP operation began in 1989 the first of the services based on clusters of micro-processors came on-line, improving price/performance by an order of magnitude compared with the supercomputers.

As the LHC experiments began to assess their computing needs we found ourselves once more in a situation in which the funding for computing at CERN could not provide the capacity required. This time, however, there was no hardware breakthrough on the horizon to solve the financial problem, but there was one big difference with LHC: this would be by far the most powerful accelerator available to high energy physicists and as a result the experimental collaborations would involve scientists from very many institutes spread around the world, many with their own computing capabilities. The challenge was how to integrate these diverse facilities to provide a coherent computing service.

3.2 Base Technologies

There are several characteristics of experimental High Energy Physics data analysis that govern the selection of the technology that is used for the computing systems providing services for data analysis today. Three of these arise from the way in which the data is organised and the analysis algorithms that have been developed. Good *floating point performance* is important, but much of the executed code consists of logical decision making and integer arithmetic, and there are few algorithms that can benefit from vector arithmetic. In fact the codes are well suited to the processors designed for general purpose PCs for office and home use. *Memory requirements* are modest. With careful design a memory size of 2 GigaBytes per programme is sufficient, another factor that makes today's standard PC a good fit. *Event-level parallelism:* analysis usually requires the processing of a large number of events (collisions), each of which can be computed independently. The application can organise the work into a number of independent jobs, each processing a certain number of events, and all of these jobs can execute in parallel. Only when all of the jobs have completed are the results merged for presentation to the physicist.

These factors set High Energy Physics apart from the classical users of high performance computing, where very high floating point performance is needed together with support for fine-grained parallel process execution in order to achieve the required performance. The ability to select essentially any level of granularity for parallel execution enabled High Energy Physics to migrate from specialised scientific *mainframe* computers towards the powerful scientific workstations that emerged during the second half of the 1980s. Using single chip scientific processors these systems were an order of magnitude more cost effective for High Energy Physics codes, and would lead to the emergence of a new class of scientific computing where throughput is more important than single task performance. The problem was how to manage large clusters of independent computers and the associated high volume of data.

> In High Energy Physics throughput is more important than single machine performance.

In the summer of 1989 Ben Segal, a network and distributed computing expert at CERN, proposed an R&D project to the Hewlett Packard (HP) company to see how far the Apollo DN10000 system could take over CERN's physics data processing load, which was at that time using Cray and IBM supercomputers. HP accepted the proposal and a project was set up by the end of the year including the OPAL experiment, one of the large experiments that was using the newly commissioned LEP accelerator at CERN. The project was rather successful as far as handling the more computation-intensive simulation work was concerned, but it brought into focus three practical difficulties with using networked workstations to provide a reliable production service:

1. Distributed data management.
2. Support for magnetic tape storage, which at that time was the only cost-effective solution for very large data collections.
3. Networking – both in terms of performance and the availability of a ubiquitous reliable file access protocol.

> The SHIFT system sets the prototype for all the High Energy Physics computing "farms" based on PC hardware.

The work led directly to the design of a simple distributed architecture ("SHIFT" – Scalable Heterogeneous Integrated FaciliTy [1]) that defined functional interfaces for mass storage, workload management, and data caching in such a way that the three components could operate in a loosely coupled mode suitable for implementation as servers on a network. The architecture also allowed each component to be implemented as a set of distributed servers. A key goal was that the service could be built up progressively using heterogeneous hardware with different performance characteristics – enabling it to integrate new technology as it became available. In fact, the most difficult component to implement was mass storage, which had to integrate magnetic tape storage with on-line disk caches. The only reliable tape hardware and software available at that time was designed for mainframes, and so a Cray supercomputer was linked into the early cluster as a tape server.

The flexibility of this architecture enabled the computing services at CERN to exploit successive generations of single chip processors, continually improving price/performance. By the time we started to design the computing services for LHC all of the computer centres used for High Energy Physics were using a similar model.

3.3 The Computing Challenge Posed by LHC

As the energy frontier is pushed forward by successive generations of accelerator, enabling the investigation of conditions closer and closer to the *Big Bang*, the complexity and cost of the accelerators has also increased. As a result LHC is the only machine in the world that will operate in the 14 TeV[1] energy range, and the wealth of opportunities for discovery that it offers has attracted thousands of physicists to form the worldwide collaborations that have built the initial four large and complex detectors that are needed to decode the results of particle collisions occurring forty million times per second. This leads to two additional characteristics, which are more problematic for the computing service to handle.

> Very large, data-centric international collaborations, PetaBytes of data to be treated and analysed.

The first of these is the very large numbers of physicists and engineers collaborating on the experiments at LHC – CMS alone has over 3,000 scientists – who are organised into many groups, studying different aspects of the detectors or different physics processes, and with independent approaches to analysis, *but all sharing the same data*. The other characteristic that is problematic for the computing service arises from the data handling needs of the experiments. The detectors include systems to reduce the data rate by five orders of magnitude, filtering out the most interesting events and compressing them to be recorded for later analysis. Nevertheless the final few hundred Hertz recording rate will generate many PetaBytes of new *raw* data each year, to which must be added a comparable volume of processed data as the various steps of the analysis proceed. All of this must be managed and made readily accessible to all of the physicists of each collaboration, but the way in which the data will be accessed is hard to predict,

---

[1] A gram of protons contains 6×10^{23} protons; a proton accelerated at the LHC energy of 14 TeV (14×10^{15} electron-Volts) acquires very approximately the kinetic energy of a fly.

especially during the early years, as this will depend on the physics that emerges and the novel ideas that develop for ways of analysing it.

The volume of data and the need to share it across large collaborations is the key computing issue for LHC data analysis. The volume is too large for conventional database management systems and so specialised data and storage management systems have been developed within the physics community [9, 19]. These implement a dynamic storage hierarchy (active data on disk, archive and inactive data on magnetic tape) and provide cataloguing and metadata facilities with the necessary performance and capacity characteristics.

3.4 Early Planning

When the project to design and construct the LHC accelerator was approved in 1996 the computing services that would be required for data analysis were not included in the financial planning. There were two reasons for this. It was considered that it would be difficult to make a good estimate of the computing requirements for experiments which were still in the design phase. Secondly, computing technology was developing very rapidly at that time, with huge improvements in performance appearing each year, accompanied by significantly lower costs. Indeed, the Pentium Pro chip had been introduced by Intel in November 1995, the first PC chip to offer a reasonable scientific (floating point) performance, and High Energy Physics was busily engaged in understanding how to exploit it effectively for large scale computing services.[2] The feeling was that we would continue to see dramatic improvements in price/performance and so by the time the accelerator would be ready the annual operating budgets for computing services at the major physics labs would be sufficient to provide for the data analysis load. There were some who considered this to be optimistic, in view of the data rates and event complexity at LHC, and they were to be proved correct.

> The LHC computing needs were such that no single computing centre could satisfy them.

Over the next 2 years, studies continued on estimating the computing requirements and guessing how computing technology and costs might evolve in the years before the LHC would be ready. We were reasonably confident that the basic architecture, which was then 10 years old, had sufficient flexibility to adapt to the

---

[2]The first full PC-based batch services at CERN were introduced in March 1997 using Windows NT, but this was rapidly superseded by the first Linux PC service opened in August of the same year.

capacity and performance requirements of LHC, and also enable us to contain costs by continuing to exploit mass market components. Nevertheless it was clear that the overall capacity required for the initial four experiments was far beyond the funding that would be available at CERN, even assuming the most optimistic scenario for technology and cost evolution. On the other hand most of the laboratories and universities that were collaborating in the experiments had access to national or regional computing facilities, and so the obvious question was: Could we in some way integrate these facilities along with CERN to provide a single LHC computing service?

At that time each of the major accelerator labs, including CERN, provided the majority of the computing and storage capacity needed for analysing all of the data from the local experiments. Only parts of the less demanding (in computer service terms) tasks were sometimes handled elsewhere. Examples included the generation of simulated data and the re-processing of raw data, tasks which can be organised as "production" activities and which have relatively modest data access requirements. Distributing the heart of the analysis, the tasks that need access to very large and dynamic data sets, would be much more difficult. This is not simply a question of access to resources – rather it would require the experiment's core data and all of the infrastructure needed to manage it to be developed to operate as a distributed system. However, the easy parallelism inherent in the analysis, together with the rapid evolution of wide area networking – increasing capacity and bandwidth, coupled with falling costs – made it look possible and a feasibility study was launched in the autumn of 1998, the MONARC project [20], proposed and led by Harvey Newman, a physics professor from the California Institute of Technology. MONARC established some fundamental principles:

- A few large data-intensive centres – with major investments in mass storage services, round the clock operation, and excellent network connectivity. These were called the "Tier-1" centres, and would provide long term data warehousing, hold synchronised copies of the master catalogues, be used for the data-intensive analysis tasks, and act as data servers for smaller centres.
- The end-user analysis tasks would be delegated to smaller centres, which would not have to make the same level of commitment in terms of data and storage management services. These "Tier-2" centres would also be responsible for some of the less data-intensive production background tasks, such as the generation of simulated data.
- CERN, as the "Tier-0", would perform initial processing of the data and maintain master copies of the raw and other key data-sets, but would rapidly push out the data to the Tier-1s and Tier-2s.

In 1988 the MONARC project defines the tiered architecture of what will be the LHC Grid.

Modelling work performed in the MONARC project indicated that the inter-site data rates would be within the capabilities of the wide area networking services expected to be available when the accelerator began operations. There was however a major problem to be solved: how to integrate many different computing centres in such a way that the physicists would see a single service, enabling them to concentrate on their analysis without being troubled by the details of where the data was located, where the computational capacity was available and how to authenticate themselves and obtain resource allocations at more than a hundred computer centres. Each of the centres was independently managed, and most of them provided services that were also used by other applications, in physics or other areas of science. It would not be possible to dictate a specific solution – rather we would have to agree on interfaces and standards that could be integrated with their general scientific services.

Around this time a general solution to the problem was proposed by two computer scientists working in the United States, Ian Foster from the Argonne National Laboratory and Carl Kesselman from the Information Sciences Institute at the University of Southern California. Foster and Kesselman had worked on problems of fine grained parallel programming and realised there was an opportunity to tackle much larger problems if the problem could be broken up and the different components computed in parallel on several supercomputers installed in different laboratories and universities. They built a prototype called the I-Way [5], which attracted considerable interest and provided the impetus to launch the Globus [25] project to develop a general purpose software toolkit – for what by this time was being called Grid Computing [12].

One of the early users of the Globus toolkit was the Particle Physics Data Grid [27], a collaboration of High Energy Physics laboratories in the United States, which was set up in 1999 to gain experience in applying this technology to the problems of particle physics. In Europe the Italian National Nuclear Physics Institute (INFN) had implemented in 1998 a distributed simulation facility interconnecting resources at 27 INFN institutes across Italy using a distributed resource management system called Condor [22]. Although Condor did not have all the features of a Grid, the INFN project gave us some confidence that distributed computing services for High Energy Physics would be feasible on the time scale of LHC.

3.5 Grid Technology

The basic components required to build a Grid are summarised below.

The security framework provides the basis for trust between the sites connected to the Grid, defining the rules for authentication of users and Grid components using digital certificates.

The compute element (CE) is a gateway operated at each Grid site that provides services for the submission and control of jobs (tasks). The CE maps universal Grid formats to site-specific entities such as authentication and authorisation credentials, resource requirements, job control functions, etc. The CE also provides visibility of the status of the job to the Grid, maintains a detailed log of progress and error conditions, and ensures that this is returned to the initiator, along with any output data when the job has completed.

The storage element (SE) provides a view of the data available on permanent storage at the site, mapping Grid names to the local namespace, and supports services for storage allocation, data access, and data transfer to and from the site, along with the necessary access controls.

The information system enables sites to advertise the services and resources available, and provides distributed query mechanisms to allow applications to locate the sites with the resources or services that they require.

Tools for managing the *virtual organisations* (VO) that use the Grid. The VO is a structure that defines a set of collaborating users. An individual user is registered with the VO, and is then able to use resources and services at any site which supports that VO. Some of the resources in the Grid may be owned exclusively by the VO, others may be shared between several VOs.

The Grid components and the services that they implement.

An application will require many additional services, built on top of the basic Grid components. In general these may be implemented in different ways according to the needs of the application and the VO, and their compatibility with the Grid sites that the VO wishes to use. The most important of these services are discussed below.

Reliable File Transfer Service. Low level network problems would normally be recovered by the file transfer protocols, but higher level incidents such as hard failures of storage devices, scheduled maintenance, or temporary unavailability of Grid services can cause file transfer failures. In a large geographically distributed service very many components may be involved, increasing the probability of failure. The Reliable File Transfer Service (FTS) hides most of these problems from the application. Requests are queued and data movement is initiated according to the request priority and the network load. In the event of a failure FTS analyses the error and decides how best to recover.

Storage Management services. Several different storage management systems are used by sites providing services for the LHC experiments, the choice of system depending on the requirements (capacity, performance, functionality) of the services that the site must provide to LHC and to other user communities. For example Tier-1 sites must provide long-term secure data archiving on magnetic tape, while some of the Tier-2s need only support data cached from other sites with no need for a local backup copy. In some cases the principal usage is high speed data analysis on read-only data-sets, where low file access overhead and single file data rates are most important. Each of the storage management systems used supports a standard set of functions for manipulating storage: specifying storage classes, allocating data spaces, naming files, initiating archive and recall, etc. The *Storage Resource Manager* [28] functions are accessible through the storage element, enabling applications and administrators to manipulate storage remotely using the same interfaces at all Grid sites.

Distributed database services. These require very close interaction between the hosting sites, which runs somewhat against the principle that the Grid is a loose coupling of services and resources. For LHC two distributed database services are supported: a system that maintains a local cache of queries made against a central master database; a service that provides asynchronous replication of a small number (order of ten) of peer databases.

File replica catalogues. These map the Grid-wide file name from the application's global name space to the Grid site and local name space for all of the replicas of the file that are present in the Grid.

Workload management systems. These perform resource brokerage services between sites and users, matching the job's requirements to available resources, according to policies set by sites and the VOs. Two classes of resource scheduler are used for LHC: schedulers that maintain a Grid-wide view of the resources available to a VO and schedulers that maintain central prioritised queues of work for the VO. The former, forward jobs to sites according to data location and available computational capacity. The jobs are then scheduled for execution by the local scheduler, which will in general be receiving work from several different Grid schedulers. The latter, submit *pilot jobs* to sites that appear to match the requirements of queued jobs. When the pilot job enters execution it calls back to the scheduler with information about the resources actually available and the central scheduler returns a suitable job which is then executed directly under control of the pilot job.

3.6 From Prototyping to Building a Service

By the end of 1999 we were beginning to have practical experience within the High Energy Physics community with general purpose software for providing straightforward distributed services, but it was clear that there was much to be done before we would be able to support reliable computing services on the scale needed for LHC, with enormous active data collections shared by the members of very large application groups.

Rather than develop a special solution for LHC it looked more profitable to collaborate with other groups that were building on the Foster and Kesselman vision of the Computing Grid. We hoped that this would in the longer term enable LHC computing to benefit from a general scientific Grid infrastructure, in the way that we were exploiting the established international research networking infrastructure operated as a general service for science. Fabrizio Gagliardi, a computer scientist who was at that time a senior member of the Information Technology Division at CERN, was given the task of exploring ways of moving forward.

> The International Research Networking Infrastructure taken as a model for the early Grid projects.

Gagliardi brought together 21 scientific and industrial organisations from 11 European countries which proposed the European Data Grid [23] project to the European Commission for funding under its *Fifth Framework* R&D programme. The objective was to build a computing infrastructure that would support intensive computation and analysis of shared large-scale databases, from hundreds of TeraBytes to PetaBytes, used by widely distributed scientific communities. The project included software development activities, particularly in areas that had not yet been addressed in depth by the Globus toolkit (e.g. data management, job scheduling, virtual organisation management) but the emphasis was on demonstrating the ability of Grid technology to provide a practical distributed computing service for the three application areas active in the project: physics, biology and earth observation.

The wheels of administration turn slowly, and it was not until January 2001 that the project was approved and funded for a three year period. In the meantime several groups had begun to prototype and develop their own software, not all of it based on the Globus toolkit, and not all of the groups could be included in the European Data Grid project. In particular, development of Grids for physics was moving ahead rapidly in the United States. The Grid Physics Network (GriPhyN) [13], which included the Globus development teams and many High Energy Physics groups in the U.S., had begun in September 2000, funded for 5 years by the National Science Foundation. Diversity and competition are of course essential for the sound development of new technologies, but with LHC in sight we were also concerned with building a solid computing operation that could serve all of the LHC experiment collaborators. Bringing together these different groups and developments at a time when there was little practical experience of Grids and before standards had evolved was going to be a major challenge.

> A difficult start. A CERN initiative to coordinate the construction of a Grid for LHC, in close collaboration with American and European Grid technology projects.

In order to start this process Manuel Delfino, the head of the Information Technology Division at CERN, made an initial proposal [6] to the CERN directorate in February 2001 for a project that would coordinate the building of a Grid for LHC, funding a core activity at CERN, integrating the resources made available to the experiments at other sites, and collaborating as far as possible with the Grid projects in Europe, the U.S. and elsewhere. An in-depth LHC Computing Review (see Chap. 1 for a description of this review) had taken place the previous year and had established an agreed set of requirements for all four experiments [2]. Delfino guided the proposal through the various administrative and scientific committees at CERN and the CERN Council agreed to the creation of the *LHC Computing Grid Project* (LCG) in September 2001 [15, 21].

The importance of the LCG project was that it placed a clear priority on building a computing service for the LHC experiments, included the experiment collaborations directly in the project management, and was open to all institutes that wished to contribute computing resources. By committing itself to use a Grid for the LHC data handling service LCG provided a large scale applications test-bed that projects developing and prototyping Grid software could use for demonstrating their tools in a critical and demanding environment. At the same time, LCG would necessarily have to place constraints on functionality, delivery schedules, reliability metrics, interoperability, and other factors that were not always compatible with the development of new technology. If the Grid projects did not deliver, the essential functionality would have to be developed by LCG itself or by the LHC experiments. The symbiotic relationship between LCG and the growing number of Grid technology projects, many of whose scientists were also involved with LHC experiments, would require a great deal of technical, organisational and managerial flexibility.

3.7 Data and Storage Management

The major technical challenge in building a Grid for LHC is the management of the vast quantity of data distributed across more than a hundred sites. The master data will grow at around 15 PetaBytes per year, but with intermediate versions and copies, the volume of disk storage that must be managed will grow at more than 45 PetaBytes per year, organised at the application level into tens of millions of files. This has all to be moved securely around the Grid, whether as part of the scheduled production processes of generating simulated data and re-processing the master data, or driven dynamically by the needs of the analysis activities. While the data rates required for the former have been estimated with some level of confidence, the latter will depend entirely on the physics that is uncovered and the imaginative ways in which the physicists and their students decide to approach the analysis. In terms of numbers, CERN must sustain about 2 GigaBytes/second to the Tier-1s, and each Tier-1 must sustain up to a GigaByte/second to serve its associated Tier-2s.

The management of data storage is a major responsibility for the Tier-1 sites, which require the expertise to operate the complex management software needed to provide the appropriate levels of performance and reliability. With the very large number of mechanical components involved in storage systems this is the single most common reason for failures. Techniques are available to minimise single points of failure but nevertheless the Tier-1s must provide (expensive) call-out services to assure round the clock operation – an essential feature of a Grid on which the sun never sets.

There are three storage management systems in use that manage pools of storage spread across nodes of the cluster: dCache and CASTOR, both of which support a hierarchical storage structure of on-line disk and offline magnetic tape, and DPM [18] which is a simpler system used at sites that have only disk storage. Each of these products supports the common Storage Resource Manager (SRM) interface. Lately, the physics community is showing a growing interest in the xrootd [30] system which is described in detail in Chap. 10.

3.8 Application-Specific Software

To complete the task of masking the complexities of the Grid from the end user each experiment has designed its data management and job submission systems to deal with the distributed environment.

> Masking the Grid complexity to the users and still delivering the full functionality is one of the greatest challenges.

For example, the Ganga data analysis framework [7], used by two of the experiments, enables the user to work in the same way whether running analysis across limited data samples on a local system or in parallel mode across very large data-sets on the Grid. The user specifies the algorithm and defines the data to be analysed (e.g. as a query on the metadata catalogue). Ganga uses the experiment's data management system to establish the files that are required and their locations. Using this information and various application specific rules it then splits the task into a number of independent jobs. When the jobs have completed Ganga merges the outputs and returns the results to the user. In this way the user can move easily from testing an algorithm to applying it to the full data-set.

The production systems for controlling the event simulation have also been developed to exploit Grid computing. These systems generate large numbers of independent jobs that are distributed to suitably configured sites and the files of generated data are automatically routed back to permanent storage in Tier-1 centres and registered in the experiment's data management system.

3.9 Operating the Service Round the World and Round the Clock

The LHC Computing Grid service opened in September 2003 [17], using an integrated set of middleware tools developed by several different groups. The basic components were taken from the Globus toolkit [11] which was integrated along with components of the Condor system [22] and some other tools into a package called the Virtual Data Toolkit [29]. The higher level components had been developed by the European Data Grid (EDG) project. These tools allowed a simple distributed service to be operated suitably for compute intensive work such as event simulation, but with limited support for data management and significant reliability problems. A specification of the basic Grid functionality needed by the LHC experiments, the *High Energy Physics Common Application Layer* – HEPCAL [4] had been drawn up to guide the middleware developers, but much of the implementation of this had not yet reached the stage of maturity required by the computing service. EDG, which had significant development resources, was reaching the end of its three-year period of funding, but a successor project, Enabling Grids for E-SciencE (EGEE) [8], was in the process of being approved. This would build on the experience of EDG and LCG, with a strong focus on production quality and extending support to other application areas. Although supported by the European Union, the EGEE project was open to other countries and in particular it was agreed that any site providing resources for LHC could participate. An important characteristic of EGEE was that it was a collaboration of national and regional Grid infrastructures. It did not replace or compete with local organisations, but it encouraged them to agree on standards and common tools and procedures that enabled them to present a single integrated Grid that could be used by international scientific collaborations.

> A 24 × 7 round-the-world service merging local and trans-national Grid infrastructures. A world première.

The operation of the LCG service outside of the U.S. and a couple of other areas was merged with the EGEE infrastructure, with a common operations management structure, giving LCG access to expertise and resources that would not otherwise have been available to High Energy Physics during the period of development of the tools and skills needed to operate a worldwide Grid. The EGEE project also provided an environment in which the High Energy Physics community was able to collaborate with computing and applications specialists from other disciplines on the development of this new approach to distributed computing. A similar process was taking place in the United States at this time, where a multi-science Grid operations project, the Open Science Grid – OSG [26], was set up with funding from the National Science Foundation and the Department of Energy. OSG, like EGEE, was integrated with the U.S. High Energy Physics Grid infrastructure that had developed through a series of projects during the preceding years.

The LCG Grid service now uses these two infrastructure Grids,[3] each of which provides services to integrate, test and distribute software packages that provide the

---

[3]During the first half of 2010 the infrastructure services of the EGEE project were absorbed into a successor project, the *European Grid Infrastructure*.

basic Grid components, and also perform Grid operations services like monitoring, trouble ticket management, quality measurement and accounting. Almost all of the sites providing resources for LHC are connected to one of these Grids.[4] It is fortunate that there has been a convergence on only two infrastructures as Grid technology has not yet matured to the point at which all of the basic services, interfaces and protocols have been standardised, and small differences between the Grids increase the complexity of higher level services. EGEE and OSG share a common heritage of Globus and the Virtual Data Toolkit, and there has been a continuous focus on interoperability, and several components critical for LCG are available on both Grids.

Nevertheless, by the beginning of 2005 there was growing concern that interfaces and services provided at sites used by LCG were drifting apart, not only at the Grid level but also services at individual sites for data management. It was agreed to define a set of *baseline services* [3], which would have to be offered by each infrastructure Grid and site connected to LCG, in order to provide an agreed level of compatibility for the applications. In some cases the services were defined in terms of a standard interface or protocol, but in most cases standards were not available and specific implementations of the service were mandated. This was an important document as it defined the basic services that must be provided in common, placing constraints on developers to maintain a compatible service environment as each component would evolve over the coming years.

3.10 Middleware in Practice

The initial set of middleware used by LCG had some deficiencies, and this was addressed by the EGEE project which had a high level goal of *re-engineering* the middleware tools, to make them more robust and address their functionality and performance limitations. This looked like the right approach, but it turned out to have a serious shortcoming from the standpoint of the LHC service. The resources available for re-engineering were sufficiently comfortable that, rather than concentrating on a solid implementation of tools to provide functionality in areas that were at that time reasonably well understood, the developers could permit themselves to include from the start new functionality in unexplored areas. As a result, initial implementations were slow in coming and in some cases did not fulfil the expectations of the experiments. For some critical functions the LCG project implemented its own solutions, and in other cases the functionality was implemented as part of the application code, in different ways by different experiments.

---

[4]Sites in the Nordic countries are connected via the Nordic Data Grid Facility [14], and sites in Canada through a local Grid infrastructure.

> Managing evolution of a brand new software on a world-wide production
> system implied a very steep learning curve and some very tough compromises.

As this was by now a production facility the service was constrained in its freedom to introduce new versions of the middleware, especially if they were not backward compatible with the previous version. A period ensued when the EGEE/LCG service was operated using one middleware package, *LCG-2*, an evolution of the initial software, while the EGEE project planned to introduce a new and incompatible version of the middleware, *gLite*, that would have disrupted the service. An agreement was finally reached on a convergence strategy whereby *gLite* [16] would be introduced progressively, on a component by component basis, avoiding as far as possible incompatibilities with the running software. While this enabled the continuous operation of the service two important opportunities were lost: the final *gLite* package contained many of the original components with their inherent reliability shortcomings; some basic functions that could have been provided as a common Grid service remained within the applications, implemented in different variations by each experiment.[5]

3.11 International Networking

Reliable high bandwidth international networking is an essential requirement in order to achieve the performance and reliability goals for LHC data transfer. The first plans developed by the MONARC project in 1999 assumed that the inter-site bandwidth would be limited by the cost of the connections. The original hierarchical architecture of the computing model was designed to minimise data transfer, concentrating the major data flows to the links between CERN and the Tier-1s. The evolution to a Grid architecture provided much more flexibility in designing the data flows, but exploitation of this would require an equivalent flexibility in the availability of network bandwidth. Fortunately the dot-com boom of the late nineties had generated a demand for ubiquitous high performance networking. New fibre infrastructure was being installed at a tremendous rate, and this would lead to rapidly falling prices.

The organisations managing the international research networking infrastructure had matured during the nineties and provided reliable connectivity between all of the sites involved in LHC computing. In the large majority of cases these organisations were quick to take advantage of the falling prices, making dramatic improvements in the available bandwidth between Tier-1 and Tier-2 sites. In a few cases, however,

---

[5]Examples include job scheduling and data catalogue management.

things are more difficult. In some countries telecom monopolies or cartels continue to operate and price Gigabit/second bandwidth beyond the reach of science research. In other cases submarine cables are required and these do not necessarily follow the topology that would be natural for the High Energy Physics community.

> International networking is a pre-condition for the Grid infrastructure.

For communication between CERN and the Tier-1s and for inter-Tier-1 traffic an optical private network (OPN) has been established [10]. This provides point-to-point connections between the sites, each implemented as a wavelength carried across an optical fibre, capable of transferring data at 10 Gigabits/second. High-speed network switches at some of the sites enable data to be switched between these optical links, providing redundant data paths that can be used in the case of link failure. The private nature of the OPN means that the bandwidth between sites is guaranteed, ensuring that the required end-to-end data rates can always be achieved.

3.12 Collaboration

The LHC Computing Grid (LCG) is a service provided for the LHC experiments but there is no formal contractual relationship between the organisations providing the many different components of the service. Rather, LCG is organised as a collaboration of the participating institutes. There is a *memorandum of understanding* that lays out the overall goals, the organisational structure, and the framework for the provision of resources and services. But these, as is often the case for scientific collaborations where many participants have insecure funding, are intentions to deliver rather than binding commitments. Each member funds its own participation, there is no central budget or authority, and agreements and decisions are made by consensus. High Energy Physics experiments have a long experience of operating through large international collaborations. In the case of LCG, however, many of the members are computing centres that had not previously collaborated in this way.

> Find the right balance between innovation and competition on one side and stability and reliability on the other.

Innovation and competition are essential to early adoption of new technologies, but in the end clear choices and decisions are required in order to run a coherent and reliable round-the-clock production service involving over 130 independently managed sites. Add to this the close inter-dependence of the LCG collaboration and the multi-science Grid projects and we have an environment in which reaching agreement by consensus was sometimes difficult. However, we had the great advantage of a single goal that everyone embraced: being ready to process the data from LHC. As the date for the accelerator start-up approached pragmatism increased and consensus became markedly easier to achieve.

3.13 Scale and Performance

In May 2008, a few months before the scheduled start-up of the LHC accelerator, the final tests of the computing Grid took place. Using a combination of simulated data and real data from cosmic rays the tests covered the full processing chain, from data acquisition at the detector to end-user analysis, demonstrating a sustained data distribution rate from CERN that exceeded the target of two GigaBytes per second. By the time that the accelerator started operating in September 2008 the LHC Grid was consistently running more than 400,000 jobs each day across 135 different sites. An analysis of the accounting data shows that less than 15% of the processing took place at CERN and more than 50% at the Tier-2 sites. There was a small number of very large sites and a large number of small sites: 50% of the load was delivered by about 15 sites, and 60% of the sites delivered only 10% of the load. But this was the aim – to enable all sites, large and small, wherever they may be located, to participate effectively in the analysis.

> One of the major objectives reached was to enable all sites, large and small to access and process LHC data.

One of the spin-offs of this highly distributed computing environment is that any of the research institutes whose physicists are taking part in the experiments at CERN can themselves become actively involved in the data processing, creating a tangible local manifestation of the experiment, and encouraging novel approaches to building computing systems for physics data analysis.

Unfortunately, after only a short period of operation, a serious incident damaged the accelerator causing it to be closed for 14 months for repair and the installation of additional safety equipment. During this time the computing Grid continued in operation at a lower level of activity, being used for simulation and processing of new cosmic ray events. This was a major setback to the whole LHC project, but the time was well spent by the experiments and the computing centres to bring the whole data processing system to a good level of maturity. By the time that the accelerator restarted in November 2009 the Grid service had been very well tested and proved itself able to handle without difficulty the first full year of data taking. At CERN up to 70 TeraBytes of data was generated each day, accumulating

5 PetaBytes by October 2010. The network interconnecting CERN and the Tier-1 centres was supporting data transfer rates up to 70 Gigabits per second, significantly higher than the original target. About 1,900 physicists were involved in processing and analysing the data, generating one million jobs per day which consumed the equivalent of a hundred thousand fully-utilised processors, 55% of this load being carried by the Tier-2 centres. Further development and evolution of the Grid will be required as the performance of the accelerator evolves, but the primary goal of the Grid project was achieved.

3.14 Future Challenges

Hardware evolution, virtualisation and "clouds" are all unknown factors which will be however determinant for the future of the Grid.

For the past 20 years High Energy Physics has been able to benefit from the development of increasingly cost-effective computing technology that is driven by the personal computer market. Whether this trend will continue is unclear as we see the evolution for many mass-market applications from desktop PCs to notebooks, netbooks, tablets and even smart-phones, with reduced energy consumption becoming more important than increased processor speed. Operation and management are of course also important factors in the overall cost of the computing service, and these would benefit from the economies of scale of a more centralised model than that currently used by the LHC Grid. It may be that the most cost-effective solutions in the future will be provided by many-core processors integrated in large power-efficient assemblies strategically sited in locations that benefit from low-cost renewable energy. An important goal of the LHC Grid project was to enable each of the many funding agencies to make its own decision on how best to provide the resources that it contributes. With the realisation of a Grid model for LHC computing we are well placed to consolidate the capacity in a few large centres, integrate resources wherever they may be located in the world, or to continue the current trend towards even wider distribution of resources.

The re-appearance of utility computing in the form of proprietary clouds, with their economies of scale, may well provide attractive solutions for High Energy Physics once the performance characteristics for large scale data access are understood, and commercial offers are adapted to the science research market. The LHC Grid model should make the integration of clouds relatively easy.

Another old idea that has re-emerged is the virtualisation of the base computing system – the definition of a *virtual machine* implemented as a set of low-level functions that can support a full-blown operating system such as Linux and that can be mapped on to a wide range of hardware. The *real* operating system is reduced to being a virtual machine manager, while the application is bundled together with a specific version of a specific operating system. This promises a number of advantages for High Energy Physics.[6]

- *Portability* is greatly improved as the experiment no longer has to adapt, configure and test its software for each version of the operating system that is used at the different sites that provide computing resources. This becomes increasingly important as the number of Grid sites grows and as powerful personal systems contribute a significant fraction of the computing resources.
- *Physics applications* depend to a large extent on functionality that is provided by common packages, shared libraries and the operating system itself. As these evolve the application developers must re-integrate their programmes with each new version. Virtualisation enables the application to manage this process at its own pace, relieved of the problem of dependencies changing under its feet as a computer service upgrades to a new version of the operating system. Conversely the computer service is free to schedule upgrades to the virtual machine manager (for example to support new hardware or resolve a security issue) without being constrained by the time-scale of the application.
- *Computer service providers* are able to operate hardware more efficiently, running many virtual machines with different applications on the same physical computer. This has been one of the main drivers for virtualisation, and indeed all of the commercial cloud offerings have adopted a virtual machine model. The computational requirements of High Energy Physics applications are rather large, and so there is generally little difficulty in one application filling a machine. Nevertheless this may become more important if systems with a very large number of cores appear.

High-end notebooks with a large screen, multi-core processor, high speed wireless connection and a TeraByte or more of disk storage will be widely used for LHC data analysis, as the user travels between home, university, CERN and elsewhere. These powerful systems will not be simple clients of computer centre services, but will themselves hold important subsets of the physics data. Integrating the mobile end-user with the Grid storage and services will be a challenge that must be addressed with urgency.

3.15 Conclusions and Lessons Learned

Grids are all about sharing. They are a means whereby groups distributed around the world can pool their computing resources, large centres and small centres can all contribute, and users everywhere can get equal access to data and computation.... without having to spend all of their time seeking out the resources.

---

[6]For a detailed discussion on virtualisation see Chap. 6.

Grids also allow the flexibility to place the computing facilities in the most effective and efficient places, exploiting funding wherever it is provided, piggy-backing on existing computing centres, or exploiting cheap and renewable energy sources.

The High Energy Physics community, in collaboration with several Grid development projects, has extended the Grid concept to support computational and storage resources on the massive scale required for LHC data analysis, and brought the worldwide LHC computing Grid into full scale round the clock operation in time to handle the first data from the accelerator. The first full year of data taking has shown that this approach can deliver a robust, high throughput and data intensive scientific computing environment.

The Grid model has stimulated High Energy Physics to organise its computing in a widely distributed way, involving directly a large fraction of the physics institutes that participate in the LHC experiments. We have built a very flexible environment that places us well to benefit from new opportunities as they arise. But we must keep our focus firmly on the goal – LHC data analysis. The Grid is merely the current means to that end.

References

1. Baud, J.P., et al.: SHIFT, The Scalable Heterogeneous Integrated Facility for HEP Computing. In: Proceedings of International Conference on Computing in High Energy Physics, Tsukuba, Japan, March 1991, Universal Academy Press, Tokyo
2. Bethke, S. (Chair), Calvetti, M. , Hoffmann, H.F., Jacobs, D., Kasemann, M., Linglin, D.: Report of the Steering Group of the LHC Computing Review. CERN/LHCC/2001-004, 22 February 2001
3. Bird, I. (ed.): Baseline Services Working Group Report. CERN-LCG-PEB-2005-09. http://lcg.web.cern.ch/LCG/peb/bs/BSReport-v1.0.pdf
4. Carminati, F. (ed.): Common Use Cases for a HEP Common Application Layer. May 2002. CERN-LHC-SC2-20-2002
5. Defanti, T.A., Foster, I., Papka, M.E., Stevens, R., Kuhfuss, T.: Overview of the I-WAY: Wide-Area Visual Supercomputing. Int. J. Supercomput. Appl. High Perform. Comput. **10**(2/3), (Summer - Fall 1996), 123–131
6. Delfino, M., Robertson, L. (eds.): Solving the LHC Computing Challenge: A Leading Application of High Throughput Computing Fabrics combined with Computational Grids. Technical Proposal. CERN-IT-DLO-2001-03. http://lcg.web.cern.ch/lcg/peb/Documents/CERN-IT-DLO-2001-003.doc(version1.1)
7. Elmsheuser, J., et al.: Distributed analysis using GANGA on the EGEE/LCG infrastructure. J. Phys.: Conf. Ser. **119** 072014 (8pp) (2008)
8. Enabling Grids for E-Science. Information Society Project INFSO-RI-222667. http://www.eu-egee.org/fileadmin/documents/publications/EGEEIII_Publishable_summary.pdf
9. Ernst, M., Fuhrmann, P., Gasthuber, M., Mkrtchyan, T., Waldmann, C.: dCache – a distributed storage data caching system. In: Proceedings of Computing in High Energy and Nuclear Physics 2001, Beijing, China. Science Press, New York
10. Foster, D.G. (ed.): LHC Tier-0 to Tier-1 High-Level Network Architecture. CERN 2005. https://www.cern.ch/twiki/bin/view/LHCOPN/LHCopnArchitecture/LHCnetworkingv2.dgf.doc

11. Foster, I., Kesselman, C.: Globus: A Metacomputing Infrastructure Toolkit. Int. J. Supercomput. Appl. **11**(2), 115–128 (1997). http://www.globus.org/alliance/publications/papers.php#globus
12. Foster, I., Kesselman, C.: The Grid: Blueprint for a new computing infrastructure. Morgan Kaufmann, San Francisco (1999). ISBN: 1-558660-475-8
13. Grid Physics Network – GriPhyN Project. http://www.mcs.anl.gov/research/project_detail.php?id=11
14. NorduGrid. http://www.nordugrid.org/about.html
15. Knobloch, J. (ed.): The LHC Computing Grid Technical Design Report, CERN June 2005. CERN-LHCC-2005-024. ISBN 92-9083-253-3
16. Laure, E., et al.: Programming the Grid with gLite. EGEE Technical Report EGEE-TR-2006-001- http://cdsweb.cern.ch/search.py?p=EGEE-TR-2006-001
17. LHC Computing Grid goes Online - CERN Press Release. 29 September 2003. http://press.web.cern.ch/press/PressReleases/Releases2003/PR13.03ELCG-1.html
18. Light weight Disk Pool Manager status and plans. Jean-Philippe Baud, CERN - EGEE 3 Conference Athens April 2005. https://svnweb.cern.ch/trac/lcgdm/wiki/Dpm
19. Lo Presti, G. , Barring, O. , Earl, A. , Garcia Rioja, R.M., Ponce, S., Taurelli, G., Waldron, D., Coelho Dos Santos, M.: CASTOR: A Distributed Storage Resource Facility for High Performance Data Processing at CERN. In: Proceedings of the 24th IEEE Conference on Mass Storage Systems and Technologies, pp. 275–280. IEEE Computer Society (2007)
20. Newman, H. (ed.): Models of Networked Analysis at Regional Centres for LHC Experiments (MONARC) Phase 2 Report. CERN 24 March 2000. CERN/LCB 2000-001
21. Proposal for Building the LHC Computing Environment at CERN. CERN/2379/Rev. 5 September 2001. http://cdsweb.cern.ch/record/35736/files/CM-P00083735-e.pdf
22. The Condor Project. http://www.cs.wisc.edu/condor
23. The European Data Grid Project - European Commission Information Society Project IST-2000-25182. http://eu-datagrid.web.cern.ch/eu-dataGrid/Intranet_Home.htm
24. The European Grid Infrastructure. http://www.egi.eu/
25. The Globus Alliance. http://www.globus.org/
26. The Open Science Grid. http://www.openscienceGrid.org/
27. The Particle Physics Data Grid. http://ppdg.net/
28. Storage Resource Managers: Recent International Experience on Requirements and Multiple Co-Operating Implementations, Arie Shoshani et al., pp. 47–59, 24th IEEE Conference on Mass Storage Systems and Technologies (MSST 2007), 2007
29. The Virtual Data Toolkit. http://vdt.cs.wisc.edu/
30. Scalla: Scalable Cluster Architecture for Low Latency Access Using xrootd and olbd Servers - http://xrootd.slac.stanford.edu/papers/Scalla-Intro.pdf

Chapter 4
The Realities of Grid Computing

Patricia Méndez Lorenzo and Jamie Shiers

For approximately one decade Grid [1] computing has been hailed by many as "the next big thing". This chapter looks at two specific aspects of Grid computing, namely the issues involved in attracting new communities to the Grid and the realities of offering a worldwide production service using multiple Grid infrastructures. The first topic is relevant in terms of the cost benefits – or possibly drawbacks – of Grid computing compared with alternatives, whereas the second goes beyond the hype to examine what it takes to deploy commission and run a 24 × 7 production service at the "Petascale" – now increasingly used to refer to 10^5 (of today's) cores.

As Cloud[1] computing is increasingly gaining popularity we compare these two approaches against the concrete requirements of the most demanding applications introduced in the first part of the chapter, namely those of the Large Hadron Collider (LHC) experiments.

4.1 Introduction

Although born in the late 1990s Grid computing burst onto the High Energy Physics (HEP) computing scene in the year 2000. In many ways a continuation and evolution of previous work, Grid computing succeeded in catalysing not only significant funding but also mobilising a large number of sites to pool resources in a manner that had never been achieved before. This chapter discusses a well established process for attracting new user communities to the Grid environment, including the technical

---

[1] See, for example, this Gartner report on the impact of Cloud Computing: http://www.gartner.com/it/products/research/cloud_computing/cloud_computing.jsp.

P.M. Lorenzo (✉) · J. Shiers
CERN, Geneva, Switzerland
e-mail: Patricia.Mendez@cern.ch

issues involved in migrating existing applications and supporting the corresponding communities.

There is no doubt that the largest user community (or communities, as there are several Virtual Organisations involved) that uses production Grid services today is that of High Energy Physics (HEP). Given the scope of this book a discussion of the requirements that have driven the various Grid projects, together with an analysis of the current state of production Grid services, is highly germane. However, technology never stands still and alternatives are being increasingly discussed – including at the major Grid conferences around the world. Although a full production demonstration of Cloud technology against HEP use cases has yet to be made, it is important to understand the pros and cons of such technology – at the very least for discussions with funding bodies!

The Worldwide Large Hadron Collider (LHC) Computing Grid (WLCG [2]) was officially declared open and ready to handle the analysis of data from the LHC on Friday 3rd October 2008. This was the result of close to a decade of research and development, service deployment and hardening and is the result of work by several hundred individuals worldwide. The LHC is expected to operate for well over a decade and will require every-increasing production computing services for this period and beyond. Whether Grid computing will still be in use in 10–15 years time (or even what Grid computing will mean in 10–15 years time) is clearly beyond the scope of this book.

4.2 Motivation

There is a wide range of applications that require significant computational and storage resources – often beyond what can conveniently be provided at a single site. These applications can be broadly categorised as *provisioned* – meaning that the resources are needed more or less continuously for a period similar to, or exceeding, the usable lifetime of the necessary hardware; *scheduled* – where the resources are required for shorter periods of time and the results are not necessarily time critical (but higher than for the following category); *opportunistic* – where there is no urgent time pressure, but any available resources can be readily soaked up. Reasons why the resources cannot easily be provided at a single site include those of funding, where international communities are under pressure to spend funds locally to institutes that are part of the collaboration, as well as those of power and cooling – increasingly a problem with high energy prices and concerns over greenhouse gases.

Whilst Grid computing can claim significant successes in handling the needs of these communities and their applications, the entry threshold – both for new applications as well as additional sites / service providers – is still considered too high and is an impediment to their wide-scale adoption. Nevertheless, one cannot deny the importance of many of the applications currently investigating or using Grid technologies, including drug research, disaster response and prediction, Earth observation as well as major scientific research areas, typified by High Energy Physics and CERN's Large Hadron Collider programme, amongst others.

> Grid computing has still a very high entry threshold in terms of application porting.

Currently, adapting an existing application to the Grid environment is a non-trivial exercise that requires an in-depth understanding not only of the Grid computing paradigm but also of the computing model of the application in question.

The successful demonstration of a straightforward recipe for moving a wide range of applications – from simple to the most demanding – to Cloud environments would be a significant boost for this technology and could open the door to truly ubiquitous computing. This would be similar to the stage when the Web burst out of the research arena and use by a few initiates to its current state as a tool used by virtually everyone as part of their everyday work and leisure. However, the benefits can be expected to be much greater – given that there is essentially unlimited freedom in the type of algorithms and volumes of data that can be processed.

4.3 First Steps in the Grid

In order to begin to use the Grid, both a user and the Virtual Organisation (VO) to which the user belongs need to be established. Today, this is somewhat akin to obtaining a visa: first, the host country has to be recognised – typically a lengthy and complex process. Once this has been achieved, individual citizens of the newly recognised country may apply for visas for other countries – somewhat quicker, but still non-trivial (in our extended analogy, Australia's Electronic Travel Authority (ETA) System is exemplary in this respect). On the other hand, a well defined and automated mechanism of new VO setup is required. Delays would be unacceptable – primarily by nature of the time required – to many times of disaster response for instance, which needs – by definition – to be rapid and may well be triggered out of hours or during holidays / weekends.

Once this has been achieved, the process of porting the application(s) to the Grid can begin. Today, this requires not only in-depth Grid knowledge, but also a good understanding of the computing model of the applications involved. Only once both are sufficiently mastered – typically by involving Grid experts in a close dialogue with those of the application – can porting to the Grid commence. Again, this is an area that needs streamlining: one cannot simply scale the support team with the number of Grid applications. As it is not reasonable to expect that any *central* authority can ever master the design and coding internals of all applications that will join the Grid, the only area where we can focus on is simplifying the

Grid infrastructure expansion and the Grid middleware evolution and deployment. Similarly, adding new applications is valuable to the Grid community, helping to motivate funding for a longer term, sustainable e-infrastructure.

> Open Grid Standards are slowly appearing, facilitating the process of enabling users on the Grid.

At the time of writing, Grids are in production use by numerous HEP experiments around the world. However, with the start-up of the LHC, one of the main challenges ahead was to extend this usage from a relatively small number of "production users", but to large numbers – possibly hundreds or more – of less specialist physicists, whose primary job is to perform analysis (and write papers). To paraphrase a senior ATLAS physicist – the Grid should not "get in the way", but rather facilitate.

This should not, however, be seen as an impossible task. By analogy, in the early 1990s, the number of people at CERN, or indeed in the world, who were able to make content available on the Web was rather small. Of course, this changed very rapidly – so much so that the Web is perhaps more of our daily life than a television or telephone. Had Mosaic and then Netscape and much more not appeared, together with web authoring tools that masked the HTML complexity and had the Apache [3] Web not come up automatically with the Unix OS installation, the Web take-off wouldn't have come so early and to such numbers. The Grid is missing such tools still today. Similar changes have also taken place regarding network and cluster computing. Again, in the early days of the Large Electron Positron (LEP) collider at CERN in the 80's, these were considered almost "dark arts", but have long since become part of the expected skill set of an even novice IT professional when they became off-the-shelf machine attributes and/or plug-and-play network components. For a detailed account of the early evolution of the Grid at CERN, see Chap. 3.

4.4 Bringing New Communities to the Grid

Based on the experience gained supporting the HEP communities on the WLCG structure and building on the reputation won through many years of effective support to the most demanding VOs several research applications – some of them quite separated from the High Energy environment – have asked for the help of the Grid support team at CERN to "gridify" their applications. United Nations Agencies such as the International Telecommunication Union (ITU) and the United Nations Institute for Training and Research Operational Satellite Applications Programme (UNOSAT [4]), medical applications, simulation tools such as GEANT 4 [5], theoretical physics studies in the Quantum ChromoDynamics (QCD) field and collimation and tracking studies of LHC accelerator have benefited strongly from the same infrastructure that is currently provided to the HEP communities in terms of support, infrastructure and level of services and operations.

> CERN has acquired a substantial experience in "enabling" new application communities to use the Grid.

In order to do this, the Grid teams at CERN have developed an effective procedure that has successfully allowed the introduction of the applications mentioned above into the Grid environment. This procedure has been based on an in-depth analysis of the requirements of each individual community together with a continuous and detailed follow up of their Grid production activities. The "gridification" procedure is based on the following 3 blocks.

At first the Grid team ensures the availability of a significant amount of stable resources and services. This Grid capacity is targeted at production work, rather than a basic training infrastructure. Based on the experience gained with several applications, the Grid teams at CERN have provided the Grid infrastructure for a generic VO that is managed by Grid members for communities to whom support has been agreed.

The second step consists in the support team then providing continuous follow-up of the application – no knowledge of the Grid environment or technologies is assumed at the outset – this should not be a barrier to using the Grid.

Finally, a set of user-friendly tools is provided in order to rapidly adapt the application to the Grid. These tools must satisfy the requirements of the application in terms of reliability, tracking of job status and fast access to job outputs. All communities have different goals and requirements and the main challenge of the support team is the creation of a standard and general software infrastructure to allow rapid adaption to the Grid. This general infrastructure effectively "shields" the applications from the details of the Grid (the emphasis here is to run applications developed independently from the Grid middleware). On the other hand, it has to be stable enough to require an acceptable level of monitoring and support from the Grid team and also of the members of the user communities. Finally, it must be flexible and general enough to match the requirements of the different productions without requiring major changes to the design of the tool. As general submission, tracking and monitoring tool the support team have chosen the Ganga [6] and DIANE [7] infrastructure as the official tool for all new "gridifications". This infrastructure is adapted to the requirements of each production with a minimum impact in the general tool. It also includes a layer to MonALISA [8] to monitor the status of the jobs at each site and keep information on processing history.

This infrastructure has been able to attract many applications to the Grid environment and these three fundamental blocks will likely be relevant for "gridifications" for current or future Grid projects. How to generalise this procedure (mostly in terms of support) to ensure a continuous flow of communities, which will be able to use the Grid infrastructure with a minimum and standard support infrastructure is an issue, which will have to be considered by any Grid project.

4.5 Non-HEP Applications and the WLCG Environment

A deep knowledge of both the Grid infrastructure as well as the requirements and expectations of new applications are fundamental to cover the needs of a new application arriving to the Grid. In this sense the roles played by the WLCG, the Enabling Grids for E-sciencE (EGEE [9, 10]) and the European Grid Initiative (EGI [18]) projects have been fundamental to establish collaborative efforts and share experiences and expectations among different research fields. In this environment once again the HEP communities have played the role of "catalysts" for new facilities and solutions that are now being used by many other communities.

In the context of the collaboration between CERN and United Nations initiatives, CERN helped actively the International Telecommunication Union. In May 2006, the ITU hosted in Geneva a world conference to establish a new frequency plan for the introduction of digital broadcasting in Europe, Africa, Arab States and former Russian Federation States. The software developed by the European Broadcasting Union (EBU), performed compatibility analysis between digital requirements and existing analogue broadcasting stations. The peculiarity of this software was that the CPU requirement of each job was not easily predictable since it depended on the details of the input data set. During 5 weeks the ITU run more than 100,000 jobs per week using the WLCG infrastructure, with the particular requirement that the full job set had to be executed during the weekends so that the corresponding results were available for the following week. DIANE and Ganga were used during this production to ensure the full execution of the whole bunch of jobs achieving a 100% of efficiency during the 5 weeks that the conference took.

In the context of the United Nations initiatives the UNOSAT agency has established a collaboration with CERN to use the computing resources of the laboratory. UNOSAT provides the humanitarian organisations with access to satellite imagery and Geographic Information System services and plays a central role in those regions suffering from disasters. Indeed all the images of the Tsunami following the Indian Ocean earthquake in 2004 were managed and stored at CERN.

> ITU and UNOSAT have already used the World LHC Grid infrastructure for major research programmes.

Due to the large number of images to be handled by UNOSAT, the agency entered the Grid environment in 2005. The project consists on providing the UNOSAT experts placed in different parts of the world with images (previously stored on the

4   The Realities of Grid Computing

Fig. 4.1 Raw UNOSAT image

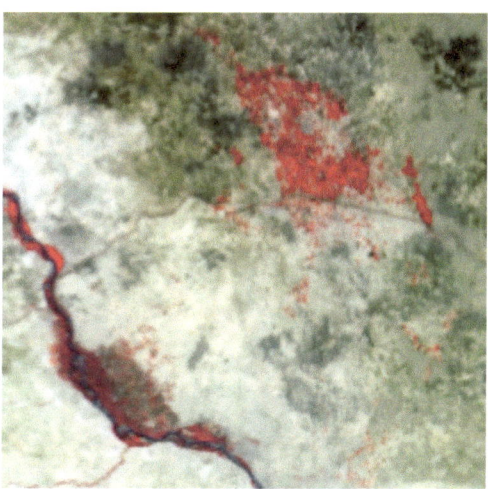

Fig. 4.2 Multi-resolution pyramid UNOSAT image

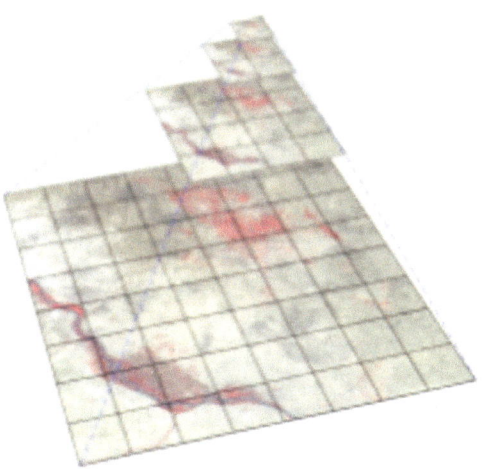

Grid) of a certain area giving as inputs the physical location of the user. Via the GPS technology this physical coordinates will be sent to the Grid, which will provide in a very short time an image of the region corresponding to these coordinates directly in mobile devices. A prototype of the project has been presented at CERN in November 2008 with very promising results. Figure 4.1 presents an example of a raw image of UNOSAT. This type of image has a size of between 200 MegaBytes and 1 GigaBytes and need to be processed before providing it to the end users. In addition the Fig. 4.2 shows a second type of UNOSAT images – the so called Multi-Resolution Pyramid Image – which consists of lot of small images (tiles) that have a smaller size and can be directly managed by the users. Both types of images are stored on Grid resources before providing them to the users via mobile devices.

UNOSAT opens a new concern regarding Grid security. It is not realistic to assign one proxy[2] per UNOSAT member since many of these users will be placed in different countries that still do not have any Grid infrastructure. Group proxies, pin mobile proxies, etc are some of the directions we will have to discuss in the future.

We present now the first community gridified by the Grid Support team beyond the four LHC experiments – the GEANT 4 community. GEANT 4 is both a general purpose software toolkit for simulating the transport and interaction of particles through matter as well as the collaboration that maintains it. The toolkit is currently used in production in several particle physics experiments (BaBar, HARP, ALICE, ATLAS, CMS and LHCb) and in other areas as space science, medical applications and radiation studies.

> The GEANT 4 community is pioneering the usage of the Grid for software quality control and non-regression testing.

The GEANT 4 collaboration performs two major releases per year: in June and December. Before providing a new GEANT 4 version the software has to be tested following a "regression" strategy: a set of physical observables are defined and compared between the new candidate and the previous released version through a large number of configurations based in different energies, particles, physics configurations, geometries and events. In total we are speaking about 7,000 jobs for each new candidate. The collaboration tests up to seven candidates before releasing a new software version. The full production has to be executed in a maximum of three weeks. This community is therefore an ideal candidate to use the Grid and they started to use the WLCG infrastructure to test their releases in December 2004. The experiences gained with GEANT 4 in terms of gridification tools, VO infrastructure, registration and recognition of this community by the EGEE project have led to the definition of the current generic gridification. Figure 4.3 shows a typical GEANT 4 image of a detector simulation.

This generic infrastructure, created in 2005 for the ITU community is ready to host any new application arriving to the Grid and willing to test the infrastructure for their own purposes. The level of support and service created for this generic Grid infrastructure is following the same rules applied to the HEP communities, ensuring in this way the highest possible level of efficiency and reliability to any community facing the Grid world for the first time.

---

[2] A proxy (server) is a server computer system or application program acting as an intermediary from clients requesting resources. The client connects to the proxy server, requesting some service, such as a file, connection, web page, or other resource, available from a different server. The proxy server evaluates the request according to its rules and grants or denies it.

Fig. 4.3 GEANT 4 visualisation of a simulated detector

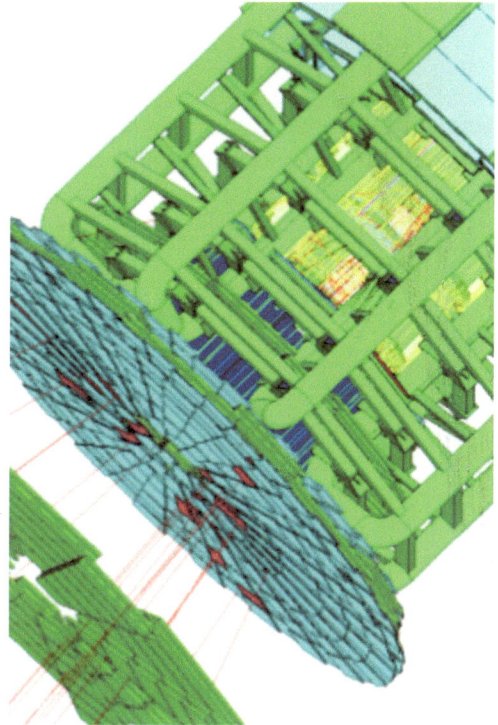

4.6 The Worldwide LHC Computing Grid

In order to process and analyse the data from the world's largest scientific machine, a worldwide Grid service – the Worldwide LHC Computing Grid – has been established, building on two main production infrastructures: those of the Open Science Grid (OSG) [11] in the Americas, and the European Grid Initiative in Europe (EGI).

The machine itself – the Large Hadron Collider (LHC) – is situated some 100 m underground beneath the French-Swiss border near Geneva, Switzerland and supports four major collaborations and their associated detectors: ATLAS, CMS, ALICE and LHCb.

Even after several levels of reduction, some 15 PetaBytes of data are being produced per year at rates to persistent storage of up to few GigaBytes/s – the LHC itself having an expected operating lifetime of some 10–15 years. These data (see Fig. 4.4) are being analysed by scientists at close to two hundred and fifty institutes worldwide using the distributed services that form the Worldwide LHC Computing Grid (WLCG) [12–14]. Depending on the computing models of the various experiments, additional data copies are made at the various institutes, giving a total data sample well in excess of 500 PetaByte and possibly exceeding 1 ExaByte.

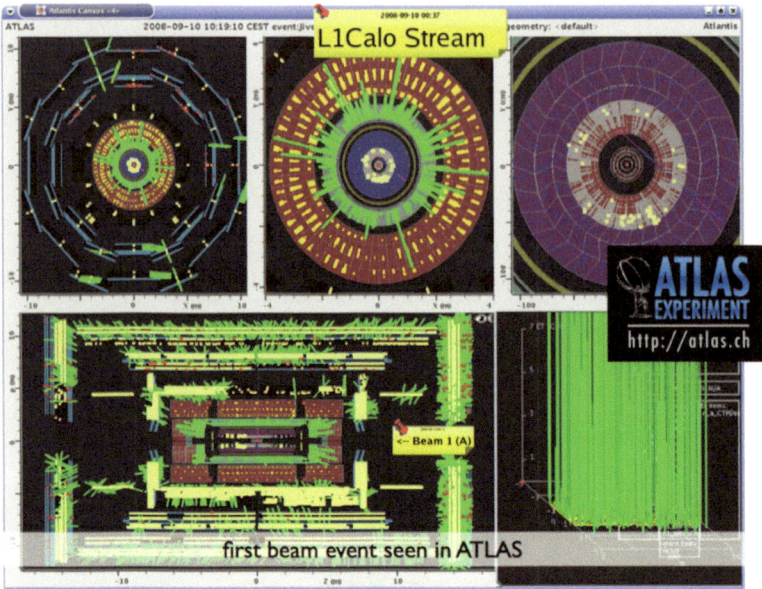

Fig. 4.4 First beam event seen in the ATLAS detector

Running a service where the user expectation is for support 24×7, with rapid problem determination and resolution targets requested by the experiments (30 min for critical problems to be resolved!), is already a challenge. When this is extended to a large number of rather loosely coupled sites, the majority of which support multiple disciplines – often with conflicting requirements but always with local constraints – this becomes a major or even "grand" challenge. That this model works at the scale required by the LHC experiments – literally around the world and around the clock – is a valuable vindication of the Grid computing paradigm (see Fig. 4.5).

However, even after many years of preparation – including the use of well-proven techniques for the design, implementation, deployment and operation of reliable services – the operational costs are still too high to be sustained in the long term. This translates to significant user frustration and even disillusionment. On the positive side, however, the amount of application support that is required compares well with that of some alternate models, such as those based on supercomputers. The costs involved with such solutions are way beyond the means of the funding agencies involved, nor are they necessarily well adapted to the "embarrassingly parallel" nature of the types of data processing and analysis that typify the High Energy Physics domain (Fig. 4.6).

The LHC Grid is a "Grand-challenge"-scale problem and an ideal test-case for Cloud computing.

4 The Realities of Grid Computing

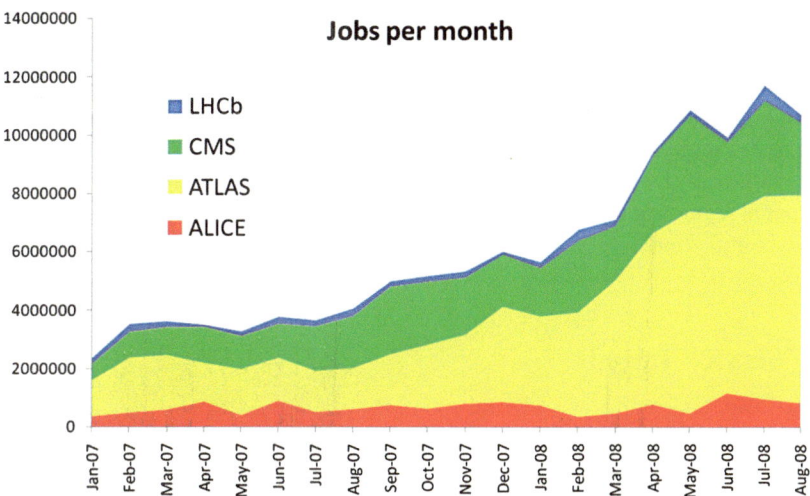

Fig. 4.5 Jobs per month by LHC VO

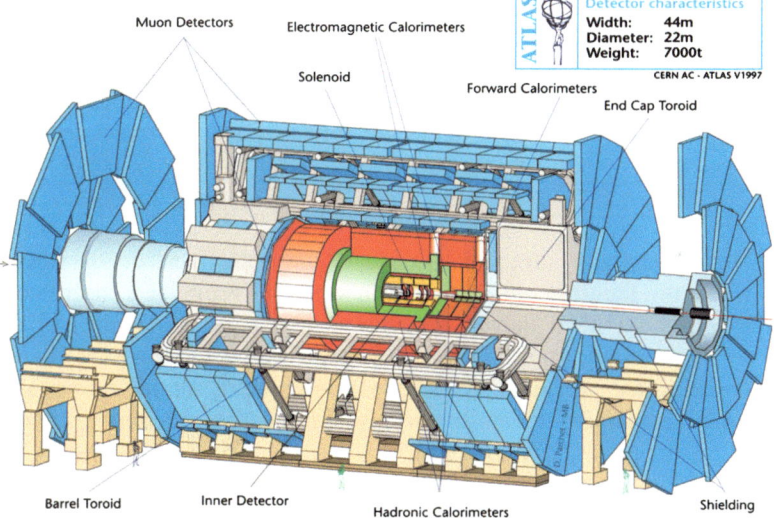

Fig. 4.6 The ATLAS detector

This makes HEP an obvious test-case for Cloud computing models and indeed a number of feasibility studies have already been performed. In what follows we will explore the potential use of Clouds against a highly ambitious target: not simply whether it is possible on paper – or even in practice – to run applications that typify our environment, but whether it would be possible and affordable to deliver a level of service equivalent to – or even higher – than that available today using Grid solutions. In addition to analysing the technical challenges involved, the "hidden benefits" of Grid computing, namely in terms of the positive feedback provided – both scientifically and culturally – to the local institutes and communities that provide resources to the Grid, and hence to their funding agencies who are thus hopefully motivated to continue or even increase their level of investment, are also compared. Finally, based on the wide experience gained by sharing Grid solutions with a large range of disciplines, we try to generalise these findings to make some statements regarding the benefits and weaknesses of these competing – or possibly simply complementary – models.

4.7 Service Targets

There are two distinct views of the service targets for WLCG: those specified up-front in a Memorandum of Understanding [15] (MoU) – signed by the funding agencies that provide the resources to the Grid – and the "expectations" from the experiments. We have seen a significant mismatch between these two views and attempted to reconcile them into a single set of achievable and measurable targets (See Table 4.1).

> One of the most crucial aspects is the agreement on a set of "metrics" to define the quality of the service and its relation with the user expectations and satisfaction.

The basic underlying principle is not to "guarantee" perfect services, but to focus on specific failure modes, limit them where possible, and ensure sufficient redundancy is built in at the required levels to allow "automatic", or at least "transparent", recovery from failures – e.g. using buffers and queues of sufficient size that are automatically drained once the corresponding service is re-established. Nevertheless, the targets remain ambitious, specified both in service availability measured on an annual basis as well as the time to respond when necessary (see Table 4.2 and Fig. 4.7).

Table 4.1 Extract of service targets (Tier 0)

Service	Maximum delay in responding to problems		
	Interruption	Degradation > 50%	Degradation > 20%
Raw data recording	4 h	6 h	6 h
Event reconstruction or distribution of data to Tier1s	6 h	6 h	12 h
Networking service to Tier1s	6 h	6 h	12 h

4 The Realities of Grid Computing

Table 4.2 Service criticality (ATLAS virtual organisation)

Criticality of service	Impact of degradation / loss
Very high	Interruption of these services affects online data-taking operations or stops any offline operations
High	Interruption of these services perturbs seriously offline computing operations
Moderate	Interruption of these services perturbs software development and part of computing operations

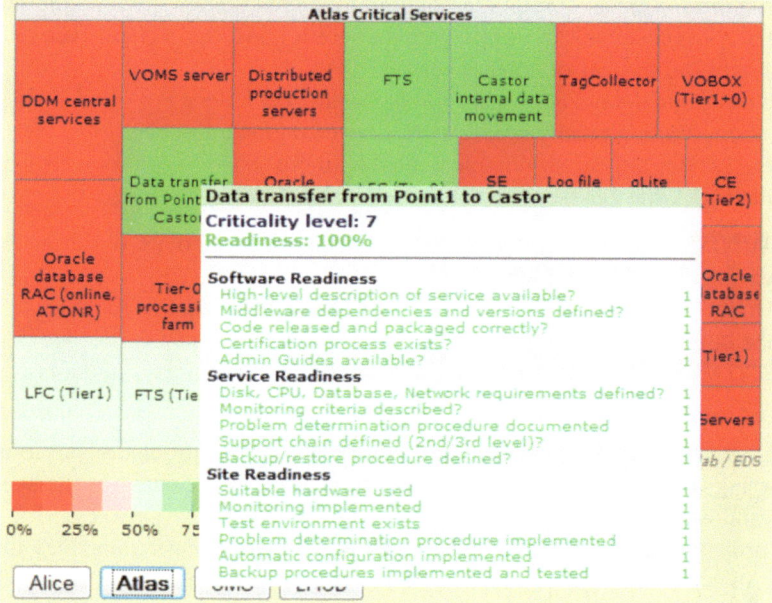

Fig. 4.7 "GridMap" visualisation of service readiness

As currently defined, a very small number of incidents are sufficient to bring a site below its availability target. In order to bridge the gap between these two potentially conflicting views – and building on the above mentioned industry-standard techniques – we observe relatively infrequent breaks of service: either those that are directly user-visible or those that cannot be smoothed over using the buffering and other mechanisms mentioned. We have put in place mechanisms whereby appropriately privileged members of the user communities can raise alarms in these case – supplementing the automatic monitoring that may not pick up all error conditions – that can be used 24 × 7 to alert the support teams at a given site.

Whilst these mechanisms are used relatively infrequently – around once per month in the most intense periods of activity – the number of situations where a major service or site is either degraded or unavailable for prolonged periods of time with respect to the targets defined in the MoU is still fairly high – sometimes several times per week. Most of these failures fall into a small number of categories.

Power and cooling – failures in a site's infrastructure typically have major consequences – the site is down for many hours. Whilst complete protection against such problems is unlikely to be affordable, definition and testing of recovery procedures could be improved – e.g. ensuring the order in which services are restarted is well understood and adhered to, making sure that the necessary infrastructure – redundant power supplies, network connections and so forth – are such as to maximise protection and minimise the duration of any outages.

Configuration issues – required configuration changes are often communicated in a variety of (unsuitable) formats, with numerous transcription (and even interpretation) steps, all sources of potential errors.

Database and data management services – the real killers. For our data intensive applications, these typically render a site or even region unusable.

These service targets are complemented by more specific requirements from the experiments. A site can be in one of the following 3 states:

1. *COMMISSIONED*: daily rules satisfied during the last 2 days, or during the last day and at least 5 days in the last 7
2. *WARNING*: daily rules not satisfied in the last day but satisfied for at least 5 days in the last 7
3. *UNCOMMISSIONED*: daily rules satisfied for less than 5 days in the last 7

The purpose of these rules is to ensure as many sites as possible stay in commissioned status and to allow for a fast recovery when problems start to occur.

The Fig. 4.8 shows a historical snap-shot of CMS Tier2 sites for the specified time-window.

In principle, Grids – like Clouds – should offer sufficient redundancy that the failure of some fraction of the overall system, with the exclusion of the Tier 0 (Table 4.3), can be tolerated with little or preferably no service impact. This is not, unfortunately, true of all computing models in use in HEP, in which for reasons of both geography and funding, specific dependencies exist between different sites – both nationally and internationally. Furthermore, sites have well defined functional roles in the overall data processing and analysis chain which mean that they cannot always be replaced by any other – although sometimes by one or more specific sites. This is due to the fact that the problem at hand is essentially data-centric

Table 4.3 Targets for Tier 0 services

When	Issue	Target
Now	Consistent use of all Service Standards	100%
30'	Operator response to alarm / alarm e-mail	99%
1 h	Operator response to alarm / alarm e-mail	100%
4 h	Expert intervention in response to above	95%
8 h	Problem resolved	90%
24 h	Problem resolved	99%

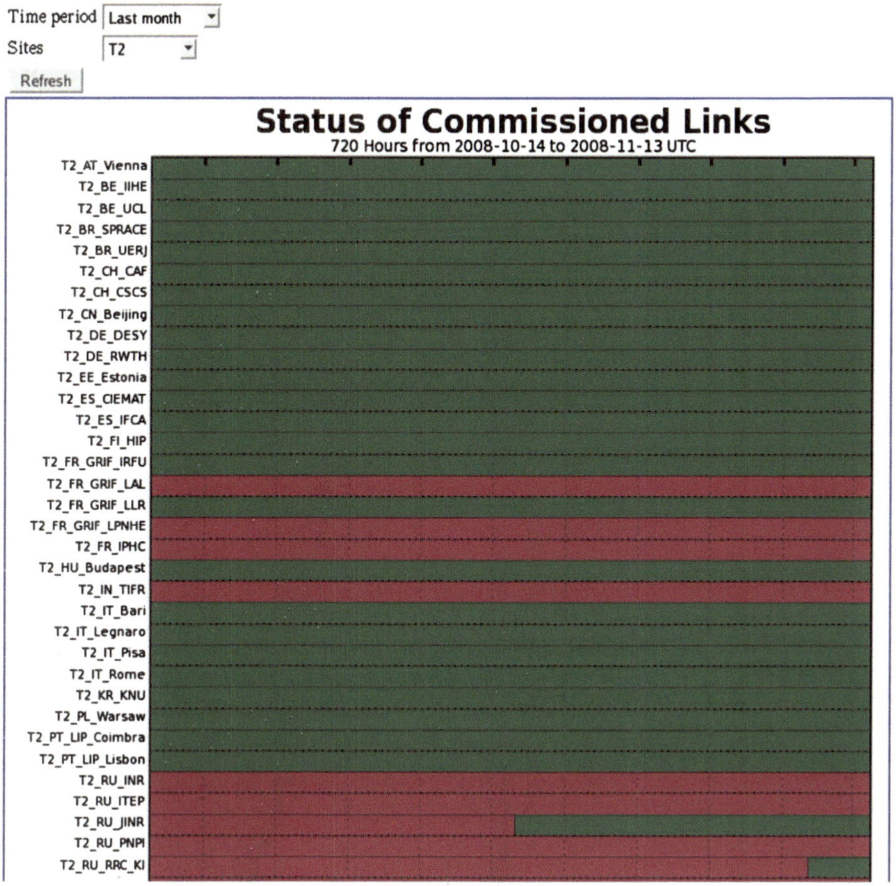

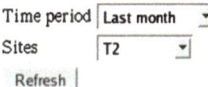

Fig. 4.8 Status of CMS links

and for financial reasons not all data can be replicated at all centres. This is not a weakness of the underlying model but simply a further requirement from the application domain – the proposed solution must also work given the requirements and constraints from the possibly sub-optimal computing model involved. The different "tiers" have specific responsibilities within the WLCG Grid. The Tier 0 (CERN) has the responsibility of safe keeping of RAW data (first copy), first pass reconstruction, distribution of RAW data and reconstruction output (Event Summary Data or ESD) to Tier 1 and reprocessing of data during LHC downtimes. The Tier 1s should provide safe keeping of a proportional share of RAW and reconstructed data, large scale reprocessing and safe keeping of corresponding output, distribution of data products to Tier 2s and safe keeping of a share of simulated data produced at these Tier 2s. Tier 2s handle analysis requirements and a proportional share of simulated event production and reconstruction.

In the considerations below, we will discuss not only whether the Cloud paradigm could be used to solve all aspects of LHC computing but also whether it could be used for the roles provided by one or more tiers or for specific functional blocks (e.g. analysis, simulation, re-processing etc.)

4.8 The Data is the Challenge

Whilst there is little doubt that for applications that involve relatively small amounts of data and/or data rates the Cloud computing model is almost immediately technically viable, large data handling is one of the largest areas of concern for our application domain. More specifically, if long-term data curation has to be the responsibility of "the user" a significant amount of infrastructure and associated support is required to store and periodically migrate data between old and new technologies over long periods of time – problems familiar to those involved with large scale (much more than 1 PetaByte) data archives. As far as data placement and access is concerned, although we have been relatively successful in defining standard interfaces to a reasonably wide-range of storage system implementations rather fine-grained control on data placement and data access has been necessary to obtain the necessary performance and isolation of the various activities – both between and within virtual organisations. Data transfer is possibly a curiosity of the computing models involved and strongly coupled to the specific roles of the sites that make up the WLCG infrastructure – bulk data currently needs to be transferred at high rates in pseudo real-time between sites. Would this be simplified or eliminated using a Cloud-based solution? Figure 4.9 shows the percentage of file transfers that are successful on the first attempt. It is clearly much lower than desirable, resulting in wasted network bandwidth and extra load on the storage services, which in turn has a negative effect on other activities (Fig. 4.10). Finally, database applications are behind essentially all data management applications even if a variety of technologies are used – often at a single site. Again, deep knowledge of the hardware configuration and physical implementation are currently required to get an acceptable level of service.

> Cloud computing might not be the silver bullet for such a heavily data-centric activity as the one of High Energy Physics.

It is perhaps unfair to compare a solution that has evolved over around a decade, with many teething problems and a number of major outstanding issues, with an alternative and impose that a well-established computing model must be supported without change. On the other hand, targets that are relatively independent of the implementation can be defined in terms of availability, service level, computational

4 The Realities of Grid Computing

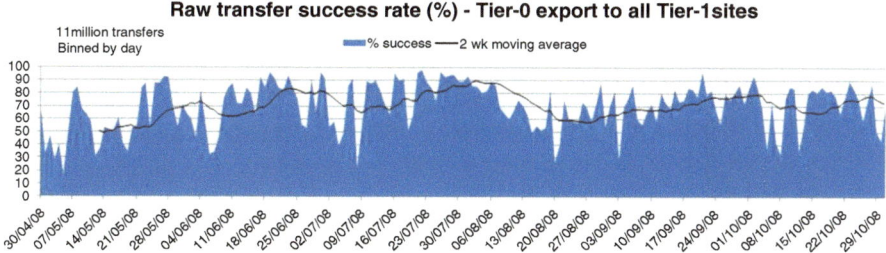

Fig. 4.9 Success rate of file transfers

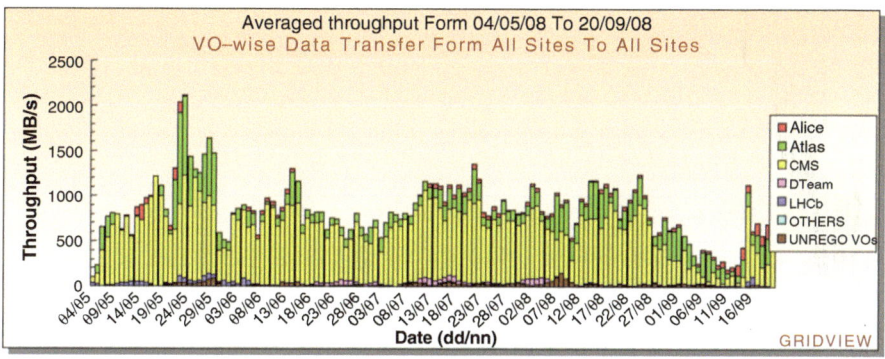

Fig. 4.10 Averaged daily data transfer rates

and data requirements. It may well be that on balance the technical and managerial advantages outweigh any as yet to be found drawbacks. This would leave unavoidable issues such as cost, together with the sociological and other "spin-off" benefits.

4.9 Operations Costs

The operations costs of a large-scale Grid infrastructure are rarely reported and even when this is done it typically refers to the generic infrastructure and not to the total costs of operating the computing infrastructure of a large-scale collaboration. At neither level are the costs negligible: the European Grid Initiative Design Study estimates the total number of full time equivalents for operations-related activities across all National Grid Initiatives to be broadly in the 200–400 range – with an extremely modest 5(!) people providing overall coordination (compared to 15–20 in the EGEE project, which just completed its 3rd and final phase). The operations effort required for a single large virtual organisation, such as the ATLAS experiment – the largest LHC collaboration, is almost certainly in excess of 100. A typical WLCG "Grid Deployment Board" – the monthly meeting working on the corresponding issues – also involves around 100 local and remote participants, whereas a "WLCG Collaboration workshop" can attract closer to 300 – mainly site administrators and other support staff.

> Total operational cost and long-term maintenance of the Grid infrastructure are very difficult to evaluate now, but they are essential parameters for the future evolution of the system.

These costs are not always easy to report accurately as they often covered – at least in part – by doctoral students, post-doctoral fellows and other "dark effort". However, any objective comparison between different solutions must include the total cost of ownership and not just a somewhat arbitrary subset.

4.10 Grid-Based Petascale Production is Reality

Despite the remaining rough edges to the service, as well as the undeniably high operational costs, the success of building a Petascale (using the loose definition of 100,000 cores) [16] world-wide distributed production facility that is built using several independently managed and funded major Grid infrastructures – of which the two main components are built out of O(100) sites (EGI and OSG) – must be considered a large success. The system has been in production mode – with steady improvement in reliability over time – since at least 2005. This includes formal capacity planning, scheduling of interventions – the majority of which can be performed with zero user-visible downtime – and regular reviews of availability and performance metrics. A service capable of meeting the evolving requirements of the LHC experiments must continue for at least the usable life of the accelerator itself plus an additional few years for the main analysis of the data to be completed. Including foreseen accelerator and detector upgrades, this probably means until around 2030! Whether Grids survive this long is somewhat academic – major changes in IT are inevitable on this timescale and adapting to (or rather benefiting from) these advances are required. How this can be done in a non-disruptive manner is certainly a challenge but it is worth recalling that the experiments at LEP – the previous collider in the same tunnel – started in an almost purely mainframe environment (IBM, Cray, large VAXclusters and some Apollo workstations) and moved first to farms of powerful Unix workstations (HP, SGI, Sun, IBM) and finally PCs running Linux. This was done without interruption to on-going data taking, reprocessing and analysis, but obviously not without major work.

Much more recently, several hundred TeraBytes of data from several experiments went through a "triple migration" – a change of backend tape media, a new persistency solution and a corresponding re-write of the offline software – a major

4 The Realities of Grid Computing   109

effort involving many months of design and testing and an equivalent period for the data migration itself. (The total effort was estimated at ∼1FTE/100 TeraByte of data migrated).

These examples give us confidence that we are able to adapt to major changes of technology that are simply inevitable for projects with lifetimes measured in decades.

4.11 Towards a Cloud Computing Challenge

Over the past few years and before the start up of the LHC a series of "service challenges" has been carried out to ramp-up to the necessary level required to support data taking and processing at the LHC. This culminated in 2008 in a so-called "Common Computing Readiness Challenge (CCRC'08)" – aimed at showing that the computing infrastructure was ready to meet the needs of all supported experiments at all sites that support them. Given a large number of changes foreseen prior to data taking in 2009, a further "CCRC'09" was scheduled for 2 months prior to data taking in that year. This was a rather different event that the 2008 challenge, relying on on-going production activities, rather than scheduled tests, to generate the necessary workload. Where possible, overlap of inter-VO, as well as infra-VO, activities were arranged to show that the system can handle the combined workloads satisfactorily. An important – indeed necessary – feature of these challenges has been metrics that are agreed upfront and are reported on regularly to assess our overall state and progress. Whilst it is unlikely that in the immediate future a challenge on an equivalent scale could be performed using a Cloud environment, such a demonstration is called for – possibly at progressively increasing scale – if the community is to be convinced of the validity and even advantages of such an approach.

The obvious area where to start is that of simulation – a compute-dominated process with relatively little input/output needs. Furthermore, in the existing computing models, at the Tier 2 sites – where such work typically but not exclusively takes place – data curation is not provided. Thus, the practice of storing output data at a (Tier 1) site that does provide such services is well established. Thus, the primary question that should be answered by such a model is: *Can Cloud computing offer compute resources for low I/O applications, including services for retrieval of output data for long-term data storage "outside" of the Cloud environment, in a manner that is sufficiently performant as well as cost-competitive with those typically offered today by Universities and smaller institutes?*

The readiness of the WLCG infrastructure has been verified by large-scale operations on simulated data, called "data challenges".

To perform such a study access to the equivalent of several hundred – a few thousand cores for a minimum of some weeks would be required. There is little doubt that such a study would be successful from a technical point of view, but would it be not only competitive or even cheaper in terms of total cost of ownership? The requirements for such a study have been oversimplified – e.g. the need for access to book-keeping systems and other database applications and a secure authentication mechanism for the output storage – but it would make a valuable first step. If not at least in the same ball-park in terms of the agreed criteria there would be little motivation for further studies.

Rather than loop through the various functional blocks that are mapped to the various tiers described above, further tests could be defined in terms of database and data management functionality – presumably both more generic as well as more immediately understandable to other disciplines. These could be characterised in terms of the number of concurrent streams, the type and frequency of access (sequential, random, rarely, frequent) and equivalent criteria for database applications. These are unlikely to be trivial exercises but the potential benefit is large – one example being the ability of a Cloud-base service to adapt to significant changes in needs, such as pre-conference surges that can typically not be accommodated by provisioned resources that do not have enough headroom for such peaks, often synchronised across multiple activities, both within and across multiple virtual organisations.

4.12 Data Grids and Computational Clouds: Friends or Foes

The possibility of Grid computing taking off in a manner somehow analogous to that of the Web has often been debated. A potential stumbling block has always been cost and subscription models analogous to those of mobile phone network providers have been suggested. In reality, access to the Web is often not "free" – there may not be an explicit charge for Internet access in many companies and institutes – and without the Internet the Web would have little useful meaning. However, for most people Internet access is through a subscription service, that may itself be bundled with others, such as "free" national or even international phone calls, access to numerous TV channels and other such services.

A more concrete differentiator is the "closed" environment currently offered as "Clouds" – it may be clear how one purchases services but not how one contributes computational and storage resources in the manner that a site can "join" an existing Grid. A purely computational Grid – loosely quantified as one that provides no long term data storage facilities or curation – is perhaps the most obvious competitor of Clouds. Assuming such facilities are shared as described above between provisioned, scheduled and opportunistic use a more important distinction could – again – be in the level of data management and database services that are provided.

> The size of the data to be handled may well be the deciding element between Grid and Cloud computing.

A fundamental principle of our Grid deployment model has been to specify the interfaces but not the implementation. This has allowed sites to accommodate local requirements and constraints whilst still providing interoperable services. It has, however, resulted in a much higher degree of complexity and in less pooling of experience and techniques than could otherwise have been the case. This is illustrated when the strategies for two of the key components – databases and data management – are compared. The main database services at the Tier 0 and Tier 1 sites (at least for ATLAS – the largest VO), have been established using a single technology (Oracle) with common deployment and operational models. Data management services are implemented in numerous different variations. Even when the same software solution is used, the deployment model differs widely and it has proven hard to share experience. Figure 4.11 shows the diversity in terms of front-end storage solutions: in the case of dCache not only are multiple releases deployed but also the backend tape-based mass storage system (both hardware and software) varies from site to site – creating additional complexity.

There is little doubt that the cost of providing such services as well as the achieved service level suffers as a result – even if more "politically correct". Any evolution or successor of these services would benefit from learning from these experiences.

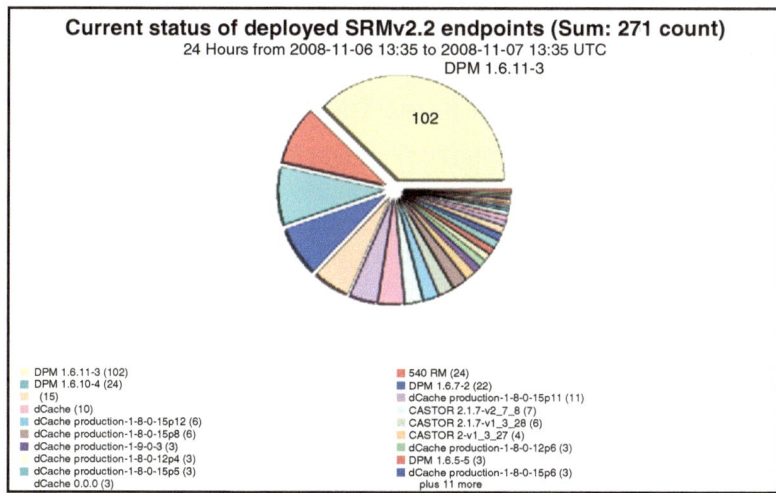

Fig. 4.11 Distribution of storage solutions and versions

4.13 Grid Versus Clouds: Sociological Factors

For many years an oft-levelled criticism of HEP has been the "brain-drain" effect from Universities and other institutes to large central facilities such as CERN. Although distributed computing has been in place since before the previous generation of experiments at the LEP collider – formerly housed in the same 27 km tunnel as the LHC today – scientists at the host laboratory had very different possibilities to those at regional centres or local institutes. Not only does the Grid devolve extremely important activities to the Tier 1 and Tier 2 sites but the key question of equal access to all of the data is essentially solved. This brings with it the positive feedback effect mentioned above which is so important that it probably outweighs even a (small) cost advantage – to be proven – in favour of non-Grid models.

4.14 Is the Gain Worth the Pain?

It should be clear from the above that some of the major service problems associated with today's production Grid environment could be avoided by adopting a simpler deployment model: fewer sites, less diversity but also less flexibility. However, much of the funding that we depend on would not be readily available unless it was spent – as now –primarily locally. On the other hand – and in the absence of any large-scale data-intensive tests – it is unclear whether a Cloud solution could meet today's technical requirements. A middle route is perhaps required, whereby Grid service providers learn from the difficulties and costs of providing reliable but often heterogeneous services, as well as the advantages in terms of service level, possibly at the cost of some flexibility, through a more homogeneous approach. Alternatively, some of the peak load could perhaps be more efficiently and cost effectively handled by Cloud computing, leaving strongly data-related issues to the communities that own them and are therefore presumably highly motivated to solve them.

For CERN, answers to these questions are highly relevant – projections show that we will run out of power and cooling in the existing computer centre on a time-scale that precludes building a new one on the CERN site (for obvious reasons, priority has been given in recent years to the completion of the LHC machine). Overflow capacity maybe available in a partner site to tide us through: do we have the time to perform a sufficiently large scale demonstration of a Cloud-based solution to obviate such a move? Is there a provider sufficiently confident of their solution that they are willing to step up to this challenge? There have been no takers so far and time is running out – at least for this real-life ExaByte-scale test-case. In the meantime our focus is on greatly improving the stability and usability of our storage services, not only to handle on-going production activity with acceptably low operational costs, whilst sustaining the large-scale data-intensive end-user analysis that has come with the first real data from the world's largest scientific machine.

4.15 Conclusions and Lessons Learned

After many years of research and development followed by production deployment and usage by many VOs, worldwide Grids that satisfy the criteria in Ian Foster's "Grid checklist" [17] are a reality. There is significant interest in longer-term sustainable infrastructures that are compatible with the current funding models and work on the definition of the functions of and funding for such systems is now underway. Using a very simple classification of Grid applications, we have briefly explored how the corresponding communities could share common infrastructures to their mutual benefit. A major challenge for the immediate future is the containment of the operational and support costs of Grids, as well as reducing the difficulties in supporting new communities and their applications. These and other issues are being considered by a project for a long term e-infrastructure [18]. Cloud computing may well be the next step in the long road from extremely limited computing – as typified by the infamous Thomas J. Watson 1943 quote "I think there is a world market for maybe five computers" – to a world of truly ubiquitous computing (which does not mean free). It is clear that the applications described in this document may represent today's "lunatic fringe", but history has repeatedly shown that these needs typically become main-stream within only a few years. We have outlined a number of large-scale production tests that would need to be performed in order to assess Clouds as complementary or even replacement technology for the Grid-based solutions in use today, although data-related issues remain a concern. Finally, we have raised a number of non-technical, non-financial concerns that must nevertheless be taken into account – particularly by large-scale research communities that rely on various funding sources and must – for their continued existence – show value to those that ultimately support them: sometimes a private individual or organisation but often the tax-payer.

In a manner that draws inevitable comparisons with the Web, the emergence of commodity computing and a convergence of technologies have made a new era of computing possible, namely that of Grid computing.

We have not yet gained sufficient experience in this environment for a fully objective analysis – this must wait another few years, including the onslaught of full LHC data taking and analysis.

References

1. Foster, I., Kesselman, C.: The GRID – Blueprint for a New Computing Infrastructure. Morgan Kaufmann, San Francisco. ISBN 1-55860-475-8
2. The Worldwide LHC Computing Grid (WLCG). http://lcg.web.cern.ch/LCG
3. http://www.apache.org
4. UNOSAT web page. http://www.unosat.org
5. http://geant4.web.cern.ch
6. Ganga web page. http://ganga.web.cern.ch/ganga/

7. Diane web page. http://www.cern.ch/diane
8. MonaLisa web page. http://monalisa.caltech.edu/monalisa.htm
9. EGEE technical information page. http://technical.eu-egee.org/index.php?id=254
10. The Enabling Grids for E-sciencE (EGEE) project. http://www.eu-egee.org/
11. The Open Science Grid. http://www.opensciencegrid.org/
12. Shiers, J.D.: Lessons Learnt from Production WLCG Deployment. In: The proceedings of the International Conference on Computing in High Energy Physics, Victoria, BC, September 2007
13. Shiers, J.:The Worldwide LHC Computing Grid (worldwide LCG). Comput. Phys. Comm. **177**, 219–223, CERN, 1211 Geneva 23, Switzerland (2007)
14. LCG Technical Design Report, CERN-LHCC-2005-024. available at http://lcg.web.cern.ch/LCG/tdr/
15. Memorandum of Understanding for Collaboration in the Deployment and Exploitation of the Worldwide LHC Computing Grid. available at http://lcg.web.cern.ch/lcg/mou.htm
16. Bell, G., Gray, J., Szalay, A.: Petascale Computational Systems: Balanced CyberInfrastructure in a Data-Centric World
17. Foster, I.: Argonne National Laboratory and University of Chicago. What is the Grid? A Three Point Checklist (2002)
18. The European Grid Initiative Design Study. website at http://www.eu-egi.org/

Chapter 5
Software Development in HEP

Federico Carminati

Software Engineering and the "Software Crisis" that it aims at solving, have been fundamental issues at the heart of the software development activity since the moment software existed. A book about software in High Energy Physics ("HEP") would not be complete without some remarks on this matter, and this chapter provides such remarks.

5.1 Introduction

Modern HEP needs to use computers to manipulate and analyse the data from the experiments. Today's experiments have software suites consisting of $O(10^7)$ lines of code written in fairly powerful (and complex!) languages, such as C++, Java, Perl and so on. In general, HEP has been quite successful at developing its own software, even if with varying degrees of success. A quote attributed, probably apocryphally, to S. Ting, 1976 Physics Nobel Prise winner, is "no experiment ever failed for its software". Whether this has to be considered as praise to HEP software developers is still an open question.

In spite of its rather honourable record, HEP has never claimed to use, nor has it used on a large scale, traditional Software Engineering methods, which have been specifically conceived in order to facilitate the production of large programs. HEP never developed an independent school of thought of its Software Engineering practises. However, some intuitive Software Engineering was applied without ever being formalised. Some of it was quite innovative, and indeed ahead of its time. The failure to recognise this led to a complicated and not very efficient relationship with Software Engineering and software engineers, that still continues today.

F. Carminati (✉)
CERN, Geneva, Switzerland
e-mail: Federico.Carminati@cern.ch

This chapter will provide some elements of explanation of how this could happen, and what are the reasons behind this apparent contradiction.

5.2 Traditional Software Engineering in HEP

HEP has written and operated its own software for years without claiming any formal method. Several generations of successful experiments have produced complex software systems. These have been maintained over the years, adapting to changes in the detector, the replacement of the population of users and developers, and the evolution of the basic technologies of hardware, operating systems and data storage media. Non-professional programmers with no formal Software Engineering training have successfully done all this. Was this madness and pure beginners' luck or was there method in it?

5.2.1 *Traditional Software Engineering*

Writing large programs has always been recognised as a problematic activity. Already in 1958 [1] papers appeared at conferences on the production of large computer programs. As early as 40 years ago a conference was held to address the fact that "Software projects finish late, they run over budget and their products are unreliable" [2]. At this conference the term Software Engineering, which could well be one of the most persistent and pernicious oxymora of computer history, was introduced.

The fundamental predicate of Software Engineering is that solid engineering principles should be applied to the production of software in order to transform it into a more predictable and efficient activity. As one can see this is more an objective than a definition, which is moreover based on a questionable assumption on the existence and field of application of "solid engineering methods". The literature on Software Engineering is vast, ranging from management and process control, to formal methods to prove programs correct, to psychology and philosophy. In very broad terms, the strategy devised by classical Software Engineering to achieve its objective of introducing predictability and efficiency in software production, has been to divide the software production process into a series of clearly defined and well-controlled steps, which could be planned in advance.

> Traditional software engineering: a solution or simply an aspect of the problem?

Unfortunately it is seems to be an unavoidable fact of life that any program, once it is tried out in reality, needs to be changed to conform to its intended usage and behaviour. Software Engineering promoters soon had to accept that no software production activity could be handled in a single pass through these steps. Confronted with change, traditional Software Engineering tried to devise a strategy consistent with the supposed similarity between the software production activity and the engineering processes that, as we have seen, constitutes its founding analogy. Change had to be avoided or, at least, disciplined and controlled via a formal process.

Given that the need to have "feedback loops" in the process is unavoidable, traditional Software Engineering sees the minimisation of these "feedback loops" as the way to streamline and optimise the software production process. To achieve this goal, proponents of classical Software Engineering suggested that software professionals start with as complete and exhaustive a design as possible. Additionally, classical Software Engineering supporters suggested the introduction of as much rigour as possible in each step, and the avoidance to returning to the step as much as possible. A typical example of this approach is the Waterfall model (see Fig. 5.1).

> The problem of software evolution: a curse or an opportunity?

One condition for the success of this strategy is to minimise the ambiguity with which information is moved from one step to the next. This has led to a considerable "formalisation" process in which terms and concepts are precisely defined within a semantic domain. Communication tends to be in the form of documents which try to be exhaustive and as detailed as possible in their attempt to cover all the aspects

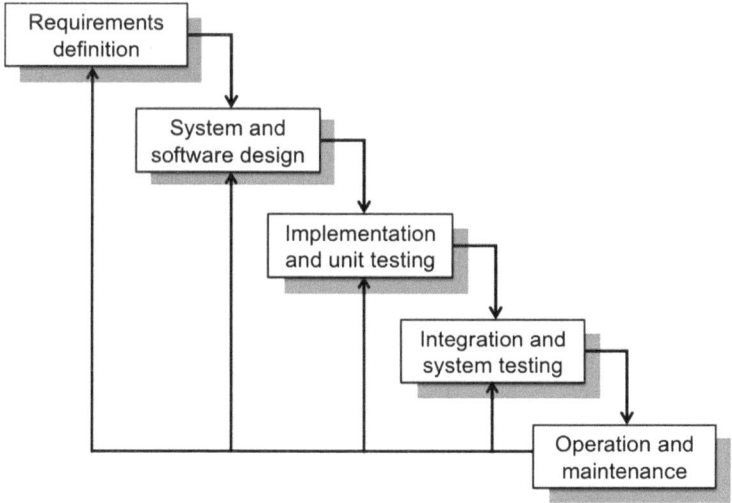

Fig. 5.1 Waterfall model

of each phase in the process in order to minimise the need of subsequent changes. It is clear that this process tends to become quite burdensome and the "overhead" with respect to the software writing activity itself may be substantial. However this could pay off if the objective, i.e. the reduction of the occurrence of change, were attained.

Unfortunately such a process tends to create a situation where the cost of change grows faster than the likeliness of change is reduced. No matter how formal or heavy is each step in the process, it is almost certain that some change in the final product will be required. But when so many formal steps are involved and so many documents have to be modified and kept consistent between each other, that even the smallest change may be the cause of substantial expenses or delays. The net result is to make change slightly less likely but much more expensive with little or no gain.

> High Ceremony Processes: cost of change grows faster than its likelihood decreases.

To be fair, this statement may or may not be true depending on the application domain. In some application domains there is a reasonable chance to capture the essence of the problem and to reach a full specification of the process with an effective reduction of late change. But these applications tend to be few and, as we will see later on, almost none are to be found in the domain of HEP research.

To try to correct this situation, several alternative models have been proposed, centred on the concept of "accepting" the recursive nature of software development. A typical example is the "spiral" model shown in Fig. 5.2.

Another aspect of classical Software Engineering is the precise classification of roles corresponding to the strict organisation of the process. This precise classification supposes that the roles are indeed distinguishable and it tends to introduce another element of rigidity into the process. A typical example can be seen in Fig. 5.3.

Due to their intended rigidity and formalism, classical Software Engineering methods became known as High Ceremony Processes ("HCP"), characterised by:

- Many formal documents
- Very detailed design models, usually difficult to read and understand
- Formal document ownership
- Rigidly distinct roles
- Communications through documents
- Formal processes to follow

HCPs are suitable for big projects with stable requirements, as the time elapsed from requirement gathering to the start of coding may be several months if not years.

HCP Software Engineering appeared in order to address the chronic problems of large industrial software projects, such as slippage of schedules, overspending

5 Software Development in HEP

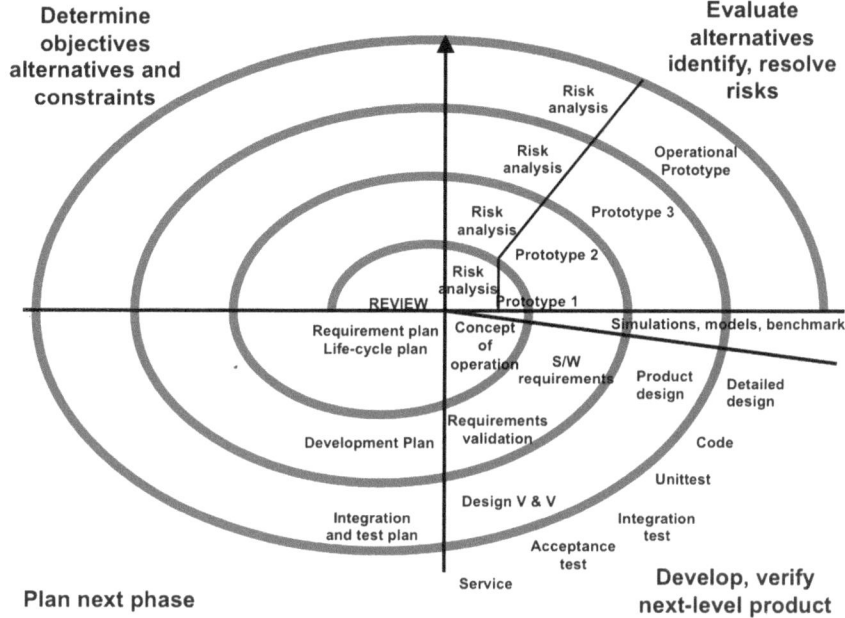

Fig. 5.2 Spiral software development

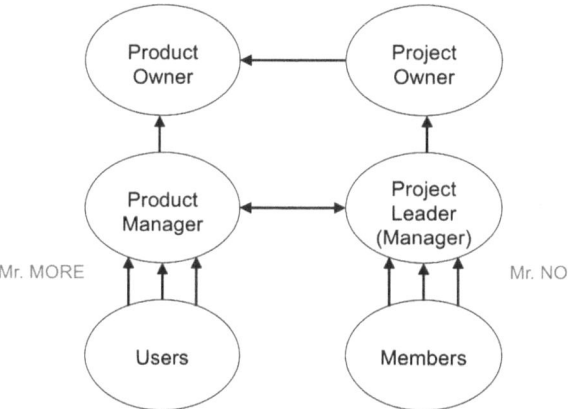

Fig. 5.3 Roles in software development

budgets, not meeting the user requirements and overblown maintenance costs. All this came to be known as the "Software Crisis".

After the Software Crisis came the Software Engineering Crisis.

It can be said that classical Software Engineering in the form of HCP did not yield the expected results. In general, the inadequacy of these methods is becoming more and more evident in the e-Business era, when change can no longer be considered an accident, but it has to be regarded as the norm. Software Engineering had its crisis too. In fact many university syllabi on Software Engineering start with the Software Engineering Crisis. As we will see later on, this is an open and active field for computer science, and new models are being proposed to overcome the difficulties of traditional Software Engineering.

One of the major dangers of the application of HCP Software Engineering to a software project comes from its tendency to adapt reality to its method rather than the opposite. The trouble with HCP Software Engineering is that the elaboration of a plan is very expensive. If reality does not follow "The Plan", the cost of the elaboration of an alternative strategy may simply be prohibitive. It is often the case that, at this point, the entire HCP construction collapses like a house of cards. All divisions of work, established roles, documents, plans, dictionaries and the like suddenly become useless and their remnants stay as a cumbersome legacy that hinders possible correcting actions. The failure of "The Plan" leaves the project with no plan and it then enters a no-man's-land where time-scales and costs-to-completion become unpredictable. The effects are the more serious the earlier this happens during the lifetime of the project. And in turn, the heavier and more rigid the project, the greater the likelihood that reality will not follow at an early stage. This has come to be commonly denoted by the sentence "too much process kills the process".

The reader at this point may expect an explanation of this failure or a moral to this story. Indeed, even if what has been said above can be shared by several professionals of the field, the opinions on the degree of success (or failure) of classical Software Engineering differ widely, and the analysis of the reasons of its achievements cover an even larger spectrum.

I have of course my own opinion, which is based on more than 20 years of development of software and management of software projects. The truth is, I believe, that although we may claim to have reached a very good level of analysis of the situation, we cannot pretend to have a silver bullet. I have led software projects since 1987 and I have never known one that was not in a crisis, including my own. After much reading, discussion and reflection on the subject, I am increasingly convinced that most of us are working with software every day for a number of reasons, without ever penetrating some of software's innermost nature. A software project is primarily a programming effort, and this is done with a programming language. Now this is already an oxymoron in itself. Etymologically, programming is "writing before": predicting or dictating the behaviour of something or someone. A language is a vehicle of communication that, in some sense, carries its own contradiction, as it is a way of expressing concepts that are inevitably reinterpreted at the receiver's end. This is no different with computers: how many times have you thought "why does this computer do what I say, and not what I want?". A language is a set of tools that have been evolutionarily developed not to "program" but to "interact". And every programmer has his own "language" beyond the "programming language". So many times opening a program file and looking at the code I could recognise the author at once, as with a hand-written letter.

> Is programming a "literary" activity as well as a "technical" one?

And if only it were that simple: If you have several people in a project, you have to program the project, and manage communication via human and programming languages between its members and its customers. And here come the engineers telling us "Why don't you build it as a bridge?". We could never build it like a bridge, no more than engineers would ever be able to remove an obsolete bridge with a stroke of a key without leaving tons of metal scraps around, as we can do regardless of how complicated and large a software project is. Software Engineering's dream –"employing solid engineering processes on software development" – is more wishful thinking than a realistic target. Software professionals all know intuitively why it has little chance of working expressed in this way, but we seem unable to explain our intuition in simple words. Again language leaves us wanting.

Trying to apply engineering to software has filled books with explanations of why it did not work and books on how to do it right, which means that a solution is not at hand. The elements for success are known: planning, user-developer interaction, communication and communication again. The problem is, how to combine them into a winning strategy. Then came Linux and the Open Source community. Can an operating system be built without buying the land, building the offices, hiring hundreds of programmers and making a master plan for which there is no printer large enough? Can a few people "in a garage" outwit, outperform and eventually out-market the big ones? Unexpectedly the answer has been yes, and this is why Linux's success has been so subversive. I think we have not yet drawn all the lessons. I still hear survivors from recent software wrecks saying: "If only we had been more disciplined in following The Plan...".

One of the keys of the problem probably lies in the deceptive ease with which software can be written, rewritten and modified. In spite of this, the intrinsic time taken by writing a large piece of code is just a small part of the cost of bringing this code into production. A rigid planning strategy is, almost by definition, poorly fit for such a "product". Any "Master Plan" is doomed to become quickly obsolete, but this does not imply that planning is "per se" useless, quite the contrary indeed. The situation is well described by the famous D. Eisenhower's sentence "In preparing for battle I have always found that plans are useless, but planning is indispensable". In order to implement this philosophy, the whole planning process has to be rethought, to make it lightweight and intrinsically adaptive. We will come back to this point when discussing about software development in HEP.

> "Plans are useless, but planning is indispensable."

Another problem lies in the fact that at the core of applying "solid engineering methods" to software development is the assumption that a personal creative activity, which often software development is, can be transferred into an industrial process. The passage art→handicraft→industry is possible only for some classes of activities. Others do not lend themselves to be embodied in an industrial process, particularly activities where the "repetitive" or mechanical parts are rather limited. As we have already observed, writing a large number of lines of code is very easy, and often the performance and success of a large system depends more on a brilliant idea than on a solid and disciplined software production activity.

At this point, however, the above account of classical Software Engineering's status should be tempered with some of its positive aspects. Information Technology has been part of the technological evolution, and in some sense it has been its motor and its fastest moving element during the latter half of the past century. It is common knowledge that the last century has changed the human condition more deeply than the previous 10,000 years. Information Technology revolution at large has changed our world in a way that was not even imaginable only 20 or 30 years ago. No method of software development could have survived this fast-paced evolution without becoming obsolete. Even if we cannot point to a single Software Engineering method that can be said to capture the essence of software development or that proposes a strategy that works, at some level, in most cases, the collective reflection on software development operated by Software Engineering has been invaluable to understand this activity which is at the heart of the Information Technology revolution and it has certainly contributed to it in many ways. Again, and in spite of its claimed engineering heritage, this resembles much more closely the situation of the human sciences than that of physics and engineering. Psychological or sociological theories can rarely be said to be scientifically true (or, better, subject to falsification), but each theory captures some of the aspects of a very complex reality that does not lend itself to be "reproduced in its simplest manifestation under controlled conditions", and they all contribute to its understanding and to the evolution of knowledge.

> In spite of its limits, the collective reflection on software development operated by Software Engineering has been invaluable.

The trouble is that Software Engineering has been offering interesting forays into a very complex field whereas the software industry needs effective strategies for software development. This complex field is at the heart of our society, constantly changing and evolving and still largely unknown; but, and for very good reasons, it has produced no practical recipe that would yield a predictable result. My own opinion is that the activity of software development is more complicated than was initially thought, and the research about it has only recently started exploring beyond the analogy with engineering. The attempt to devise practical recipes was both necessary and premature with respect to existing knowledge.

5.2.2 Classical Software Engineering in HEP

The introduction of Software Engineering methods to the industry eventually attracted the research sector's attention as well. HEP has not been immune. At CERN, for instance, every new method and formalism that appeared over the years has been tried out (Yourdon's SASD [3], Entity-Relationship (ER) [4], Booch's OOADA [5], Rambaugh's OMT [6], Shlaer-Mellor's OL [7], ESA's PSS-05 [8], UML [9]) up to the latest fancy object-oriented methods. The tools supporting these methods (ADAMO [10], I-Logix Statemate [11], OMW [12], OMTool [13], StP [14], Rational Rose [15], ObjecTime [15]), have periodically raised interest and then fallen into oblivion. Scores of computer scientists have tried to apply these methods with no noticeable impact on HEP software at large, in spite of the time and resources invested. None of the results obtained in this way have come even close to the success and robustness of CERNLIB [16] (and in particular PAW [17] and GEANT 3 [18]), developed over the years with "amateurish" methods, or of ROOT [19], the spiritual heir of CERNLIB.

In practise, all the classical Software Engineering methods introduced so far in our environment have not delivered the expected benefits. Of course people who have invested money, effort and good will, have had some reticence to admit the truth. In some cases, the simple fact that a Software Engineering method had been used for part of a project was considered a proof of success in itself, no matter how marginal the real contribution to the software of an experiment has been.

> Also in HEP, Software Engineering has not "worked as advertised".

As we have seen, it does not come as a surprise that classical Software Engineering has not worked as advertised for HEP. It can hardly be claimed that it has worked anywhere as advertised. However what is perhaps surprising is the fact that the software projects where traditional Software Engineering has been adopted, or at least attempted, are amongst the less successful ones of the history of HEP software at CERN and elsewhere. If we believe, as many did, that HEP has always developed its code without any method, then we may consider it strange that the introduction of some methodology, even if perhaps not the most suitable to our environment, did not yield any positive results.

Our interpretation of this phenomenon is that, over the years, HEP had developed a methodology for software development that was both adequate to its environment and reasonably effective. What these projects did was not to introduce some methodology in the development of HEP software, but rather to replace an existing methodology, which, as we will see, was never formalised or properly described, with one, well represented in the literature, but rather inappropriate for our environment. This chapter is, in some way, trying to reveal some of the main elements of the "HEP software methodology".

5.3 The HEP Software Development Environment

The problems that software designers are called upon to solve in HEP have some specific features that differentiate them from those of an industrial environment. The purpose of industry is to produce goods, products and services and bring them to market. The technologies involved are, in most of the cases, mature and well understood. Marketing and selling implies knowledge of the customer needs and the capacity of maintaining a quality standard for all goods that are sold. The output is measurable in terms of quantity and revenue. Late discovery of flaws is expensive and may seriously harm the image of the manufacturer.

In the research environment in general, and in HEP in particular, nearly all these points are overturned. Our general objective is to acquire knowledge. Not only is what we research still unknown, but means to achieve our objectives may not be available yet, and their acquisition may be part of the research activity itself. We are constantly required to innovate, in order to provide tools and methods that current technology cannot supply. Some of the specificity of our development environment is detailed here.

5.3.1 Evolving Requirements

The object of our research is usually very well defined. HEP projects, particularly the very large ones that characterise modern HEP, are regularly monitored by scientific and financial peer reviews. However, while the objective is clear, often the way to achieve it is not entirely specified, particularly in the software arena. This is not out of incompetence, but because of the very nature of the problem at hand. Research proceeds in steps, and each step opens up a new horizon. New ideas on how to continue and progress are brought up, which refine, or even modify, the initial requirements. Innovation and creativity are at the very heart of what we do. It is therefore impossible to define a comprehensive and stable set of requirements from the beginning. Room for progressive refinements of the requirements must be allowed for. Sometimes, it may even happen that at a given step a completely new field of investigation appears, requiring new tools and developments, which were not foreseen in the original plan.

Software in HEP is not only an instrument for pursuing the research objectives, but it is often the subject of research and development activities of its own, albeit ancillary to the main research subject. This is a further reason the requirements are not entirely known at the beginning of the activity. The software system used to analyse the data evolves together with the experiment itself, and therefore the requirements depend also on the functionality of the software and on its architecture, which change as the R&D activity proceeds. Note that this is true both for functional requirements and for those related to performance and scalability.

> High Energy Physics unique features call for a specific software development methodology, different from industry.

5.3.2 Forefront Technology

Advanced research methods move at the same speed as the technology "boundary", and actually push it. The objectives of the research are chosen so that the technology will just about allow them by the time the experiment starts. Technologies are often still relatively immature and their products do not have a level of "quality" comparable to those commonly available on the market. In addition they are not stable. Important performance enhancements are likely to happen during the life of an experiment, and new technologies may quickly become obsolete and be abandoned, if the market does not follow. Also very important are the financial aspects. Research is often constrained by its budget and by its objectives, but it is free to take substantial risks with new and unstable technologies provided they hold the promise of the results within budget.

5.3.3 One-off Systems

The software systems used in HEP are, to a large extent, unique. They are developed for one specific experiment and often their lifetime corresponds to that of the experiment. Therefore tailoring them specifically to the application is an advantage, not a penalty. Features not required are not welcome. Given their specificity, rapidly changing requirements not only do not involve the overheads incurred in the industry, but are a necessity.

5.3.4 Non-hierarchical Social Relationship

The links among HEP developers and users have a peer-to-peer nature, as opposed to the hierarchical organisation of commercial companies. The roles of user, analyst, designer, programmer, documentation writer etc. are shared among all the developers in the team, each person taking several roles at different times. In general there is limited specialisation of roles and small independent units participate in the projects, with no higher authority. Decisions are taken by consensus. The traditional distinction between designers and customers is also blurred, since often the same persons belong to both categories. We will return to this issue later in the chapter.

5.3.5 Assessment Criteria

The performance evaluation of a given HEP project can neither be based on revenue nor on marketing success. The criteria that apply are in general difficult to measure, since they concern the quality of science produced, or on subjective aspects such as user preference and satisfaction. Also, because these systems are "one off", any market study is impossible and it is impossible to answer whether things would have been better with a different system.

5.4 Software Engineering in HEP

The heart of any Software Engineering method is the attempt to control the impact and the cost of change. Common wisdom holds that changes in design are less costly at the beginning of development than later on. Therefore, it seems wise to concentrate change at the beginning, when it is easy, and to avoid as much as possible late changes. The road to this Eldorado of stability is sought in a complete, detailed and exhaustive design of the system before development starts, or at least before the system is deployed. Surprisingly enough the idea that all software systems can be designed first and then implemented has seduced several people and the very fact that it is merely a mental model implying a gross simplification of reality seems to escape the majority.

5.4.1 Who is Afraid of Change?

Customers, requirements, technology, market, people, science are all constantly changing. The faster and less predictable change is, the more irrelevant the initial design will be, and the more we will have to accept to change it. It is illusory to believe that the design will survive unscathed the effect of change, no matter what amount of effort you have invested into it. This may have been less clear at the time of the mainframes, but it is an inescapable fact now. Until very recently the reaction has been to include the option of change in the initial design, with the result of making the method heavier, more complicated, and in the end, less susceptible to unforeseen change.

> Managing software development starts and ends with managing change.

In some sense HEP has always lived in a situation where change was unavoidable at any moment. As we have explained before, some requirements evolve and become clear only during the deployment of software and its use. New ideas are born, which could not have existed if physicists did not experiment with the software. The drive to buy the cheapest possible hardware and the necessity to use existing computing equipment in the collaborating institutions, as they join the experiment, impose a continuous evolution of the computers and operating systems. Once the detector is built, change comes from the need to experiment new ideas and to adapt to unforeseen experimental conditions and to the modifications and upgrades of the apparatus.

If change cannot be avoided, then the economy of change must be reconsidered. The axiom that late change is a disaster needs to be challenged and a methodology that does not allow the cost of change to increase exponentially with time needs to be implemented. If this approach is taken, then sometimes even deferring implementation, and thus creating the possibility of a future change, can be the winning strategy, as the time to implement a given feature may never come. In this case, starting with a minimal design and then evolving it may be much more effective than spending a large amount of time designing the system while the requirements are still unclear.

5.4.2 The Hidden Methodology

Over the years HEP has developed a self-styled method that, within the constraints and relative to its objectives, has been surprisingly successful. What computer scientists have had difficulties appreciating is that HEP is an environment where a well adapted method is applied *instinctively* but successfully to the projects. What is peculiar of HEP is that this methodology is transmitted via "oral" tradition and it has never been formalised. This situation is far from ideal, as knowledge must be abstracted and formalised in order to be properly communicated and to evolve. We will come back to this point later on.

> Does a High Energy Physics Software Engineering method exist?

HEP development strategy manifestly can cope with change. The Large Electron Positron (LEP)-era experiments started when the dominant computing platforms were mainframes, and the total computing needs estimated for LEP were less than the power of a modern laptop. Yet the LEP software has been ported and adapted over all the technology revolutions intervened over 20 years, and the user investment has been protected. HEP has taken advantage of every piece of inexpensive hardware that hit the market with no disruption in the production of results. The same code that was running 20 years ago on mainframes and on the first VAXes is now running on Pentiums with little modification. We believe it is important to analyse and understand the reasons for this undeniable success to see how these can be leveraged for the new generation of experiments.

5.4.3 The HEP Development Cycle

The main feature of the HEP development method is the apparent absence of design. Indeed the most successful HEP projects started with a very precise initial idea guiding the development. Developers and users "knew what they wanted" because they knew what they missed in the past and what they would have liked to have. This is not very precise but to any Software Engineer's despair, it is as precise as you can make it without imposing useless constraints and without inventing details that most probably you will have to throw away, without wasting time. The objective or "plot" is clear, although sometimes difficult to express and formalise.

> The development is guided by a common "plot" that finds its origin in the shared experience of the HEP community and not by a formal "design".

This initial plot, or metaphor, becomes common ownership of the community of physicist users, and guides the evolution of the system. The actual code is written usually quickly in accordance with the objectives of the project, and it is quickly delivered to the community of users, who are very active in testing it because it answers their immediate needs.

Once started in this way, the software naturally goes through development cycles. The most important features are the ones on which the developers work first, because they are the ones most in demand by the community. Somehow, in the best cases, it all happens in the right way because there is a common culture of the product, the moral of the initial plot, which is shared by the community. Users usually complain vocally, but constructively, and they insist on using the product. Developers implement the most wanted features. Advanced users provide pieces of code, following the style and the philosophy of the product, that most of the time can be just "slotted" in place and become part of the common code-base.

The community that feels collectively responsible drives the code development. Parallel developments often merge to give birth to a single product. When necessary, re-factoring or even massive redesign happens. New releases appear quite frequently, and the version under development is constantly available to guinea-pig users or to those who desperately need the new features.

> The community feels responsible for and guides the software development.

Progress is guided by mutual respect, dialogue and understanding. Most of the communication happens via e-mail and it is not infrequent that some of the major developers meet only a few times in their life. Major redesigns are carried out with the responsibility to protect the investment of existing users, while paving the way for further evolution. This is not the description of a dreamy future. This has been the day-by-day story of the CERN Program Library from the late 1970s to the early 1990s when it was frozen, and it is the day-by-day story of ROOT since then. Other products, where the design-first method has been used, either have never seen the day (E. Pagiola's GEM in the 1970s) or had difficulties entering production (e.g. GEANT4 [20]) or have been rejected altogether by users (e.g. LHC++ [21]).

In what follows we will see some of the most important aspects of the HEP software development cycle in greater detail.

5.4.4 The Fine Art of Releasing

One of the most delicate points of the development of a software project in HEP is the process of delivering (*releasing*) new versions of a product (including the first one!) to the users. HEP has developed a rather sophisticated procedure for doing this, that in many ways is central to its own development process and that is instructive to examine in more depth.

The reason that the release process is of paramount importance in the HEP software development cycle is that, as we have seen before, requirements are not all and not completely defined at the beginning of the development. Therefore, as the understanding of the requirements becomes clearer and new requirements are added, new versions of the software, sometimes with substantial changes, have to be brought to the users. The acceptance of change in the development process which characterise the HEP development strategy, as opposed to the traditional HCP Software Engineering, implies that any product will go through several releases during its life-cycle, some of them carrying substantial changes in the code. The release procedure is therefore central to the success of any development strategy adopted by HEP. It is the way developers bring to the users the change that they expect when existing requirements evolve and new ones get added.

When confronted with a new entity, human beings tend to feel both positive anticipation and defiance. The trial and eventual adoption of a new entity marks a suspension, at least temporary, of defiance. The exploitation of this window of opportunity, which consists in a "suspension of disbelief", is central to a successful release process. It is here where the apparent "informality" of the HEP software development comes to an end and a predictable and dependable process has to be put in place.

> The first trial of a new piece of software requires a temporary grace period, a "suspension of disbelief".

It is perhaps quintessential to the understanding of the HEP software development process to see how the stability and dependability are transferred from the "product" itself to the "process" that provides the product, i.e. users do not expect the product itself to be frozen and to respond to all their requirements at once, but instead they come to expect that the process by which their requirements are progressively satisfied is "predictable" and, as much as possible, uniform. As a matter of fact, and we will come back to this, users do not even expect the product to be flawless, and they are ready to accept a certain amount of "frustration" due to bugs, as long as their feedback is rapidly and properly taken into account. This is in some sense a "virtualisation" of the software product, where the focus moves from the actual software to the process that produces it.

The advantage of this evolution is to add flexibility to the strategy used to achieve the final goal of providing a product satisfying the user needs. In particular this opens the way to the introduction of "competition" between different solutions in response to an emerging requirement within the project, as long as these parallel solutions are framed in a "predictable" release life-cycle management process. The most successful HEP projects have profited from this "Darwinist" approach where different solutions can coexist for a given period until one of them is privileged. The ample possibilities offered by C + + and other modern languages to abstract the functionality from the actual implementation have made this process more "respectful" of the user code investment.

Here again HCP Software Engineering is at disadvantage, because the existence of a rigid plan tends to reduce the space for emerging alternatives. So when the "chosen" solution turns out to be not the right one, the chances that a better one could find its way into the mainstream are slim. In this case it is not uncommon to see an HEP software project using HCP Software Engineering prefer to invest in Public Relation activities rather than in reworking the planning to recover from the situation. In more crude terms, failures are hidden and reality denied, while user satisfaction is sacrificed. This may seem a suicidal attitude doomed to a rapid failure. Unfortunately this can indeed be a successful strategy for a software project to survive, and even see its funding and support continued.

> The user expectation of stability and predictability is transferred from the product to the process that produces it.

The backdrop of this kind of situation is set by the fact that user satisfaction and scientific output are the only measurable return of HEP software. If the source of financing or the management is far enough from the actual users, it may have difficulties in evaluating these two factors. Moreover, admitting a divergence from the initial plan may be perceived by an outsider as a setback and may have an image cost for the project. This especially tends to happen when projects are started by "managerial decision" without a close interaction with the users of the kind described above.

Sometimes the "collateral" benefits that the user community, or a part of it, can reap by acquiescing to this policy are non-negligible. For instance some of the users may profit in their scientific activity from the funding of the project in question and so have an interest in its continuation irrespectively of the quality and usability of the software produced. Under these conditions, this strategy may turn out to be applicable. Of course users will have to find an alternative product to do their work, and this is usually produced by a "successful" HEP software project, usually developed along the lines described in this chapter.

Fortunately these situations are quite rare, but some did emerge in the recent years. One very telling variable that can help detecting them is the actual usage of a product by the users. It is significant that these projects tend to present the most strenuous resistance to the simple proposal of "counting" the users that actually use their product.

Coming back to release strategy, the first release has special importance. Usually a product is developed when there is a need for it, therefore the user community has expectations and a genuine need to use it in its day-to-day work. Delaying the release too much may force the user community into using some alternative product and then becoming entrenched in it. Early release is important: there is a window of opportunity that should not be missed due to a late release. Also, as changes will certainly be requested, early release when the code base is still limited will make substantial changes of direction easier. Late releases tend to make developers less receptive to requests for major changes in architecture and direction, and we have often seen such "original sins" having long-haul consequences on the life-cycle of the software product.

> The timing of the release process is one of the most important factors for a successful software development.

However, there is a danger in early releases as well. If the product is too immature or offers too little functionality with respect to the intended use, it risks disappointing the users who therefore will try it only once. Finding the right moment for releasing a product depends on a number of factors whose appreciation has a great influence on the subsequent success or doom of the product. These include the user community's expectations, the flexibility and "courage" of the developers in modifying the initial design to follow users' feedback and the expected rate of evolution of the requirements.

Subsequent releases follow a similar logic with the distinctive feature of being rather frequent. "Release often, release early" is the punch line, and the necessity of this follows directly from the dynamics of the development that we have explained above. As requirements "trickle" in during the project development, frequent releases are necessary to make sure that the production version of the product satisfies a large fraction of users. Also, in the absence of a "master plan", constant "user verification" of the changes is needed in order to keep corrections small and not disruptive.

Failure to follow this simple recipe has usually been the symptom of a malfunctioning project. Moreover, this is a self-catalysing process. The longer the release time, the larger the number of changes. In such a situation it quickly becomes impossible for the users to simply "replace" the working version with the new one, because the impact of a large number of changes cannot be easily estimated, and substantial backward incompatibilities may have crept into the system. This tends to require extra checking and tests, and may lead to the necessity to expose the system to the users *in parallel* with the current version for a long time. This makes administration heavier and user interaction more complicated. In a fixed-size developer team, the necessary resources are subtracted from development. Moreover, if the differences between the current and the new version are significant, few users will have the time or willingness to test the new version, with a reduction of valid feedback. This in turn slows down the rate at which the new version improves to meet user needs, which creates a vicious circle.

When this situation arises, there are essentially five major handles on which to act in order to break the stalemate and recover, already described by Beck [22].

Money: Sometimes buying better and faster machines, rewarding developers with cool laptops or installing fancy junk food vending machine in the developers' corridor might help. In most cases however the time derivative of the improvements that these measures yield is smaller (if positive!) than the degradation of the situation due to slipping release dates. Given the resource-constrained nature of most HEP experiments, depending essentially on public funding and usually chronically understaffed and underfunded, this option is not readily available to ailing HEP software projects.

Time: If the release date cannot be met, then delaying it is the most natural reaction. However very often it is the worst option too, because as time passes, new requirements arise which, *because the release is delayed anyway, and because it would look very bad if they were not introduced by the time finally the code is released* tend to be introduced. Unfortunately this delays the release even further and it is often the starting point of the vicious cycle. We have seen this happen very frequently in HEP, where even mission-critical products have seen substantial delay in their release. The peculiarity of HEP in this respect is that timing is governed by the experimental apparatus (accelerator and detector). By the time the RAW data start flowing, users will make use of whatever product is there, or will *force* the project to deliver the product. In the best case this will yield a working product with a reduced scope (see last bullet), in the worst, a partially working product which will address the user requirements only partially.

People: It is well known by now that *adding people to a project which is late usually makes it later* (Brooks [23]). Nevertheless also this seems to be the most common corrective action taken when a project is in crisis. In HEP this action can be particularly damaging because of the financial and sociological constraints of most HEP experiments and projects. The professionals that are most readily available for their "low cost" and because usually research institutions "by mandate" encourage their employment, are young students and fellows. Irrespective of their quality, which is often very high, these people need to be trained and therefore for some time are more a hindrance than a help. Moreover this investment is, most of the times, of little return, as their permanence in the project is usually limited to a couple of years at most. So I argue that HEP development environment makes even more acute the fact that adding people to a project in crisis almost always results in a deepening of the crisis. This is of course not to mean that (wo)man-power is not an essential element in the success of a software project. The point is that the addition of people to a project requires an investment from the members of the project to integrate the new resources, and in time of crisis these resources may not be available to be mobilised.

> Money, time, people, quality and scope are the parameters that can be tuned to control the software development process.

Quality: It can be very tempting to control a project via the quality of the code delivered. However writing low quality code is only marginally faster than writing good code. On the other hand, debugging low quality code can be much more time and resource consuming. Any compromise on code quality will result at the end in a net waste of resources, if not in the dooming of the entire project. Of course aiming at the highest possible quality (whatever this means) can also be a reason for failure. The definition of the "right" quality for a given code, and of the means of implementing it is one of the hardest tasks for a project manager, and we will come back to this point later on.

Scope: This is perhaps the least used parameter in software projects, and this is probably a legacy of the Software Engineering HCPs where the scope of the software is deeply embedded into the original planning. When properly employed, this is a very effective handle to control the evolution of a software project. Whenever a feature risks delaying a release, it should be seriously considered whether to re-schedule it for a subsequent release. In fact even *anticipating* a release when in a difficult situation could be a good decision, as this allows developers to focus on the really problematic items while reducing user pressure. As explained above, the natural deadlines that accompany the life of an experiment have forced HEP software development to adopt this strategy, sometimes just as an *emergency* measure without perhaps realising that this could be turned into a software development policy tool that, if judiciously employed, could increase the flexibility of the software development process. The establishment of an a priori release schedule, irrespective of requirements and development times, has often been proposed as an effective way of managing a software project. Indeed, the experience of the ALICE Offline Project, where this method is applied, although not too strictly, is quite positive in this respect.

As we have seen above, the release strategy is central to the success of a software project. As with many other aspects of HEP software development, the release strategy has evolved to adapt to the peculiar HEP environment. However most of the distinctive features of the HEP release process are common to modern Software Engineering and in particular to the Open Source community. Here, once again, we are faced with one more case of "independent discovery". Again, we can only regret that more self-consciousness of our own procedures was not present, which could have led to a constructive dialogue with modern Software Engineering experts and Open Source project managers.

5.4.5 *Testing and Documentation*

It is common knowledge that the importance of testing and documentation cannot be over-emphasised in a software project. However both come at a price, and in a resource-strapped project these may well be in competition with the writing of the actual code and meeting the project deadlines. HEP software projects have developed specific strategies to cope with these problems.

Developing an exhaustive test suite represents a substantial amount of work, often comparable to the effort invested in writing the code itself. Moreover, the relation between the effort invested in a testing suite and the number of bugs that effectively "escape" it is not linear. While a relatively simple testing suite is invaluable at detecting a large percentage of the flaws of a new piece of code, the reward in terms of problems detected versus effort deployed in the improvement of the tests decreases rapidly. The other element that has to be taken into account is the sheer (elapsed) time taken by the test suite. Given the rapid development cycles, a test suite that requires several hours to run would not be adequate to be run frequently.

One specific feature of HEP applications is that any single program can fail without dramatic human or economic consequences, unlike for instance the control system of an Airbus. This is not to say that failure of an HEP software project as a whole has no consequence. First of all there is a loss of image, which can be quite serious, and then, if the failure results in a serious damage to the discovery potential of an experiment, this may affect further funding, and ultimately damage scientists' careers and lives. However, no crash of a single program leads immediately and inevitably to a catastrophe.

> Testing and documentation are important, but in HEP the resources spent there are in competition with the actual code development and maintenance.

In this situation, the nimble release cycle effectively turns the user community into a large and exhaustive non-regression testing system. Given that the difference between releases is usually small, users can move quickly from one release to the next with their production code and therefore submit the newly released code to the most stringent test of them all, the full spectrum of real use cases. Of course this technique only works if some conditions are met.

The quality of the release must be indeed rather high. A new release that immediately breaks most user applications would simply block the development process. This means that there is a non-regression testing procedure that, however minimal, weeds out all the "trivial" bugs. Problems must be addressed quickly and professionally. The physics user community expects that programs may fail as a "fair" flip-side of seeing their requirements quickly addressed by the developers. However they also expect problems to be addressed quickly, either by workarounds or by the issue of a patched release. The code of the program must be easily available for inspection. This had often the positive effect that expert users may report not only the problem, but also a thorough diagnosis of its causes and, in several cases, a proposal for fixing it, if not the directly the corrected code!

As a matter of fact, this can also help in writing the test suite itself. It is often the case that some (or all) of the code used to test the package comes directly from the users. Usually the code is chosen so that it provides exhaustive checks and can be run quickly. Because of the general Open Source policy of the user community, there is usually a large choice for this kind of code which makes very good test cases.

Documentation poses another problem, which is different in nature but not less serious. It is clear from what has been said that writing the documentation beforehand is not an option for a project where requirements, and hence features and functionality constantly evolve. Also, the usual problem of being resource-constrained forces hard choices between the writing of the documentation and the development of the code.

> The fact that HEP code is documented only when it reaches maturity does not mean that documentation is missing or of low quality.

The lack of documentation of HEP codes is often taken as a fact, however a closer examination of the major programs shows that this is not necessarily the case. GEANT 3, GEANT 4, FLUKA, CERNLIB, PAW, ZEBRA, HYDRA, ROOT and so on all have extensive and well-written user manuals. Moreover the code of packages is usually rather well commented in-line. The fact is that this documentation is usually created only when the program reaches a certain maturity and therefore stability. The best example of documentation are those developed by a collaboration of one or more users with the authors of the packages. The authors are not always in the best position to document the system themselves, as their intimate knowledge makes it difficult for them to appreciate the points that need explanation for users.

For modern systems such as ROOT, the code is available on-line and it is transformed into an hypertext with links between different parts of the code. The in-line documentation then becomes of paramount importance to describe details of usage and also some algorithmic details that may not be found, yet, in the "printable" user guide. This is achieved either via commercial tools (e.g. DoXygen [24]) or via home-grown products like the THhtml class in ROOT.

Another tool that is worth mentioning in this respect, is the *reverse engineering* tool developed by the Fondazione Bruno Kessler institute [25]. This tool analyses the code and extracts UML diagrams representing the current status of the class structure. While UML has proven to be rather unsuccessful as design tool in our environment, it has shown a certain utility to illustrate the code to discuss code design once it is implemented. The other advantage of this tool is that it understands dependencies with polymorphic containers and therefore it can detect and display dynamic dependencies between classes.

5.4.6 Spaghetti and Heroes

During the development of a large package over a few years it often happens that some parts of the code become quite untidy, and the elegance of the initial design (if any!) is progressively lost. The resulting code tends to hide away the logic and flow of the algorithms and to become difficult to maintain. This is what has come to be known as "spaghetti code" in the programmers' jargon. "Spaghetti code" is usually considered the prime symptom of the improper application of "solid engineering principle to software production", and therefore a serious indication that there is something wrong with the corresponding product.

It is our experience that reality is, unfortunately, much more complicated, and that not all pieces of code that look "dirty" are indeed a problem to be fixed. Almost every piece of code tends, over its lifetime, to become more complex. The principal cause for this is the discovery of conditions of use that were initially not foreseen or anticipated. When more robustness and flexibility are built into the code, inevitably alternative paths and conditional branches also appear, which make the code less easy to interpret and more difficult to maintain.

> Complicated and possibly untidy code is often the embodiment of complicated problems.

However this complexity is also a richness, because it embodies the experience of the developers and the feedback from the user community. Two are the main reasons for which a complex code can be regarded, in this case, as a problem. The first is the difficulty to maintain it and the second the loss in performance that conditional branches may introduce.

Before considering any correction to the so called "spaghetti code", it should be seriously considered whether the code is indeed over-complicated, or just a mere reflection of the actual complexity of the problem at hand. In the latter case the best solution could be to find developers able to deal with this complexity. If the code has really lost performance due to the increased complexity, then action may be necessary, and we will come back to this further on.

The second cause for some code to grow complex with time is the necessity to run it on different platforms than those initially foreseen. Here design can help to minimise the problems of code porting. HEP has a long tradition of writing very portable code because of the need to exploit the less expensive combination of hardware and operating system on the market at any given time and also because each institution participating in an experiment has its own purchasing policy that cannot be influenced by the experiment itself. Machine specific parts of the code usually introduce compile time branches rather than execution branches, which also contribute to making the code less readable and more difficult to maintain.

At this point I will make a small digression on the importance of porting code to different platforms, even in the absence of an immediate "economical" reason to do this. First of all it is more than likely that, over the lifetime of an experiment, its code will have to be ported to different platforms for the economical and sociological reasons explained above. The importance of application porting as a way to improve the code quality is well known and cannot be over-emphasised. Some bugs or program flaws, possibly causing wrong results, may remain hidden forever on one architecture and immediately provoke a crash on a different one, forcing their debugging and correction.

> Writing portable code is one of the highest return investments in software development.

In HEP we have repeatedly made the experience that if a code has not been designed for portability from its inception, and, perhaps even more important, ported on different architectures since its first release, it may be very difficult or even prohibitively difficult to actually port it later in its life-cycle. However, even the best portable design does not help if the code is not actually ported, run and tested on more than one system. System and hardware dependencies tend to creep into the code unknown to the developers, and these require very heavy surgery when the necessity to port the code finally becomes inescapable. A glaring example of this has been the code of the European Data Grid (EDG [26]) and Enabling Grid for E-sciencE (EGEE [27]) Grid projects. While in the Technical Annex of the proposal to the European Commission it was foreseen to port the code onto different platforms from Linux to Windows, the developers concentrated their efforts on the

Linux version. This *original sin* caused the code to be so dependent on the Red Hat Linux version on which it was developed that even the transition to successive versions of Red Hat Linux involved a full-porting effort, and caused substantial friction with the computing centres who wanted to upgrade their machines long before the corresponding version of the Grid software was available. Moreover the porting from 32 to 64 bits architectures also took a very long time, with the result that some computing power provided in the form of 64 bits machines was not exploited, and it was still difficult to exploit ten years later.

At the other extreme the ROOT example could be cited, which was designed for portability since the beginning, and is now simultaneously released and running on a very large range of hardware architectures and operating systems. A somewhat intermediate example is the AliEn system. The framework has been designed for portability, however some of the Open Source elements have a very limited degree of portability. Thanks to the portable framework, and to the effective help and feedback of the user community, the system now runs on several different platforms, even if with a rather large number of "custom" patches to some of its Open Source components.

The example of the ALICE experiment software suite, the AliRoot system, may be instructive in this sense. At the beginning of the development in 1998 it was decided to have the system running also on a Digital AlphaServer workstation with Tru64 Unix. Although this has been the only AlphaServer which ever ran AliRoot, the exercise has been very useful, as it was one of the few 64 bit operating systems available at the time. When the 64 bits HP Itanium and the AMD Opteron systems appeared, ALICE was able to port its code on those machines very rapidly, with minor adaptations of the Tru64 version. For some time ALICE was the only experiment able to run on these platforms, before the other experiments could catch-up and port their code.

A code designed for portability concentrates the "system dependent" functionality and code constructs in a number of code elements (classes, methods or functions) that are intrinsically "clean", as each of them is specifically written for one system. Moreover they abstract the system dependencies, and therefore most of the time the porting effort is limited to implement those functions for a new system, leaving the rest of the code untouched. Thanks to this, the rest of the code is not littered with obscure system calls or with endlessly growing compilation-time system-dependencies, and the dependencies on the system are easy to find and document.

Whenever the necessity arises to correct a piece of code that has grown complex and untidy, the developers find themselves in front of a difficult choice. A very natural reaction from most developers would be to leave the code as it is and to rewrite it from scratch. This is a very tempting option as in principle it offers the chance to avoid the old mistakes and to only retain the experience of what worked embodied into a clean and efficient design. In this approach, the developers imagine to deploy a concentrated and "heroic" effort that would lead, in a short time, to a new product, free from the previous sins and embodying all virtues learnt from the past mistakes. Here again Brooks [23] explains how the second system designed by an engineer will be his worst ever. We have seen this several times in recent HEP developments, and developers seem to be so attracted by the perspective of a "new start" that they usually forget the traps lurking along this path.

> "Heroic" re-write of large codes often end up in a stalemate leaving users and developers in the middle of the wade.

The old code, while "spaghetti" or poorly performing, was nevertheless working at some level. Particularly in HEP, manpower is never redundant, and therefore the effort devoted to the development of the new "clean" project has to be inevitably subtracted from the maintenance and further development of the previous version. Whatever the ambition, capacity and commitment of the developers, the time taken by the rewrite will most probably be longer than the average time between releases. So the release cycle of the product will have to be paused, with an accumulation of bug reports and requirements, some of them new, that will have to be embedded in the redesign and rewrite. This tends to slow down the "straight shot" that was planned at the beginning. Users grow impatient with bug fixes and "simple" requirements being put on hold for the old version while the new one is worked on. At some point pressure mounts high and therefore either a new release of the old code has to be issued, further slowing down the development of the new one, or the new code is released in haste. The quick release is achieved via a compromise either on quality or scope of the new code. However reductions of scope are difficult to achieve in this case because the new rewrite started with a plan to implement "at least" the functionality of the old one, which by definition was already the result of a long evolution.

So the poor users end up with two versions. The following dialogue is a summary of endless mail threads ensuing from this situation: U(ser): "I have problem A." D(eveloper): "It is fixed in the new version." U: "But the new version crashes immediately." D: "We are working on that, use the old one for the moment." U: "Yes, but it does not solve my problem A." D: "That is solved in the new version." It looks like a bad existential play, indeed it is a typical stalemate situation we have witnessed many times. From what has been said it is clear that once the project is in this situation, the outcome depends strongly on how far the new version is from being capable of completely replacing the old one, albeit at the price of some degree of user unhappiness. The refusal to cut the Gordian node and the continued maintenance of two versions may doom the entire project, or just the new version, losing all prospects to address the structural problems of the old one. But the old one has been meanwhile worsened by hasty patches because it was supposed to be soon replaced by the new one.

> Courageous and gradual re-factoring is often the most difficult, but also most effective way to improve and evolve old code.

While every project is different, and every "spaghetti" dish is also different, the best practises observed in HEP in these cases follow the path of a progressive "re-factorisation" of the deprecated features or pieces of code within the normal development and release cycle. The necessity of limiting each step to what can reasonably be done between two releases, which, as we have seen, tend to be rather frequent, is a good guide to gain a comprehensive, in-depth and pragmatic view of what really has to be done and what the priorities are. This need for a continuous output avoids "throwing away the baby with the dirty water", and preserves the parts of the code that are not on the critical path of maintainability or performance, while having a working system that can flexibly respond to the user bug reports and new requirements. This approach is perhaps less narcissistically gratifying for developers, as they can implement their vision only piece-wise, but is much more "safe" and efficient. It is also much more difficult intellectually and professionally, as it requires an in-depth knowledge of all aspects of the systems and of the user expectations to decide what to address first and how. However, contrary to how it may seem, it does not limit the scope of the re-factoring to small interventions. Programs like PAW, ROOT, GEANT 3 or FLUKA have seen very massive re-factoring during their life-cycle, with no "messianic" introduction of an entirely new version. The progress has been gradual but constant, with no major disruption of the service provided to the users, even if some releases have been more successful than others.

Also, the timescale of such a process is comparable with, if not faster than, a complete rewrite. If judiciously managed, it tends to avoid being bogged down in vicious circles, and is less vulnerable to change in manpower situation and user requirements, which are quite frequent in HEP. As most of the other concepts described in this chapter, it is also more difficult to package into a precise recipe and most is left to the appreciation of the people who manage the project.

5.4.7 The (Wo)man of Destiny

It is clear from the above that the management of a large HEP software project requires making several choices which are highly critical for the success of the endeavour, such as the release schedule and scope, the re-factoring strategy and prioritisation and many others. All these choices do not follow directly from the technical competence of the developers, either as programmer or scientists, and they depend on a large number of factors that can be described but not easily quantified.

This is perhaps one of the weakest points of our story. While it is important to analyse the circumstances that led some projects to success and that doomed others, some uneasiness remains in the difficulty to indicate a "fail-safe" process or a silver bullet. Most of the merit or fault with the fate of a software project seem to lie with the manager of the project and with his or her ability to assemble a team of capable developers sharing the same vision and working harmoniously together, to heed the message coming from the users, and often even anticipate it. Somewhat unfortunately the personality of the manager seems to play a central role, which cannot be reduced to a set of rules to be applied.

> Even taking all the "best practises" into account, many of the reasons of the success of a project seem to hinge on the qualities of the project leader and of the developer team.

If we consider that each large software package is a one-off system, and in HEP it is meant to accompany and enable experiments to break new ground and push the knowledge boundary, then we may refine the comparison with engineering taking as terms of reference the projects which built really innovative systems, or to the work of an architect and his or her team in designing a public library, a museum or a opera house. "Solid engineering methods" are applied, every day and in all instances. But these are necessary not sufficient conditions. Projects have indeed failed because "solid engineering methods" were not applied, the code was not portable, the release schedule was not flexible, the design was wrong from the start and so on. But even if all these pitfalls are avoided, and the "best practises" are used, the difference between a successful and a failed project may lie somewhere else. It depends on the creativity of the men and women working on it, on their degree of coherence and on the skill of those who make the decisions.

Again, this is not to say that the reflection on these issues is not important, but perhaps the main failure of HCP Software Engineering is exactly that it has tried to marginalise the human factor in the production of Software, trying to minimise the aleas by ignoring the part of imagination and intuition in the whole process. True, if this is the conclusion, our hope to find a set of precise rules to be applied for the success of a software process may go frustrated. However this makes software creation a much more challenging activity, akin to other highly motivational activities and of great intellectual content, such as fine engineering, industrial design or scientific research, where mastery of the technical aspect is necessary, but rarely sufficient. I do not believe that this should be regarded as a negative conclusion.

5.4.8 The People

One of the peculiar features of the development process of HEP software is the characteristic "sociology" of a physics experimental collaboration. The coherent development of the software is usually managed by a project, often called *Offline* in contrast to the real-time data reduction which is performed *On-line* while the data is being acquired.

The scientists participating in this project work for the different laboratories and universities which collaborate at the construction and exploitation of the detector. One of these institutions, usually a large national or international laboratory, actually hosts the experiment. Usually (but not necessarily) the host laboratory has the responsibility of coordinating the Offline project, and to actually provide elements of the "framework" into which the groups responsible for the building and operation of the different sub-detectors composing the experimental apparatus will accommodate their code.

> The "sociology" of HEP experiment is quite peculiar and has an important impact on the Software Development process.

Each sub-detector is in effect a small experimental collaboration in itself, with a separate Offline project that has to work in collaboration with the overall one. The important point to appreciate here is that the different groups are geographically spread in different institutes, countries and continents. Modern HEP experiment are composed of one or two thousand physicists coming from one hundred institutions located in tens of countries in different continents. The "project" relationship is rather informal, the different institutes being bound by a Memorandum of Understanding that cannot be legally enforced. An experimental collaboration itself has only a moral but not a legal status. The real link between the physicist participating to an experiment is the common goal of building and operating a detector for fundamental research.

In this situation the management of the Offline project has to proceed by consensus, and all decisions and agreements have essentially only a moral value. Not only most of the people participating to the Offline project from different institutes have no hierarchical relationship between them, but the project itself is defined and mandated by an entity (the experiment) that has no legal existence. This introduces a vast heterogeneity of relationships among individuals. At one extreme some of the researchers belong to the same institutions and have a precise hierarchical relationship between them. Others work in different institutions of the same national research organisation, and therefore their careers and titles are comparable. Others work in different continents, and their only relationship is the affiliation to the same experiment, which is in yet another continent hosted by an institution with totally different positions and careers. The situation is even more complex for software projects that serve multiple experiments, such as ROOT or GEANT. This may seem the recipe for an utter managerial chaos, but, in the best cases, it offers the opportunity for an efficient and flexible organisation of the work.

Within the project each group enjoys substantial freedom of initiative and development. This avoids the risk that a wrong decision taken centrally drives the whole project in the wrong direction. Of course avoiding divergence and duplication of work is a constant preoccupation. However this is achieved via discussion and consensus building, as there is no practical way to "impose" a decision. Again, much depends on the personality of the Offline project leader, the Computing Coordinator, and his or her capability to reach effective compromises and build consensus.

> The loose coupling between developers is certainly a challenge, but it may turn out to be a fantastic opportunity.

The other effect of this non-uniform organisation is that communication tends to be naturally structured. Researchers belonging to the same institute or the same country tend to form a common position or at least to meet in person more frequently, and the same is true for different institutions, even across national boundaries, participating in the construction of the same sub-detector element. These communities, often partially overlapping, provide a natural structure of communication which favours the emergence of partial agreements preparing wider consensus. This avoids the risk of everybody discussing with everybody, which would result in having to find a compromise for thousands of different points of view. Of course this organisation also carries with it the danger of the formation of different "parties" which can oppose on national or sub-detector boundaries, with a breakdown of communications at some given level of granularity. Here again the role of the project leader in fostering the communication at all levels without losing its structure is paramount.

5.4.9 Why Does It Work At All?

We believe that HEP software development works out because there is a method behind it, well adapted to the environment. As we said the method is not formalised, and therefore it suffers from all the limitations of a tradition compared to a methodology. In HEP there is a continuous spectrum between users and developers, very few individuals belonging completely to one or to the other category. Requirements and conditions change very quickly and are never clear in their entirety at any given moment. The only clear item is the list of most important features for the current work, which could not possibly have been foreseen at the moment the product was first started.

A traditional computer scientist may claim that there is no design. We maintain that design is done continuously, even if implicitly. The open discussion with all the users is the design activity that influences the evolution of the product. The new code is put into production very quickly: most of the time the users have direct access to the development repository. This in turn means that testing is continuous. The functionality of the product is defined, again implicitly but exhaustively, by the use made by the user community. At the end, very few features are developed but not used, and very few of the most important desired features are not implemented while the project is active.

A common criticism is that the direction of the project is not clear. In reality the direction of the project is continuously redefined on the basis of the customer feedback: the project "goes where it needs to go", independently from, and most of the time beyond, what was decided at the beginning.

> HEP software development works out because there is a method behind it, well adapted to the environment but not explicitly formalised.

Critics claim that the code developed in this way is of low quality, difficult to maintain and to modify, not modular etc. The quality of the code developed in HEP varies largely. However most of the HEP products have served their purpose and a few products have been used by generations of experiments over tens of years and vastly changing technologies.

The downside of all this is the lack of formalisation of the process. Given the absence of a description and of manuals or other teaching materials on the software development practise, it takes time to train programmers in the HEP style of work, and people who do not have a natural inclination for it have bigger difficulties. The best developers are those who have a feeling for the user community and what it wants, and there seems to be no way to learn or transmit this without finding an equally talented disciple. This lack of formalisation turns out to be rather inefficient, and the advent of very large collaborations and a higher programmers turn-over can put the system into crisis.

A formalisation of this process would reduce these drawbacks. A clear, and possibly consensual, description of the best practises of HEP software development in the language and formalism of Software Engineering could help young researchers to become active more quickly and with less coaching. It could also be the basis on which to build a dialogue among HEP software professionals and with software engineers, leading to more fruitful exchanges and collaboration. In spite of all evident advantages, this step has not been taken formally. We believe that this happened mainly because software developers in HEP were themselves unaware to be doing Software Engineering, and also because such an activity would have required an investment of resources, particularly from the main actors of these developments, and resources are notoriously in short supply in HEP software projects. More subtly, writing such a "HEP Software Engineering Manual" is an activity that would yield only medium-long term benefits, although we are convinced, important ones. As we have explained, this timescale and perspective is at odds with the typical paradigm and working habits of HEP software professionals, where even long-term projects are driven by short-term feedback and rapid response to user feedback. We think however that this is a necessary step, and perhaps now, with the increase in size and timescale of HEP software, times are ripe for it.

5.5 Modern Software-Engineering Trends

It may now be instructive to give a quick overview of the modern Software Engineering trends to compare them with what has been said above about HEP software developments. As we have seen, HEP software developers are able to produce results under the following conditions:

- Short time-to-market
- Changing requirements
- Changing user community
- Changing hardware and Operating System base
- Changing developer teams

These features have always been specific to HEP, but they are now also becoming commonplace in the Information Technology industry, particularly in the area of e-Business. These are the features that modern Software Engineering tries to capture and satisfy.

5.5.1 The Open Software Revolution

The first hint of a real revolution in software development was the blooming of the Open Source [28] culture. Open source is much more than the availability of the source code. It includes many other specific features that characterise it.

Free Redistribution: Do not throw away long-term gains in exchange for a little short-term money. Avoid pressure for cooperators to defect.

Availability of source code: Evolve the programs allowing users to modify them.

Permission of derived works: For rapid evolution to happen, people need to be able to experiment with and redistribute modifications.

Integrity of The Author's Source Code: Users should know who is responsible for the software. Authors should know what they support and protect their reputations.

No Discrimination Against Persons, Groups or Fields: Insure the maximum diversity of persons and groups contributing to open source, allow all commercial users to join.

Distributable, non-specific and non-restrictive License: Avoid all "license traps", let distributors chose their media and format.

> Open Source is much more than the availability of the source code, it is an entire philosophy of scientific collaboration.

Many of the basic elements that have guided HEP development can be found in Open Source. Active feedback from the user community, trust and respect between the members of the user community, advanced users providing large portions of code to the system and rapid response to user feed-back. This very closely resembles the conditions of development of major codes such as GEANT and PAW since the eighties. The importance given to e-mail communication and mailing lists between colleagues that may never meet in their life is very similar to the situation of HEP developers, who often meet only at experiment weeks at the main labs. ROOT has profited largely from the formalisation of this culture and its success is also due to the conscious recast of the old CERNLIB practises and culture into the new Open Source paradigm.

5.6 Agile Technologies

Modern Software Engineering is trying to respond to the relative inadequacy of classical methods and to the rapid evolution of software development conditions. The response to High Ceremony Processes are the so-called "Agile Methodologies" [29]. These methodologies tend to adapt to change rather than predict it, and they focus on people interaction rather than on process definition. The whole process is kept as simple as possible, in order to react quickly to change. This allows the introduction of an incremental and interactive development based on short iterations of the order of weeks, in constant dialogue with the customers and users. The work is steered by coding and testing rather than by analysis and design. These technologies are uncovering better ways of developing software by valuing:

- Individual and interactions, more than processes and tools
- Working software more than huge documentation
- Customer collaboration more than contract negotiation
- Responding to change over following a plan

The value of the last items is not ignored, but the former are chosen every time there is a conflict. In analogy with High Ceremony Processes, agile technologies are also known as Low Ceremony Processes.

These new development methods, involving Low Ceremony processes, are becoming commonplace in modern Software Engineering. One of the most successful, eXtreme Programming (XP) [22], has been recently described and represents one of the first Software Engineering methodologies that tries to reduce the cost of change, reducing initial design and centring the method on the process of change itself. XP is built around four simple principles:

1. Communication: A project needs continuous communication, with the customer and among developers. Design and code must be understandable and up to date.
2. Simplicity: Do the simplest thing that can possibly work. Later, a simple design will be easily extended.
3. Feedback: Take advantage of continuous feedback from customers on a working system, incrementally developed. Programming is test-based.

4. Courage: The result of the other three values is that development can be aggressive. Every time a possible improvement in the system is detected, the system can be re-factored mercilessly.

XP is based on small, very interacting teams of people working in pairs. Testing is practised from the start, and every piece of code has its frequently run test suite. System integration is performed daily under continuous test. The development is use-case driven, with specific techniques to estimate time and cost of the project. Programs are continuously re-factored to follow the evolution of user requirements. Written documentation besides code is kept to a minimum until the program is released. The main achievement of XP is perhaps to have re-cast the traditional paradigm of Software Engineering (waterfall and spiral) into a much shorter timescale that is better adapted to present day software production in industry, which is very close to the normal practise in research (see Fig. 5.4).

Agile Technologies and eXtreme Programming are the answers to the ailing High Ceremony Processes of classical Software Engineering.

Here again the similarities with HEP are very strong, even if the conditions in which XP has been developed are somewhat different from the environment of a typical HEP experiment. Evolving design, accent on most wanted features first, very short time-to-market, general availability of unreleased code for testing, i.e. constant Quality Assurance verification by the user community are all XP features that are commonplace in HEP development. There is however an important difference: HEP

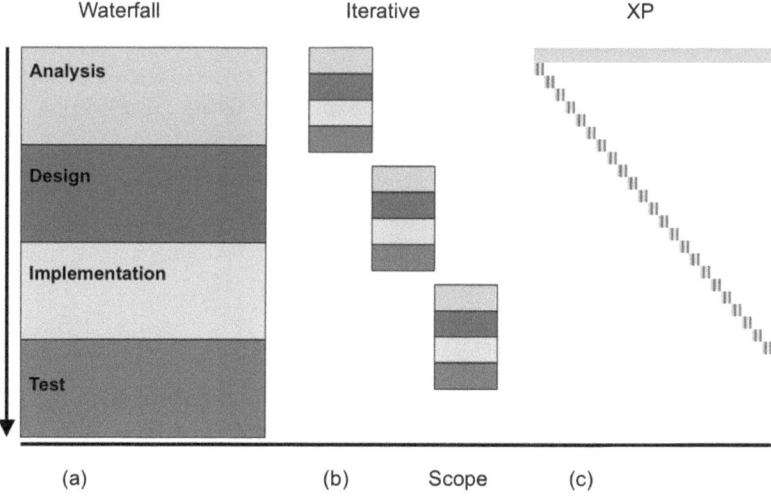

Fig. 5.4 XP programming synopsis

systems are built by distributed teams and not by small concentrated teams of programmers working in pairs as is prescribed by XP. However this needs not be a major obstacle. As with classical Software Engineering, agile methods are also models that need to be adapted to the specific software development situation. In this case the adaptation of agile technologies to HEP could give new insights into both fields.

It is very encouraging that modern Software Engineering trends seem to explain and justify what we have been doing for so many years, often feeling guilty about it. We finally start to understand why things worked in spite of the curses of generations of software pundits. This understanding, if properly exploited, may help us in doing things better, formalising and improving our process.

Now that a new culture develops and agile methodologies begin to be recognised as a perfectly honourable way of working, we feel some pride of having been precursors on this way, when the road was still full of obstacles.

5.7 One Point in Case, the Software Development in ALICE

5.7.1 The Off-Line Software

The ALICE Software Project performed the transition from FORTRAN to C++ in 1998. Once this decision was taken, a community of non-expert C++ programmers developed a new system from scratch. The development philosophy moved along the lines described above. The software development process is now taken as a whole: analysis, design, implementation, testing and documentation are all carried out together, but they are applied to solve only a fraction of the problem, the one that is known and urgent at any one time. Partial requirements are used to design a partial system that is allowed to evolve. The project progresses through subsequent refinements with the cyclic implementation and testing of ever more complete prototypes. Requirement evolution and bug fixing have become integral parts of the process, not traumatic accidents to be avoided.

A set of simple rules guide the whole development process:

- Base the development on working prototypes.
- Implement the most wanted features first.
- Get user feedback and plan development according to it.
- Release as frequently as needed.

An important element in the development of the ALICE Offline project was the early adoption of ROOT, a robust and friendly framework and an interactive analysis tool that could replace its predecessor, PAW, still connected with the FORTRAN environment. ALICE has been the first LHC experiment to officially adopt ROOT in 1998, when ROOT was still a "forbidden program" at CERN (see Chap. 1). The reaction of the CERN management was at the time rather critical, but thanks to this decision, the ALICE experiment was the first to move entirely to C++ and abandon FORTRAN. The ROOT system has been developed very much following the philosophy described above. AliRoot [30], the ALICE Offline framework, is implemented as a set of modular C++ libraries. The code, including the development versions, is contained in a central subversion ("svn") server that is publicly accessible. Developers check-in their code in this server, and every night all new code is integrated, i.e. compiled and tested, so that integration of new code is never an issue.

Documentation is kept to a minimum. Every night a special tool, which is part of the ROOT system itself, analyses the whole code and produces a hypertext-enabled version of it that is published as a web page. Another tool is also run every night to check the compliance of the code with the ALICE coding conventions [31]. The results are published on the web in form of a table where the violation to the different rules are listed for each code unit. Another tool [32] produces a set of hypertext-enabled UML diagrams of the code via reverse engineering. In this way the latest code and code design can be accessed and consulted, and this design is never out of date.

The design of the code evolves according to the users most urgent needs. A large usage is made of abstract classes that define the architecture and the design of the code in the different sub-detectors. The code for each sub-detector is kept as modular as possible. 70% of the code has been developed by almost 100 external contributors, while less than 20 people compose the core Offline group at CERN. When a new functionality is needed, the software groups working for one or more sub-detectors usually implement this independently, and the best implementation is then generalised via abstract classes. Code design proceeds in cycles. Micro-cycles happen continuously, but constant integration and testing ensure that they do not disrupt operation, while Macro-cycles, involving major re-design, happen during "Offline weeks", when most contributors come to CERN to discuss software development, two or three times per year.

> The ALICE Offline Software Framework (AliRoot) is a typical example of a system developed in the "HEP style".

ALICE has developed a Grid system called AliEn ([33]) developed with the same philosophy. Thanks to fast prototyping, intense user interaction and large use of Open Source components, the AliEn system is now able to provide a distributed system for production and analysis deployed at 80 sites in four continents. AliEn is in continuous development and is being interfaced to major Grid projects, but it also offers a solid production platform where ALICE has already run more than ten million production jobs. The development of AliEn by a small team at CERN closely follows the criteria suggested by agile methods.

Since its beginning in 1998, the ALICE Offline project has been able to provide to the ALICE users a system responding to their more urgent needs while continuously evolving toward the requirements of the final system. Most of it is due to the adoption of a flexible development philosophy. This originated from an analysis of the software development models that proved to be very effective for other HEP products such as CERNLIB and ROOT. This strategy seems now to be confirmed by the latest Software Engineering developments.

5.7.2 The Data-Acquisition Software

The ALICE Data-Acquisition (DAQ) group has developed a very similar strategy. The development group was small and composed of highly skilled people. This allowed adopting an agile way of working, where informal communication was privileged. To reduce overheads, bulky internal documentation was avoided, privileging readable code. During the execution of the project, the development plan underwent frequent revisions and modifications. One additional complication in the development of the data acquisition system was that traditional Software Engineering methods have never been able to cope satisfactorily with the complexity of real-time problems. This alone justified the introduction of a new method.

The ALICE DAQ development process is based on practises specifically developed for it and based on several years of experience of writing real time acquisition systems for HEP experiments. At first a functional DAQ prototype is designed and constructed, running in the laboratory, responding to the current requirements of potential users. This is done making use of technologies currently available which do not involve complicated Research and Development activities. The design must accept the complexity in the general architecture, while keeping the components as simple as possible.

Every software package is the responsibility of one person. The prototype is then delivered to pilot users, who accept to test it in production-grade applications. During this phase the development team must provide a close follow-up on the installation and a short response time for bug-fixing and modifications. All this development is done in an Open Source policy and encouraging the users to implement new facilities. A necessary requisite for this is to maintain a user-friendly software-distribution system and issue an up-to-date user manual. It is also necessary to maintain at all times one working version, while a new version is being designed, implemented, and tested in the lab. Backward compatibility of new releases is not a must, but the planning is made in consultation with the users to balance the needs for progress with the stability of existing users.

> Traditional Software Engineering methods have never been able to cope satisfactorily with the complexity of real-time problems.

The project started in 1996 and was named DATE (which stands for Data Acquisition Test Environment). A running experiment (NA57 [35]) asked to test the prototype and they received a limited but complete system. The system contained only the facilities requested, while any missing facility was developed upon request. The advantages of having the software running in the field were immediately clear: it is not affordable to have a test system in the laboratory that is as large as that of a real experiment.

Laboratory tests are not able to cover all the situations that can occur in the experiment. The success of the runs at the NA57 experiment gave confidence on the correctness of the development method chosen. When another starting experiment, the COMPASS [36] collaboration, asked to use DATE the experience was repeated with added advantages. COMPASS requirements were more advanced than those of NA57 and closer to ALICE ones. COMPASS programmers provided enhancements subsequently retrofitted in the code.

DATE was subsequently used by other experiments, such as HARP [37] and NA60 [38]. The acquisition at CERN of shared computing farms (involving hundreds of PCs) made it possible to have access to large test beds. They were used to run joint on-line/off-line tests, called "Data Challenges" [34], which allowed thorough assessments of the complete systems in quasi-real conditions.

The program of development of DATE has continued at a sustained pace, with at least one release per year, leading to the final system now in place. This system has now been used since three years for the commissioning of the ALICE detector with cosmic rays and the recording of the first LHC collisions, while it keeps evolving to respond to the needs of the users.

5.8 Conclusions and Lessons Learned

HEP has developed and successfully deployed software for many years without using classical Software Engineering. Market conditions are now more similar to the HEP environment, and modern Software Engineering is confirming some HEP traditions and rituals that have proven successful even though they were never codified or otherwise justified.

These developments may be important for HEP software as they may allow us to formalise our software development strategy. This could improve our planning capability for distributed software development and it may help to transmit and evolve our software practises, reducing the lead-time for developers to be productive in our field. This is particularly important now that it is difficult to find long-term positions in research, and most of the developments are due to young scientists who, after a stay in research, will look for further employment in the industry.

Our environment adds complexity to the one foreseen by agile methods, as we work in large and distributed teams, with little or no hierarchy. Therefore the challenge is now to move agile technologies into the realm of distributed development. The authors believe that this could be a worthy goal for Software Engineers working in HEP, and an occasion to collaborate more closely with advanced Computer Science and Industry.

References

1. In: Proceedings of symposium On Advanced Computer Programs for Digital Computers, sponsored by ONR. June 1956 Republished in Annals of the History of Computing, Oct. 1983, pp. 350–361. Reprinted at ICSE'87, Monterey, California, USA, March 30-April 2, 1987
2. Naur, P., Randell, B. (eds.): Software Engineering: Reports of a conference sponsored by the NATO Science Committee, Garmish, Germany, 7–11 October, 1968
3. http://www.yourdon.com
4. Chen, P.P.: The entity-relationship model - toward a unified view of data. ACM Trans. Database Syst. **1**(1), 9–36 (1976)
5. Booch, G.: Object-Oriented Analysis and Design with Applications. Benjamin–Cummings, 2nd edn. Redwood City, CA (1994)
6. Rumbaugh, J., Blah, M., Premerlani, W., Eddy, F., Lorensen, W.: Object-Oriented Modelling and Design. Prentice Hall, Englewood Cliffs (1991)
7. Shlaer, S., Mellor, S.J.: Object lifecyclesmodeling the world in states. Yourdon Press computing series. Yourdon, Englewood Cliffs, NJ (1992)
8. Guide to the software engineering standards. Eur. Space Agency, Noordwijk, **1** (1991)
9. http://www.uml.org
10. http://adamo.web.cern.ch
11. http://www.ilogix.com
12. http://sdt.web.cern.ch
13. http://omtools.com
14. http://www.aonix.com/stp.html
15. http://www.rational.com
16. http://cernlib.web.cern.ch/cernlib
17. http://paw.web.cern.ch/paw
18. http://wwwasdoc.web.cern.ch/wwwasdoc/geantold/
19. http://root.cern.ch
20. http://geant4.web.cern.ch/geant4
21. http://lhcxx.home.cern.ch/lhcxx/
22. Beck, K.: Extreme Programming Explained. Addison-Wesley, Boston (2000)
23. Brooks, F.: The Mythical Man-Month, Addison-Wesley, MA 1975 (1995). ISBN 0-201-83595-9 (1995 edn.)
24. http://www.doxygen.org
25. Tonella, P., Abebe, S.L.: Code Quality from the Programmer's Perspective. In: Proceedings of Science, XII Advanced Computing and Analysis Techniques in Physics Research, Erice, Italy, 2008
26. http://eu-datagrid.web.cern.ch
27. http://www.eu-egee.org
28. ttp://www.opensource.org,http://www.sourceforge.com,http://www.gnu.org
29. http://agilemanifesto.org
30. http://aliceinfo.cern.ch/Offline
31. Tonella, P., Abebe, S.L.: Code Quality from the Programmer's Perspective. In: Proceedings of Science, XII Advanced Computing and Analysis Techniques in Physics Research, Erice, Italy, 2008

32. Tonella, P., Abebe, S.L.: Code Quality from the Programmer's Perspective. In: Proceedings of Science, XII Advanced Computing and Analysis Techniques in Physics Research, Erice, Italy, 2008
33. http://alien.cern.ch
34. Anticic, T., et al.: Challenging the challenge: handling data in the gigabit/s range 2003
35. http://wa97.web.cern.ch
36. http://wwwcompass.cern.ch
37. http://harp.web.cern.ch
38. http://na60.web.cern.ch

Chapter 6
A Discussion on Virtualisation in GRID Computing

Predrag Buncic and Federico Carminati

In attempt to solve the problem of processing data coming from LHC experiments at CERN at a rate of 15 Petabytes per year, for almost a decade the High Energy Physics (HEP) community has focused its efforts on development of the Worldwide LHC Computing Grid.

This generated large interest and expectations promising to revolutionise computing. Meanwhile, having initially taken part in Grid standardisation process, the industry has moved in a different direction and started promoting the Cloud Computing paradigm which aims to solve problems on a similar scale and in equally seamless way as it was expected in the idealised Grid approach. The key enabling technology behind Cloud Computing is server virtualisation.

Virtualisation is not a new concept in Information Technology and it can be contended that the whole evolution of the computing machines is accompanied by a process of virtualisation intended to offer a friendly and functional interface to the underlying hardware and software layers. At age of the Grid and the Cloud, however, the concept of virtualisation assumes a whole new importance due to the difficulties inherent in the project of turning a set of in-homogeneous hardware and software resources geographically distributed into a coherent ensemble providing high-end functionality. This chapter discusses the opportunities and challenges of virtualisation in the Grid world.

6.1 Introduction

Building on the meta-computing paradigm [47] and emerging from a successful computing science experiment to link high-speed research networks to demonstrate a national Grid back in 1995, Ian Foster and Carl Kesselman published a first paper [12] in 1997 and successfully began evangelising the idea of the Grid as a

---

P. Buncic (✉) · F. Carminati
CERN, Geneva, Switzerland
e-mail: Predrag.Buncic@cern.ch

common infrastructure by publishing articles and a highly referenced book [46] which had a profound impact on the direction in which much of computing science in academic circles developed in the following years.

> ...a hardware and software infrastructure that provides dependable, consistent, pervasive and inexpensive access to high-end computational capabilities...

In the 12 years since the book was published, it achieved a biblical status in the growing community of followers who enthusiastically embraced the idea of "a hardware and software infrastructure that provides dependable, consistent, pervasive and inexpensive access to high-end computational capabilities" as a solution to a wide range of demanding scientific and societal computing problems.

To understand the path that lead in few years from the Grid "vision" to the emergence of Clouds and virtualisation, we will need to recollect briefly the history of Grid computing from an High Energy Physics point of view. Some of the content of this chapter overlaps partially with Chap. 3, but with a different accent, making indeed these two descriptions complementary.

6.2 A Brief History of the Grid from a High Energy Physics Perspective

The concept and vision of the Grid met with a world-wide success, before any production Grid was deployed and used on a large scale. The enthusiasm was fuelled by a common belief that the technology to build effective Grids existed in the form of a Globus Toolkit, released in 1998 [27], and that only an incremental effort was needed to adapt it to specific needs in various fields of science.

The phrases defining the goals of the Grid were repeated and quoted by the members of the Grid community in new proposals for funding research to further develop a Grid software infrastructure for applications in traditional domains of science that require high performance and high throughput computing (High Energy Physics, Astronomy) as well as promising to bring the Grid into our daily life (biology, medicine, Earth sciences, various societal applications). Together with these promises, the share of funding available to Grid projects grew to such an extent that any project in which computing was involved in any form in order to "survive" and secure funding had to have a Grid flavour.

> Riding on apparently unlimited funding and a growing community, Grid projects began to mushroom, but the very definition of what a Grid remains unclear.

The idea of a new, pervasive and inexpensive computing platform available to everyone had equal appeal to computing scientists in need of finding a new subject of research, researchers in domains of science traditionally hungry for computing resources and funding agencies who saw it as an opportunity to provide computing for the masses at a reasonable cost and that did not want to "miss the train" of what seemed, for once, a quite safe technological bet. It was the combination of all these elements that created the Grid movement. Riding on a wave of seemingly unlimited funding and a growing community, Grid projects began to mushroom at the dawn of a new millennium.

As the Grid became an umbrella movement for so many different scientific communities trying to solve different computational problems, this inevitably led to a diluted understanding of what exactly is meant by the term Grid. Consequently, a number of projects that were aiming to solve a problem by running a particular application for a given user community in a distributed manner claimed to be Grid implementations.

To clarify the definition, the authors of the Grid blueprint published an article [11] where they tried to redefine the Grid as a platform that allows "direct access to computers, software, data, and other resources, as is required by a range of collaborative problem solving and resource-brokering strategies emerging in industry, science, and engineering" but under the condition that such sharing remains "necessarily, highly controlled, with resource providers and consumers defining clearly and carefully just what is shared, who is allowed to share, and the conditions under which sharing occurs". Furthermore, they introduced a name "virtual organisation" to denote "a set of individuals and/or institutions defined by such sharing rules".

> Globus was designed to be a one-stop solution providing the Grid building blocks.

The message was that not every distributed computing system could be called a Grid and that, to qualify as a Grid, access to resources must be strictly controlled. At the same time, Globus Toolkit 2.0 was promoted as the solution providing all required protocols and components to build a Grid and in a way that offers the highest possible security based on the Globus Security Infrastructure (GSI) [7].

6.2.1 *The Evolution of the Globus Toolkit(s)*

To integrate distributed and heterogeneous computational resources in a large, "virtual" computer that can be used to solve a single problem at a given time, a Grid should have a middleware, i.e. a suite of services that presents to the users and the applications a common and global view of the actual computing resources.

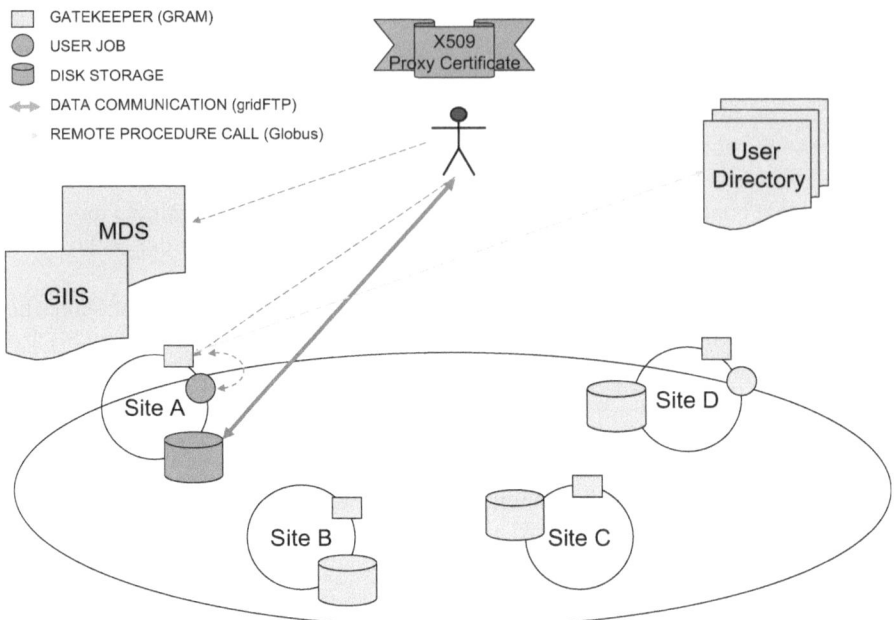

Fig. 6.1 Simplified Globus Grid model

The Globus Toolkit was internally made of a set of components (some of which are shown schematically in Fig. 6.1) that were meant to be used for developing Grid middleware.

Unfortunately, beneath the surface, the implementation of these components lacked the robustness needed by the largest communities that embraced the Grid as a promised solution for their problems. This is a problem that has plagued most of the developments around the Grid. While a large quantity of excellent work has been done in this area, the deceptive simplicity of the Grid vision, and the dire competition for the large funding available, forced many projects to declare themselves as the "final solution" rather than a prototype, in some cases very good, innovative and promising. This has had the adverse affect of confusing potential users, and sometimes even experts, and to stifle progress, as once presented as *the* complete solution, the freedom to change a given piece of middleware was considerably restrained by communication and image considerations.

> In 2002 I.Foster introduces the Open Grid Service Architecture and Infrastructure.

At the fourth Global Grid Forum (GGF) in Toronto in February 2002, Ian Foster presented plans to develop the Globus Toolkit 3.0, an implementation of the Open

Grid Services Architecture (OGSA) based on the Open Grid Services Infrastructure (OGSI). Later that year, the Globus Project announced the creation of the Globus Alliance by forming a partnership with the University of Edinburgh in Scotland and the Swedish Centre for Parallel Computers (PDC) as well as sponsorship from industrial partners (most notably IBM and Microsoft Research). The goal was to come up with a standardisation effort as a GGF activity, leading to a reference software implementation by the Globus Alliance.

While most of the projects that were officially carrying the Grid label in their name tried to build on a middleware stack provided by the Globus Toolkit, some dissident voices emerged, who claimed that working Grids could be deployed using available network protocols and open source tools well established in the Web (see Chap. 2) context, thus fully conforming to the above definition of a Grid but not using any Globus Toolkit components.

In 2004 actually the Globus Alliance announced that it would abandon work on OGSI and instead rely on the industry standard Web Services Resource Framework for implementation of OGSA in Globus Toolkit 4.0.

6.2.2 The Developments at CERN

In the wake of the success of the World Wide Web, and prompted by the ever increasing projection of the LHC computing needs, CERN took an early lead in development of the Grid. In 2001, the E.U.-funded Data Grid project started with CERN as the leading partner, aimed at addressing the challenging use cases of the handling of the data that were going to be produced by Large Hadron Collider (LHC) experiments [9]. This project did not quite achieve the initial objectives, and it was followed-up by projects aiming to re-engineer and correct the problems observed after the Data Grid experience (EGEE I, II, III [26]).

> The CERN High Energy Physics experiments pioneered the concept of "Data Grid" where data locality is the main feature.

The data-centrality of the LHC experiments led to the formulation of a "Data Grid" concept where solving the data part of the problem proved to be the hardest job, accounting for most of the development work over the following years. In particular it was felt that an extra high-level component called a Resource Broker was needed. Unfortunately the chosen architecture led to scaling issues in all the components, such as the security and the information system.

While many people put a considerable effort into building and testing individual components of the system, when it was all put together it still did not work reliably enough to fully satisfy the LHC user community.

The perception of the CERN experiments was that the Grid middleware was changing much too frequently. Therefore individual experiments, while remaining committed to test and use Grid middleware developed by LHC Computing Grid (LCG) and European Union sponsored projects, gradually took steps to protect themselves, their users and application developers against these changes in the Grid environment by developing their own distributed computing frameworks and defining their own interfaces to the Grid (AliEn [19], Dirac [48], Panda [33]).

> High Energy Physics created an "upper middleware" layer to be protected from middleware changes and to implement additional functionality needed by their "data Grid".

Additionally, the experiments were facing the reality that the dominant Grid middleware in the U.S. (OSG [30]) was different from the one used in Europe (LCG/gLite), and that even in Europe, northern countries were running separate NorduGrid [31] middleware (called ARC) while some sites simply were not yet part of any Grid infrastructure. Ironically OSG, LCG and ARC were all based on Globus middleware, although on different versions. The logical move was to create an abstraction (virtualisation) of each individual Grid software component and provide an implementation or adaptor to the underlying Grid middleware functionality. Some implementations allowed powerful generalisations such as the creation of a "Grid of Grids" self-similar hierarchy of Grid resources [6], giving these homegrown initiatives the status of competing Grid implementations.

It is interesting to note here that, even if at the time this was not explicitly recognised, virtualisation moved from ensuring seamless access to non-homogeneous computing resources to providing seamless access to non-homogeneous middleware packages. In hindsight the trend was clear.

In an attempt to resolve the confusion the LCG project at CERN appointed in 2003 a Requirement Technical Assessment Group (RTAG) called ARDA (Architectural Roadmap towards Distributed Analysis) charged to draw up a blueprint for the Grid components needed to effectively run the analysis, the most challenging of the HEP use cases. This RTAG came up with a report [2] which was based broadly on the architecture of the AliEn system. One of the main feature of AliEn is that it was, from its very beginning in 2001, developed around the Web Services model, something that was initially considered oddly different from the prescription of what Grids should look like, but which in 2003 became the blessed model.

Moreover the very blueprint of the AliEn system had the concept of virtualisation built in. See for instance in Fig. 6.2 a graphics used since 2001 to explain the ability of the system to provide a seamless interface to different Grid middleware stacks.

The plan elaborated by the RTAG is at the origin of the package, the principal components of which are shown in Fig. 6.3.

6 A Discussion on Virtualisation in GRID Computing

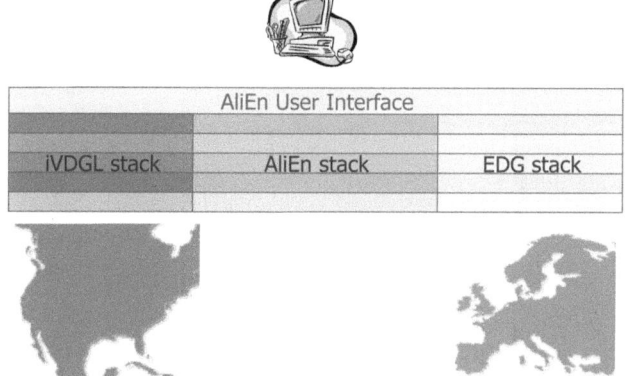

Fig. 6.2 The AliEn system as an abstraction layer to the different middleware stacks

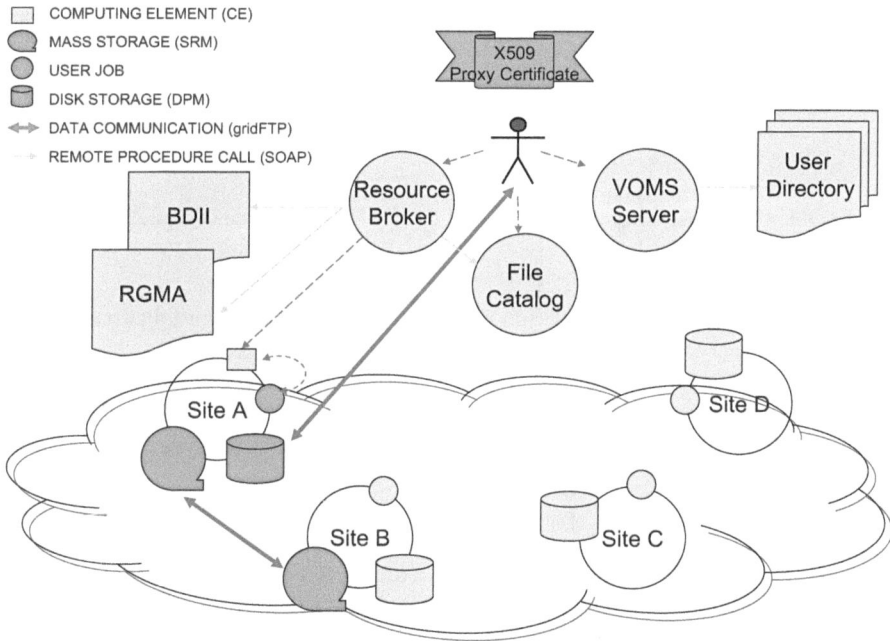

Fig. 6.3 LCG/gLite Grid model

> Workload scheduling with "pilot jobs" offers a very resilient and efficient way to distribute work on the Grid.

Meanwhile, in order to deal with the inevitable instabilities in job execution in the different centres, all experiments converged to a model where they first populate

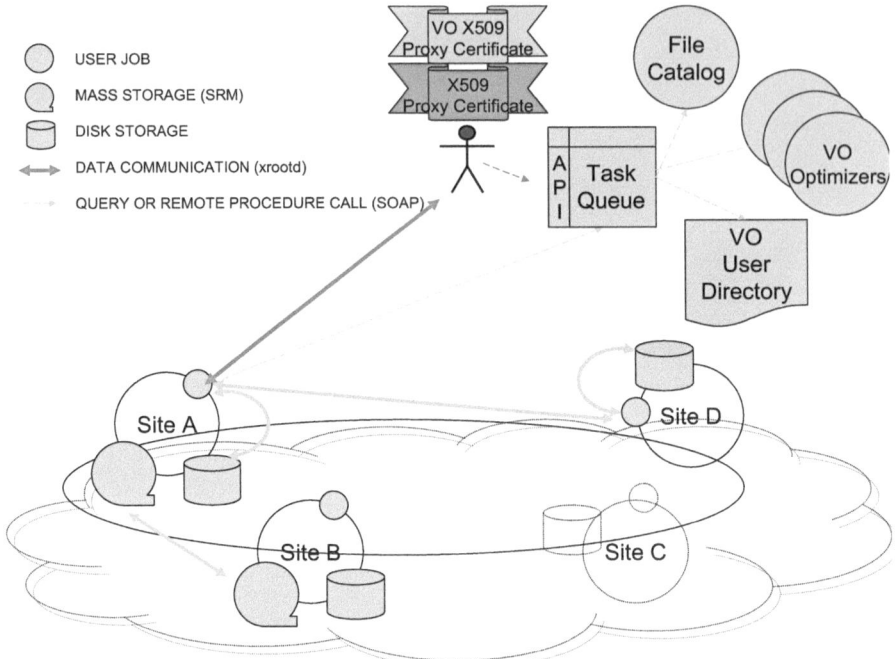

Fig. 6.4 Pilot Job model – creating a virtual Grid overlay for each Virtual Organisation

the Grid with "pilot jobs": simple job placeholders which upon landing on a worker node first inspect the environment and then contact a Virtual Organisation (VO) service known as a Task Queue which performs "just-in-time" scheduling of a real job. In contrast to underlying Grid middleware in which scheduling services rely on the information from the monitoring and information systems to push a task onto a particular site, the components of the pilot job systems operate in a pull mode.

It is interesting to note how this too can be interpreted as a "virtualisation" of a service. The pilot job is indeed a virtual job that is independent on the workload that will be assigned to it, and it is instantiated into a real entity only when the environment is sane. Here virtualisation takes a new dimension, where the connection between the virtual entity and its ontology is established "just-in-time" via an optimising scheduling procedure. Using pilot jobs, experiments create their own Grid overlays on the subset of the resources available to them in the common infrastructure, implementing their own internal, just in time scheduling and scheduling policies. This model is now called the Pilot Job model. It is typically created on top of the basic LCG/gLite Grid by VO managers while users interact with it by means of a VO middleware stack (shown in Fig. 6.4).

6.3 Grids, Clouds and Virtualisation

Although the very idea of Grid is little more than ten years old, it has undergone a profound evolution and it is already being replaced by the concept of Clouds that seems to attract the attention of the Information Technology world. In this section we will give a brief account of this evolution and we will explain how both vision have at their root the idea of virtualisation.

6.3.1 Is the Grid Fading Away?

The Grid computing model implies that the resource providers host a set of services, which, in conjunction with the services hosted centrally, are exploited to enable the work of Virtual Organisations. Since the beginning of 2000 there have been numerous national, regional and international Grid initiatives, which varied in scale and which adopted different technological approaches for the solution of similar problems. Currently there are several large-scale Grid infrastructures deployed. The biggest of them, the Worldwide LCG Computing Grid [43], comprises resources provided by more than 140 institutions and is extensively used for production, analysis and storage of data belonging to the LHC experiments. In his article "What is the Grid? A three point checklist" [13,14], Foster defines the Grid as a system that "coordinates resources that are not subject to centralised control using standard, open, general-purpose protocols and interfaces to deliver nontrivial qualities of service".

As we have seen, the nature of these services, often called Grid middleware, should be generic, i.e. their work should be based on a set of standard protocols developed with the involvement of members of the scientific community and of representatives from industry (e.g. OGF [39]), and thus they should not have any specific affiliation to a particular scientific or industrial domain. The middleware should expose to applications an interface providing uniform access to heterogeneous physical resources. It should also ensure access control, enforce resource usage policy, as well as provide the means for monitoring of and accounting for resource utilisation. In other words, the middleware deals with all the complexity associated with the integration of a vast pool of globally distributed heterogeneous physical resources into a large virtual computer that can be collaboratively exploited for carrying out computations.

Nowadays, for the computing industry, customers and software providers alike, the term Grid often refers, in contradiction with earlier Grid definitions, to a large number of private resources, hidden behind a corporate firewall and most likely belonging to a single administrative domain. These resources could possibly be physically distributed on a couple of sites and build on top of heterogeneous hardware platforms. The most frequent challenge for such Grids (that can however span tens of thousands of CPUs, e.g. in case of a large bank) is speeding up the response for standard office applications (like spreadsheets and databases). To achieve that, the applications have to be equipped with extra libraries that partition the workload, send it to the back-end for processing, and collect the results. Ideally, all this should happen within a fixed time interval and the back-end should absorb increased workload by adding more processing power on demand.

> While the Grid is entering operation and providing large amount of computing resources, the hype seems to be already moving to other buzzwords.

While the Grid has entered operation providing large amount of computing resources, the hype seems to be already moving to other buzzwords. Even two of the early flagship supporters of Grid technology, Platform Computing [32] and DataSynapse [25] no longer use "Grid computing" in their marketing materials, preferring the trendier "application virtualisation", which puts them and their products outside the scope of the Globus Grid definition which explicitly excludes "application Grids".

A quick check on Google Trends for the term "Grid computing" [28] reveals a worrying loss of interest for that keyword, indicating that the days of big hype and the hope that the Grid will indeed provide universal solutions for the computational problems of modern science, might be over.

6.3.2 Virtualisation Reborn

To distinguish themselves from the Grid, the business market is inventing more new words like Software as a Service (SaaS) and Computing Cloud, and reinvents some old ideas that were used in computer science more than 40 years ago.

In the quest to find a solution to speed up and dynamically provision resources to a specific task or application, the Information Technology (IT) industry has reinvented virtualisation, an old idea that IBM had been using successfully within its mainframe environments for many years, notably in the VM/CMS operating system later adopted by CERN.

Note that virtualisation technology is much more than virtual memory and it means different things in different contexts.

Today we talk about CPU virtualisation, network virtualisation, file virtualisation and storage virtualisation. We can use the term virtualisation to describe the representation of a physical CPU by many virtual CPUs or to describe the aggregation of many physical disks into a large logical disk.

Server virtualisation enables the application view of computing resources to be separated from the underlying implemented infrastructure view. To achieve this, software is used to create a virtualisation layer (also referred to as a *hypervisor*); this virtualisation layer enables the host platform to execute multiple virtual machines

(VMs). Each VM can run either its own copy of a different operating system, or multiple versions of the same operating system, but, importantly, it supports the pooling of resources (processor, disks, and memory) across a server, based on policy rules.

> Virtualisation is now present at several levels and it is moving "down" into the hardware layer.

Server virtualisation provides an insulation layer offering both diachronic and synchronic decoupling of the user application from the underlying hardware. The synchronic aspect facilitates physical server consolidation, reduces the downtime required for physical system maintenance, whether planned or unplanned, and increases flexibility by decoupling data processing from physical hardware, allowing computing to scale to rapid changes in demand. It also simplifies deployment of otherwise complex applications.

The diacronic aspect provides a way to support legacy systems on new hardware without disruptive upgrades and allows the application life cycle to be decoupled from the life cycle of the underlying infrastructure, which is a major benefit when the infrastructure has to serve different user communities.

From an IT perspective, all this translates into savings in terms of maintenance and support effort, planned and unplanned downtime, manpower, as well as electrical power. With all these benefits, it is easy to understand why industry is embracing virtualisation technology.

These developments resulted in a flood of software solutions allowing for Operating System and server virtualisation [22] is opening a door for physical server consolidation and improved server utilisation, service continuity and disaster recovery.

Looking at the evolution of the Grid middleware, as it has been described above, it is easy to see how at the heart of the realisation of the Grid, and now the Cloud paradigm, lies the concept of virtualisation. "Dependable, consistent, pervasive and inexpensive access to high-end computational capabilities" can only be ensured by making abstraction of the real nature of and differences between the actual resources, and the common feature of all virtualisation technologies is exactly the hiding of technical details via resource abstraction and encapsulation.

From the software perspective, the Open Grid Services Architecture glossary defines virtualisation as "making a common set of abstract interfaces available for a set of similar resources, thereby hiding differences in their properties and operations, and allowing them to be viewed and/or manipulated in a common way." The common theme of all virtualisation technologies is the hiding of technical detail, via abstraction and encapsulation.

This leads to the new definition of a Grid from an industry perspective: Grid computing enables the "virtualised", on-demand deployment of an application on a distributed computing infrastructure, sharing the resources such as storage, bandwidth, CPU cycles, etc.

Incidentally, another look at Google Trends for comparison of "Grid computing vs virtualisation" [29] confirms that the Internet spotlight has recently definitively moved from one to the other buzzword.

> Grid middleware is in itself all about virtualisation, trying to hide resource location, status and details to present a unified view to the users.

Indeed, the virtualisation of distributed computational resources has always been the goal that Grid middleware tried to accomplish. We tried hard to hide the physical location of files and created the concept of Grid Storage Elements, replicas and File Catalogues that presented the user with logical file names aiming to achieve storage virtualisation. We developed Computing Elements to abstract and hide physical details of how jobs are handled on individual sites and tried to tell users not to worry about the location where the job will actually run since the Resource Broker will take care about that and provision the user application with everything it needs in order to run on some Worker Node somewhere on the Grid. Obviously, all this had to be done observing the sometimes contradictory confidentiality requirements of the user and the need for a strict auditing trail on the part of the resource owner concerned by the site security.

6.3.3 A Virtual Path from Grids to the Clouds

However, even the virtualisation level described above may not be enough. Provided that everyone in the world would agree to use the same Grid middleware, made of the same components of the same toolkit of the same version, and maintain 100% availability of all resources, this could perhaps be made to work. Or, to make it even more difficult, if everyone would agree on what the standards were, and the different tool-kits and middleware implementations could talk to each other, we could perhaps again have the Grid we dreamt of 10 years ago.

Unfortunately this is not the world we live in. In the Internet-fuelled information technology age, a good idea has to be converted into a product that can be sold to customers on a time scale that is considerably shorter than one year, and on such a scale there is very little time left for a long standardisation process that would be fully agreed upon by all players in the field. Instead, we are often seeing claims of "de-facto" standards that are based on an individual player's self-assessed or actual market dominance. The trouble is that in a situation where technology is quickly changing and where all major software vendors are ultimately seeking such market dominance we are usually left with several competing "de-facto" standards. In that respect, the story of Grid middleware is not an exception. While still continuing to work on standardisation in the context of the Global Grid Forum, the vendors began to develop and sell end-to-end solutions sometimes bundling hardware, software and support together in order to achieve a homogeneous platform that could support the Grid paradigm.

Given all the problems with reliability of traditional Grid middleware, deployment problems in a heterogeneous environment, and the difficulty to keep up with changes and developments of operating systems, virtualisation might seem to be an obvious way forward to solve or ease the problems of Grid middleware which in turn aims to provide virtualisation of resources at different level. By offering more reliable service at the level of individual building blocks of the Grid (Storage, Computing Elements...) we could end up with a much better end-user experience without the need to redraw the architecture of current Grid implementations.

Even if it is presented as the successor of the Grid paradigm, at present, Cloud Computing has no widely accepted definition. In their paper "A Break in the Clouds: Towards a Cloud Definition" [49], Vaquero et al. give 20 definitions of the term by different authors and propose the following one themselves: "Clouds are a large pool of easily usable and accessible virtualised resources (such as hardware, development platforms and/or services). These resources can be dynamically reconfigured to adjust to a variable load (scale), allowing also for an optimum resource utilisation".

While in Grids users are provided with a full-featured batch-system-like environment that must be used "as is", in Cloud-like systems they are expected to create for themselves an infrastructure specifically tailored for a solution of a given problem using a few basic services. In the Grid case, the complex Grid middleware plays the role of a virtualisation layer, aiming to provide a homogeneous view of the resources to the end user, while in case of the Cloud, it is use of the virtualised infrastructure that allows end users to create a uniform view on a subset of resources that is appropriate to the scale of their problem. By introducing the virtualisation at the infrastructure level we have a much easier task when it comes to developing Cloud middleware suited to given Virtual Organisation as compared to the demanding task of writing and maintaining middleware that has to deal with complexity, heterogeneity and scale of the Grid.

6.4 Common Use Cases for Virtualisation in HEP

6.4.1 *Software Build and Test Systems*

Virtual machines can host various operating systems (or several flavors of the same operating system) on a single physical machine and use them for building and testing software releases. Testing software releases in different environments helps to uncover bugs, and virtualisation technology in conjunction with continuous integration tools (e.g. Buildbot [4], Hudson[34], ETICS [37]) can facilitate the process of testing during the integration stage of the development cycle.

6.4.2 Critical and Dedicated Services

The use of virtual machines for running critical experiment-specific services (the so-called VO boxes), which are typically hosted in the centralised IT infrastructure, can help to consolidate underutilised physical nodes. In some cases virtualisation technology can be used to provide automatic fail-over functionality for high availability services, reducing downtime and improving overall quality of service.

6.4.3 Virtual Batch Environment

The difficulties experienced by experiments in moving from one major version of the operating system to the next have been all too evident. As a consequence, moving a part of the batch capacity to a new Operating System flavour would often lead to wasted capacity as users would never be fully ready to move to the new platform. By virtualising the batch environment one could dynamically provision standard batch nodes of a given Operating System flavor according to demand [35]. With such an approach, efficient utilisation of computing resources could be achieved by dynamically consolidating idle virtual worker nodes on a single physical machine and by powering-off unused physical machines.

6.4.4 Private Clouds

Although already proven to be valuable on their own, the use cases mentioned above develop their full potential when combined in a fully virtualised infrastructure. A virtualised infrastructure is composed by powerful tools (e.g. [38, 44] that allow virtual appliances, storage, and network to be managed centrally, based on bare hardware resources. Such a unified infrastructure provisions resources for all required services, and the total capacity can be extended just by adding more hardware. In contrast to public and commercial Clouds like Amazon Elastic Compute Cloud (EC2) [1], we call such an infrastructure a Private Cloud (as shown in Fig. 6.5).

When building Private Clouds, it is important to stress the need for a common Application Programming Interface (API). It is certainly possible to imagine a Hybrid Cloud scenario where the Private Cloud at peak time uses resources from the Public Cloud in a fully transparent manner for the end user, so that the actual API exposed by the Private Cloud is not so important. However, if the API were identical to that of the Public Cloud, the end users would benefit from freedom of choice. While not standard in the traditional sense, the Amazon EC2 API has established itself as a de-facto standard simply by leading the market and having a strong developer community behind it. Indeed, there are already several examples of HEP

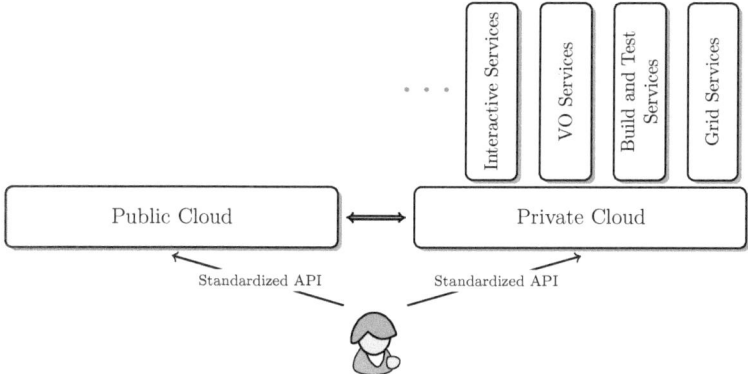

Fig. 6.5 Bringing virtualised services together in a unified infrastructure

experiments reporting successful use of Amazon EC2 in running typical workloads like Monte Carlo simulation [42] indicating that Public Cloud infrastructure is sufficiently flexible to be used for that purpose.

6.4.5 Portable Analysis and Development Environment

Given the variety of the operating systems that physicists run on their laptops and desktops (Windows, MacOSX, Linux) and the fact that most computing capacity provided by the Grid tends to be locked to a single version of the Linux operating system for considerable time, it quickly becomes impractical and often impossible to install, develop and debug locally the experiment software before eventually submitting jobs to the Grid. This is evidently one more use case where virtualisation technology can come to rescue, in the form of a Virtual Machine that would run the Linux operating system compatible with one available on the Grid and independent of underlying hardware or software platform.

6.4.6 Cloud and Volunteer Computing

We have described previously the pilot job technology. By consolidating the basic functionality (e.g. delivery and execution of complete frameworks on worker nodes) into a common reusable toolkit, the pilot job model opens the possibility of transparently merging Grid infrastructures with Clouds.

Users can run their programmes in the same environment where they develop them. Resource providers can operate a basic and VO-agnostic set of resource provisioning services. The homogeneity ensured by virtualisation allows to eliminate from the pilot framework VO-specific features, improving long term maintenance and support. This approach could unveil a potentially large amount of computational capacity so far untapped by LHC experiments by exploiting Volunteer Computing resources [41].

An example of implementation is the CernVM Co-Pilot framework [5, 16], used to prototype the integration of virtualised resources into the Grid frameworks of ALICE [3, 17] and ATLAS experiments [36].

6.4.7 Long Term Data Preservation

Long term preservation of detector data is a common problem for High Energy Physics experiments that have completed their life cycle. This includes of course preservation of the software infrastructure required to reprocess the data. This is important not only as a legacy to the future generations, but also in case new phenomena are found at the LHC, which may prompt to look for similar signals in legacy data.

Such a preservation is a challenging task due to the continual evolution in computer languages, hardware and software which makes older programmes quickly obsolete unless they are ported and kept up-to-date [10]. A specific study group on Data Preservation and Long Term Analysis in HEP [45] was formed at the end of 2008 with the objective to address and clarify the management of persistent data and of experiment-specific software over long time periods [10].

Virtualisation could achieve such preservation. Using CernVM, software releases could be archived on a simple Web server and a description of the VM needed to run it could be stored in a database along with all components that are needed to re-build the image. A generic tool like rBuilder [40] and its database could accommodate changes in the computing environment over many years and allow rebuilding images suitable for some future hypervisors. Such an approach would make HEP applications and the corresponding virtual machine image independent of change of Operating System version, hardware or hypervisor programme and would shield them from incompatibility issues in the long run.

6.5 Virtualisation Benefits and Pitfalls

Virtualisation should not be seen as the solution for all problems and it certainly comes with some drawbacks. Understanding them and learning how to work around potential pitfalls is an important step in adapting to the Cloud Computing paradigm. Introducing a virtualisation layer will inevitably result in some performance loss. Measurements have shown that the performance penalty for a typical HEP application can vary between 5% and 15%, depending on the type of workload (CPU or I/O intensive), on the chosen hypervisor and on the CPU capabilities available to support the latest instructions aimed at improving virtualisation performance [15, 50]. If significantly worse performance is observed in a virtualised environment, one should inspect the system configuration for possible problem sources, in particular the host and guest kernel options, the hypervisor configuration, and the use of paravirtualised [8] drivers. These drivers, when installed on a guest machine, have the ability to communicate directly with the hypervisor without the need for device emulation leading to the increase in performance. Currently, the focus is on the paravirtualisation support for the network and block devices, the two most I/O intensive types of devices on a virtual machine.

Once many VM instances are run on a single physical host, additional problems start to appear. Proliferation of VM instances requires extra system management effort and results in the need for extra network addresses. This may indeed pose serious problems in environments where the number of public IP addresses is limited. To mitigate this one could run VM instances on a private network and channel all outbound traffic via Network Address Translation (NAT) and caching network proxies. As a side effect, this may actually improve overall security and even scalability of the virtualised infrastructure. To manage growing numbers of VM instances, tools (e.g. [38, 44]) have been developed by the hypervisor vendors and Open Source community. Since such tools typically control VMs via some sort of API, they can do much more (e.g. suspending, restarting, migrating to another host) than what could be done by running on the physical host. Again, this results in a high degree of automation and a more resilient infrastructure.

From the security perspective, running the hypervisor on a physical node actually weakens security as it opens another attack vector from VM to hypervisor. While this is possible in principle, hypervisor exploits are in reality very rare and, unless compromised, the hypervisor acts as an additional isolation layer between host and guest Operating System.

One of the biggest concerns of the resource providers is the image generation process. For VMs to be accepted for deployment on the site they must be trusted [18]. Creating a trusted image may be a lengthy process as it requires several parties to audit and sign the image before it can be approved. The process also requires the integrity of such an image to be guaranteed when it gets transferred across the network to different sites. The image size (typically 2–4 GigaByte) may pose a problem on its own and if we consider a more straight-forward approach and add to such an image the experiment software (2–8 GigaByte per release) we are easily getting over 10 GigaByte per image. In this situation, point-to-point (P2P) transport methods may be required to move such an amount of data across the distributed infrastructure [35] and even that may not be enough to cope with frequent release cycles e.g. more than once per week. These issues have a large impact on the practical use of virtualisation in distributed environments.

Virtualisation technology also opens a possibility for Virtual Organisations and even individual users to create ad-hoc, overlay Grids capable of efficiently running specific applications by deploying a pool of virtual machines on top of physical hardware resources. In doing so, they would create a virtual Grid or virtual cluster which would have a much smaller scale of physical Grid (corresponding to a subset of resources available to each Organisation at a given point in time), a scale on which many common solutions (such as Condor [24], PROOF [23], Boinc [20]) developed over the years to support high performance and throughput computing, would just work, without the need to develop any new middleware. This could give Virtual Organisations and users a complete freedom of choice when it comes to selecting the middleware that is best suited for their applications while enhancing security on the resource owner's side by completely isolating a user application running on the site's resources.

In many ways this is similar to current Pilot Job frameworks developed by the LHC experiments (see Fig. 6.4) and could be easily achieved if the resource providers would agree to move away from the traditional model of running batch jobs and simply host VO crafted virtual machines which would in turn run jobs for users within the VO.

> Moving from "middleware virtualisation" to "server virtualisation": the Amazon Cloud.

The result of this could be something completely different from what was initially understood by the Grid idea, but could actually work; by abandoning the concept that Grid middleware is in charge of the virtualisation of resources and instead using server virtualisation to reduce the complexity of the Grid problem to a level where existing, proven solutions could work. By removing the need for complicated middleware that has to scale infinitely and support drastically different use cases, we could concentrate on building a large physical pool of resources with a simple interface to allow the user to create his or her overlay. Today, this concept is fashionably called a "Computing Cloud" and there is already an implementation of a commercial service by Amazon that provides all three important blocks to construct ad-hoc Grids. The *Simple Queue Service (SQS)* that provides a reliable communication channel between components, the *Scalable Storage Service (S3)*, which is a location-independent redundant storage system and the *Elastic Compute Cloud (EC2)*, a pool of physical servers on top of which users can instantiate an arbitrary number of virtual machines based on custom images.

> While some say that "all clouds are Grids", the middleware is playing a different role in the two scenarios.

While some are already arguing that "all clouds are Grids" [21] it is important to note the different roles that "Grid" middleware has to play in the two scenarios: while in the case of a traditional Grid, the middleware is given the tasks of resource virtualisation and resource management, and has an often complex user interface, in the case of the Cloud model, middleware is simply an interface which allows for dynamical provisioning of computational resources that can be linked together using a messaging protocol and can use storage irrespective of location.

In the case of a typical Data Grid like the one being developed at CERN to fulfil the needs of the LHC experiments, the storage is localised since the need to store a massive amount of data mandates the proximity of Mass Storage Systems which are located in typically very large computing centres. Unless we can one day move away from that model towards a "storage cloud" where data would be stored on spinning media at multiple locations, it will be difficult to benefit completely from the elegance and simplicity of the Cloud model.

If this does happen, we would be in the position to use server virtualisation technology not only to improve the quality, reliability and availability of Grid services but also to construct a simple, thin middleware that can be used to create ad-hoc Grids for a specific application in a given time window, irrespective of user location.

6.6 Conclusions and Lessons Learned

Today, we call the vehicle that moves primarily on the road, has four wheels and carries people an automobile – just as we called the first one constructed about 200 years ago. The technology has changed but the name has survived. Similarly, as technology changes, the Grids of the future are likely only to share a name with the Grids of today. The emerging virtualisation technology has the great potential to offer a clear separation between the physical infrastructure (in the hands and under the responsibility of its owners) and many middleware implementations tailored to the specific needs of a user or a group of users (and under their responsibility).

In HEP virtualisation technology offers an opportunity to decouple infrastructure, operating system and experiment software life-cycles as well as the responsibilities for maintaining these components. While in the traditional Grid model, modifying any one of them affects all of the others, virtualisation allows for the different parts to evolve independently. The site operators have the responsibility to run virtual machines with minimal local contextualisation, the experiment collaborators maintain the necessary libraries and tools, and the end-user receives a uniform and portable environment for developing and running experiment analysis software on both single desktops and laptops as well as batch nodes in computer centres and computing clouds.

As a result, proliferation of Virtual Machine images can be avoided as well as the need to frequently update and distribute them to many sites along with all the related security issues is eliminated.

Evidently, there will be always some price to pay for using virtualisation in terms of performance, but if we measure the overall project time leading to publication of results (time-to-solution), rather than just comparing benchmark results on a single node, small performance penalties will be seen to be quickly offset by the convenience and improved service quality that virtualisation can provide. This will open the possibility for much faster development and favour evolutionary cycles that will ultimately lead to better, more comfortable and economical, faster and yet safer Grids of the future.

References

1. Amazon Elastic Compute Cloud (EC2). http://aws.amazon.com/ec2
2. ARDA: Architectural Roadmap towards Distributed Analysis. http://lcg.web.cern.ch/LCG/SC2/cern-sc2/cern-sc2-2003/CERN-LCG-2003-33.pdf
3. Bagnasco, S., et al.: AliEn: ALICE environment on the GRID. J. Phys.: Conf. Ser. **119** (2008) 062012
4. Buildbot system. http://www.buildbot.net/
5. Buncic, P., et al.: A practical approach to virtualization in HEP. Eur. Phys. J. Plus **126**(13) (2011) DOI: 10.1140/epjp/i2011-11013-1
6. Buncic, P., Peters, A.J., Saiz, P.: The AliEn system, status and perspectives. In: The Proceedings of 2003 Conference for Computing in High-Energy and Nuclear Physics (CHEP 03), La Jolla, California, 24–28 Mar 2003, pp MOAT004 [arXiv:cs/0306067]
7. Butler, R., Welch, V., Engert, D., Foster, I., Tuecke, S., Volmer, J., Kesselman, C.: A National-Scale Authentication Infrastructure. Computer **33**(12), 60–66 (2000)
8. Campbell, S., Jeronimo, M.: Applied Virtualization Technology, Intel Press, Hillsboro (2006)
9. Common Use Cases for a HEP Common Application Layer (HEPCALII). http://lcg.web.cern.ch/LCG/SC2/cern-sc2/cern-sc2-2003/CERN-LCG-2003-32.doc
10. Dphep Study Group. Intermediate report of the ICFA-DPHEP Study Group **arXiv:0912.0255v1** (2009)
11. Foster, I., Kesselman, C., Tuecke, S.: The Anatomy of the Grid: Enabling Scalable Virtual Organizations. Lect. Notes Comput. Sci. **2150**, 1 (2001)
12. Foster, I., Kesselman, C.: Globus: a Metacomputing Infrastructure Toolkit (1997). ftp://ftp.globus.org/pub/globus/papers/globus.pdf
13. Foster, I.: What is the Grid? A Three Point Checklist (2002). http://www-fp.mcs.anl.gov/~foster/Articles/WhatIsTheGrid.pdf
14. Foster, I.: What is the grid? a three point checklist. Grid today **1**(6), 32–36 (2002)
15. Ganis, G., et al.: Studying ROOT I/O performance with PROOF-Lite. In: Proceedings of XVIII. International Conference on Computing in High Energy Physics (2010)
16. Harutyunyan, A., et al.: CernVM CoPilot: a Framework for Orchestrating Virtual Machines Running Applications of LHC Experiments on the Cloud. In: Proceedings of XVIII. International Conference on Computing in High Energy Physics (2010)
17. Harutyunyan, A.: Development of Resource Sharing System Components for AliEn Grid Infrastructure, Ph.D. thesis, State Engineering University of Armenia, CERN-THESIS-2010-084 (2010)
18. HEPiX working group. http://www.hepix.org
19. http://alien.cern.ch
20. http://boinc.berkeley.edu/
21. http://en.wikipedia.org/wiki/Grid_computing
22. http://en.wikipedia.org/wiki/Virtualization
23. http://root.cern.ch/twiki/bin/view/ROOT/PROOF
24. http://www.cs.wisc.edu/condor/
25. http://www.datasynapse.com
26. http://www.eu-egee.org
27. http://www.globus.org
28. http://www.google.com/trends?q=grid+computing
29. http://www.google.com/trends?q=grid+computing+virtualization

30. http://www.opensciencegrid.org
31. http://www.opensciencegrid.org
32. http://www.platform.com
33. https://twiki.cern.ch/twiki/bin/view/Atlas/PanDA
34. Hudson, Extensible continuous integration server. http://www.hudson-ci.org/
35. Integration of VMs in the Batch System at CERN. Proceedings of XVIII. International Conference on Computing in High Energy Physics (2010)
36. Maeno, T.: PanDA: distributed production and distributed analysis system for ATLAS. J. Phys.: Conf. Ser. **119** (2008) 062036
37. Meglio, AD..: ETICS: the international software engineering service for the grid. J. Phys.: Conf. Ser. **119** (2008) 042010
38. Nurmi, D., et al.: The eucalyptus open-source cloud-computing system. In: Proceedings of the 2009 9th IEEE/ACM International Symposium on Cluster Computing and the Grid, pp. 124–131 (2009)
39. Open Grid Forum, http://www.ogf.org
40. rPath rBuilder. http://www.rpath.com
41. Segal, B., et al.: LHC Cloud Computing with CernVM. In: Proceedings of XIII International ACAT Workshop Pos(ACAT2010)004
42. Sevior, M., Fifield, T., Katayama, N.: Belle monte-carlo production on the Amazon EC2 cloud, J. Phys.: Conf. Ser. **219** 012003 (2009)
43. Shiers, J.: The worldwide LHC computing grid (worldwide LCG). J. comput. phys. comm. **177**(1-2), 219–223 (2007)
44. Sotomayor, B., et al.: Virtual infrastructure management in private and hybrid clouds. IEEE Internet Comput. J. **13**(5), 14–22 (2009)
45. Study Group on Data Preservation and Long Term Analysis in High Energy Physics. http://www.dphep.org/
46. The Grid: Blueprint for a New Computing Infrastructure. http://portal.acm.org/citation.cfm?id=289914
47. The Metacomputer: One from Many, NCSA (1995). http://archive.ncsa.uiuc.edu/Cyberia/MetaComp/MetaHome.html
48. Tsaregorodtsev, A., et al.: DIRAC - Distributed infrastructure with Remote Agent Control. In: The Proceedings of 2003 Conference for Computing in High-Energy and Nuclear Physics (CHEP 03), La Jolla, California, 24–28 Mar 2003, pp TUAT006 [arXiv:cs/0306060]
49. Vaquero, L.M., et al.: A break in the clouds: towards a cloud definition. ACM SIGCOMM Comput. Comm. Rev. **39**(1), 50–55 (2008)
50. Yao, Y., et al.: Performance of ATLAS software on virtual environments. In: Proceedings of XVIII. International Conference on Computing in High Energy Physics (2010)

Chapter 7
Evolution of Parallel Computing in High Energy Physics

Fons Rademakers

Computing in High Energy Physics (HEP) has always required more computing power than could be provided by a single machine. In fact, HEP has been pushing the capabilities of computers since the first mainframes were produced in the early sixties. Every CPU cycle and every last bit of memory and disk has always been used, running data acquisition, simulation, reconstruction and analysis programs.

7.1 Introduction

In this context, the idea of using a group of CPUs or computers in parallel to increase the combined CPU, memory and disk available to a program, is only natural. In addition, in HEP data processing *events*, i.e. particle collisions recorded by detectors, can be processed easily in parallel as they are all independent. A problem that can be easily parallelised in large chunks is called *embarrassingly parallel* and most HEP processing falls in that category. In theory running an embarrassingly parallel program on multiple machines looks fairly trivial, but the practise is much more complicated.

In this chapter I will concentrate on tracking, for a large part, the evolution in parallel computing in HEP by describing the evolution of two major parallel data analysis systems have been working on over the past 15 years. But first, what makes efficient parallel computing so hard? For that, lets have a look at Amdahl's law.

7.2 Amdahl's Law

Amdahl's law [1], named after computer architect Gene Amdahl, describes the relationship between the expected speedup of parallel implementations of an algorithm relative to the serial algorithm. For example, if a parallel implementation

---

F. Rademakers (✉)
CERN, Geneva, Switzerland
e-mail: Fons.Rademakers@cern.ch

of an algorithm can run 15% of the algorithm arbitrarily fast, while the remaining 85% of the operations are not parallel, the law states that the maximum speedup of the parallel version is $\frac{1}{1-0.15}$ = 1.18 times faster than the non-parallel version. This can be written as:

$$Speedup = \frac{1}{1 - P} \qquad (7.1)$$

If none of the code can be parallelised, $P = 0$ and the *Speedup* = 1 (no speedup). If all the code is parallel, $P = 1$, the *Speedup* $\to \infty$ (in theory). If 50% of the code can be parallel, the maximum *Speedup* = 2, meaning that the code will run twice as fast.

Parallel computing performance is governed by Amdahl's law.

In the case of parallelisation, Amdahl's law states that if P is the fraction of the program that can be made parallel, and $(1 - P) = S$ is the fraction that cannot be parallel (remains serial), then the maximum speedup that can be achieved by using N processors is:

$$Speedup = \frac{1}{S + \frac{P}{N}} \qquad (7.2)$$

In the limit, as N goes to infinity, the maximum speedup goes to $1/S$. In practise, price/performance increases rapidly as N is increased once there is even a small component of S. See Fig. 7.1 for speedup versus number of processors plots for different values of P. Notice for example, if P is 90%, i.e. there is only a 10% serial part, the program can be sped up by a maximum factor of 10, no matter how many processors are used. For this reason, parallel computing is only useful for either small numbers of processors, or problems with very high values of P, so-called *embarrassingly parallel* problems. A great part of the craft of parallel programming consists of attempting to reduce S to the smallest possible value.

A great part of the craft or parallel programming is reducing the serial part to a minimum.

7.3 Some History

As most computing in HEP concerns event processing, and since events are independent, they can be trivially processed in parallel. However, most computer centres had only one or at most two large mainframes, and thus there was no

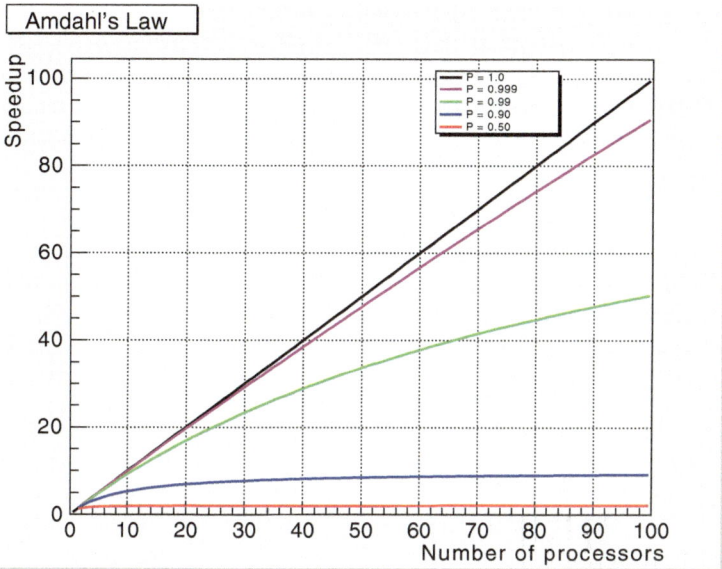

Fig. 7.1 Plot showing speedup as function of number of processors for different values of P

infrastructure to run programs in parallel. However, with the introduction of powerful mini-computers, like the Digital VAX series, in the middle of the 1980s and the even more powerful and cheaper workstations, like the HP PA-RISC machines, in the early 1990s, affordable quasi-parallel processing became feasible. This was typically done by running the same programs on different parts of the data under the control of a batch system. These programs all ran independently and there was no communication between them. Real parallel computing in HEP is a relatively recent activity. The first serious efforts were made only in the mid 1990s, when in addition to the processors also the networking became faster and more affordable (FFDI, HiPPI and fast-Ethernet). However, the still relatively slow communication could introduce a large serial component into the parallel programs and limit their efficiency to only a few processors. There were two main solutions to this problem, using low volume, and therefore expensive, networking technologies like Myrinet, HiPPI or SCI, or design algorithms and programs with very low communication overhead. The latter was the preferred and most economic solution used in the early parallel data analysis systems.

7.4 Parallel Computing Opportunities in HEP

HEP computing, being centred around event processing, allowed for many parallel computing opportunities. Over the last three decades parallel solutions were developed in some areas related to HEP.

Theoretical QCD lattice gauge calculations: several characteristics of lattice calculations make them simpler than other large problems. For instance, researchers can easily divide the uniform Grid evenly among the processors of a parallel computer, in such a way that individual processors rarely need to trade information. Also, lattice calculations don't require much input and output, and compared to other calculations, need relatively little memory. The trend used to be to build custom parallel computers optimised for floating point calculations, however currently, clusters made of off the shelf components are favoured.

On-line triggering, filtering and data recording systems: traditionally this field has used multiple machines to parallel the detector read-out. In the LHC era, due to the complexity of the events and the high rates, large clusters are being used to run the event trigger algorithms in parallel.

Monte-Carlo simulations: the most embarrassingly parallel area of HEP computing. Very large numbers of simulated events are needed for detector and model studies. The events are typically produced by running a large number of batch jobs in parallel. There is no need for finer-grained parallelism as there is no real interest to get quickly a single event, but rather to maximise the number of events produced in a given time.

Event reconstruction: in event reconstruction the raw detector data of the millions of recorded events is reconstructed to physics data. Also here, as for Monte-Carlo simulations, one needs all events reconstructed before continuing to the data analysis phase, so batch jobs running on large clusters is the preferred solution.

Data analysis and data mining: during data analysis, physicists mine the reconstructed events to find patterns hinting at new physics or to make precise measurements of specific quantities. This is the field where fast response times are of importance as during an analysis many different hypothesis have to be tried. Here finer-grained parallel solutions will provide a real benefit.

> HEP presents several parallelisation opportunities which have been exploited over the years.

In the rest of the article I will focus on parallel data analysis, parallel hardware architectures, parallel programming techniques and the parallel data analysis systems we have developed in the last 20 years.

7.5 Parallel Data Analysis

Data analysis, also called data mining outside the field of HEP, is the process concerned with uncovering patterns, associations, anomalies, and statistically significant structures in data. One of the key steps in data analysis is pattern recognition

7 Evolution of Parallel Computing in High Energy Physics 181

or the discovery and characterisation of patterns in data. Patterns are identified using features, where a feature is any measurement or attribute that can be extracted from the data.

In general data analysis is an interactive, iterative, multi-step process, involving data preparation, search for patterns, knowledge evaluation and possible refinements based on feedback from one of the steps. In HEP data preparation is done by first processing the raw detector data to reconstruct the event topology and dynamics, then the event data is reduced to key quantities (features) that will be used in the data analysis, like number of particle tracks, particle types, momentum distribution, energy depositions, etc. During the analysis, the key quantities are processed for all events in the data sample and interesting patterns are identified using classification and clustering algorithms. Finally, the patterns are presented to the user in an easy-to-understand manner, often in the form of histograms, graphs and other types of plots. As the various steps in this process are refined iteratively, the model that is build from the data is validated (or invalidated).

Data analysis is a multi-disciplinary field, combining ideas from diverse areas such as machine learning, statistics, high performance computing, mathematical programming and optimisation, and visualisation.

As the data sets to be analysed are becoming increasingly larger and more complex, it is becoming clear that we require a process that is scalable. In a parallel environment, by *scalable* we generally mean the ability to use additional computational resources, such as CPUs, memory and disks, in an effective way to solve increasingly larger problems.

> The reason to develop parallel data analysis systems is that the wall-clock time needed to process the data is a major factor that can affect the ability of a user to explore many different hypotheses.

The fundamental reason driving the development of scalable data analysis systems is simple – we need to build accurate models quickly. This can be beneficial in several different ways. Firstly, faster turnaround times tend to yield better models. The wall-clock time needed to process the data is a major factor that can affect the ability of a user to explore many different hypotheses. Fast processing allows multiple hypotheses to be explored quickly. It will also allow the exploration of a larger set of models and thus be more likely to find a better model. Secondly, there can be great value for analysis of data as it is being acquired. This prompt analysis is important in cases where the transient effects are the important ones (triggering on event types in HEP, or on-line fraud detection in business transactions), the processing must keep up with the rate of new data acquisition.

There are several key issues that must be considered in order to make data analysis scalable.

The scalability of the algorithm. Here scalability is a description of how the total computational work requirements grow with problem size, an issue that can be discussed independent of the computing platform. Ideally, we would like the computational requirements for an algorithm to grow slowly as the problem size increases. While this may not always be possible, the payoffs can be great, even on a single processor.

Ease of data movement. Each step of the data analysis process is performed many times over the course of one complete analysis. It is therefore important that we manage the data appropriately between the various steps. For example, the output of the reconstruction step should be in an appropriate format so that it can be directly used by the analysis program. Another important concern arises if the data is being generated at a location different from where it is analysed. Data sets are growing much faster than increases in network and disk I/O bandwidths. As the size to bandwidth mismatch continues to grow, the choice of where and when to process and analyse the data will be crucial to the overall success of the project.

Fast data access. Since most data sets are much larger than what can be stored in core, it is important to create fast, parallel, specific access methods that *forgo* the access methods that are invoked by the query optimiser of a standard relational DBMS. In typical DBMS, the overhead coming from transaction semantics, locking, logging, and other features that are unnecessary for analysis application can slow the entire process down immensely. In HEP we have developed data storage systems that are optimised for *write-once, read-often* access that employ data layout schemes that are specifically optimised for analysis. These systems typically cluster same kind attributes for all events in large buffers stored consecutively on disk (*column-wise* storage), instead of having the attributes grouped per event (*row-wise* storage). This is hugely beneficial as during analysis one is typically interested in only a limited number of attributes of an event instead of the full event. Column-wise storage speeds up access by reading only of the desired attributes from disk.

7.6 Parallel Architectures

An overview of parallel data analysis would be incomplete without a brief overview of parallel architectures and the issues to be considered in the efficient implementation of parallel algorithms. Programming for a parallel machine is significantly more complex than programming for a serial machine and requires a good understanding of the underlying machine architecture.

Parallel computers have had a long history, mainly as research efforts in scientific computing (e.g. for QCD calculations). However, in the last two decades, as the need for computing power has increased, there has been greater acceptance of parallel computers in both the scientific and commercial domains.

> Programming for a parallel machine is significantly more complex than programming for a serial machine and requires a good understanding of the underlying machine architecture.

Traditionally, the distinction between parallel architectures was between "single instruction stream, multiple data stream" or SIMD machines and "multiple instruction stream, multiple data stream" or MIMD machines. In the former, thousands of relatively simple processing elements, with limited local memory, executed the same instruction on different data. MIMD systems on the other hand, had fewer, but more powerful processors, each executing potentially different programs. Several commercial systems were built, exploiting both SIMD and MIMD architectures. However, except for specialised applications such as image processing, MIMD architectures currently dominate the market; the distinction today is not between SIMD and MIMD systems, but between shared-memory and distributed memory architectures.

7.6.1 Shared-Memory Architectures

Shared memory machines, also known as symmetric multiprocessors (SMP), have multiple processors accessing the common memory through a single shared bus, see Fig. 7.2. Examples of such systems include various Cray X-MP and Y-MP vector/multiprocessor systems, the HP/Convex and the SGI Power Challenge.

The programming model for SMP systems is based around a single address space. Data are shared by processes when they directly reference the address space. The programmer must properly identify when and what data are being shared and synchronise the processes explicitly to ensure that shared variable are accessed in the proper order.

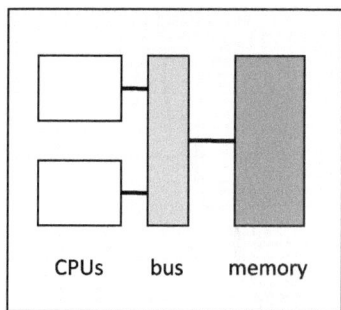

Fig. 7.2 SMP architecture

In practise, software is written for SMP systems using either *pthreads*, which is an IEEE Posix standard, or the *OpenMP* standard for portable shared memory programs. Both approaches require the programmer to parallelise the code. The pthread standard uses function calls, while OpenMP uses compiler directives.

SMP systems typically consist of 2 to 16 or more CPUs or cores. As the number of processors increase beyond 16 or so, they have to be connected by buses that are increasingly wider, with a higher bandwidth and lower latency. Unfortunately, as the number of processors sharing the bus increases, the bandwidth available per processor decreases, resulting in substantial performance degradation. The lack of scalability of bus-based architectures can be overcome either by using a different system interconnect such as crossbar interconnects or by moving to distributed memory architectures.

7.6.2 Distributed-Memory Architectures

High-performance systems that can scale to large number of processors are designed from the start to avoid bottlenecks of shared interconnects. Such systems, called distributed memory machines, "shared nothing" systems or massively parallel processors (MPP), consist of separate computing nodes with no shared components other than the interconnect, which is usually a switched or crossbar connection that scales with the number of processors, see Fig. 7.3. As a result, the interconnect bandwidth available per processor remains relatively constant across configuration sizes. Examples of this kind of system include Cray T3D and Intel Paragon.

Distributed memory systems use the message passing programming model to communicate among processors. The memory of the entire system is partitioned

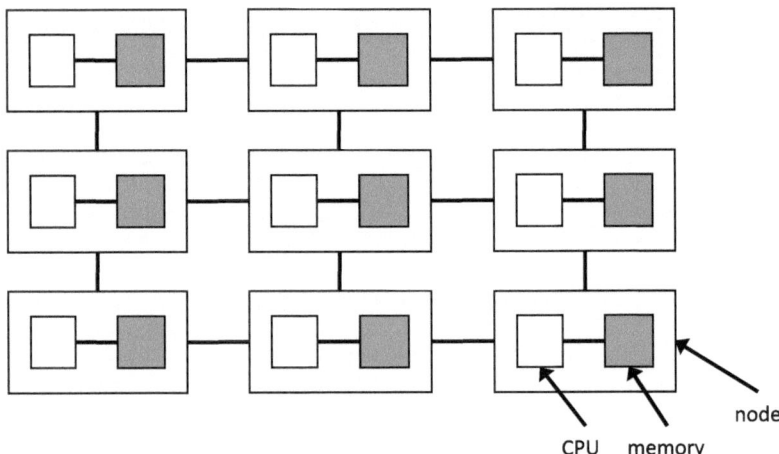

Fig. 7.3 MPP architecture

in distinct address spaces and each processor has only access to its local memory. The program on each processor must be aware of when and what data it needs to send to, or receive from, another processor, and must include the code that handles this communication. As a result, MPP systems are typically harder to program than SMP systems. In addition, as communication is done through a software layer, the overhead costs can be sizeable.

The two most popular message passing software packages currently available are the Message Passing Interface (MPI [2]) and the Parallel Virtual Machine (PVM [3]). While the two packages are very similar, a richer set of communication features and improved communication performance tend to make MPI the preferred choice in most cases.

7.6.3 Hybrid Architectures

As the need for processing power increases, an obvious solution is to use an SMP system as the basic node in an MPP system, see Fig. 7.4. This approach, sometimes referred to as "clusters of SMPs", can provide a significant increase in the peak performance at a cost that is not much higher than the cost of a uni-processor with the same amount of storage. Such hybrid architecture machines include commercial machines as the IBM SP-2, SGI Origin and the HP/Convex systems.

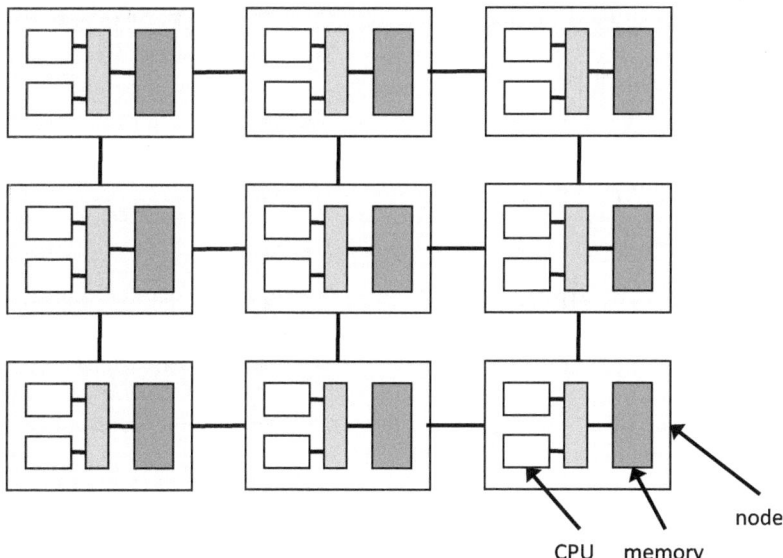

Fig. 7.4 Clusters of SMPs

A major, and yet unsolved, problem with clusters of SMPs is that the programmer now has to manage two levels of parallelism, with big differences in performance and function between the inter-node and intra-node communication. One approach, using software is to use OpenMP within the node and MPI access across the nodes.

A somewhat different approach is used in the distributed shared memory architectures, where the memory is physically distributed, but the system implements a single shared address space, thus retaining the shared memory paradigm for communication and programming.

The current fastest supercomputers are all clusters of SMPs, like the Blue Gene from IBM.

7.7 Data Analysis Programming Issues

Obtaining high performance from a data analysis application using fine-grained parallelism is not a straightforward task. Broadly speaking, parallelising code can follow *task parallel*, *data parallel* or *hybrid* approaches. Task parallel approaches attempt to break the work down so that each processor is working on an independent piece of the problem space, with coordination at the end to form the global result. This can require data being replicated to all processors, or clever schemes for partitioning the data across processors, without incurring large communication costs. Data parallel approaches involve partitioning the data across nodes and having all nodes participate in expanding each part of the problem space by working together. This can involve large synchronisation costs due to poor load balancing. Synchronisation costs occur where processors idle-wait for others to catch up and pass relevant information on before proceeding. As in HEP data parallelism is the natural way to partition problems, solving these synchronisation issues were the most important challenges. In the sections on the PIAF and PROOF systems we describe how we minimised these issues.

Several factors can affect the performance of a parallel application, we discuss a few of them briefly here.

Single Processor Performance. While much of the work in parallel data analysis focuses on the parallel performance, we must also ensure that the single processor implementation of the analysis algorithm has been fully optimised. This is especially true as the newer processors all have deep memory hierarchies and care must be taken to exploit the differences in data transfer rates between the different levels.

One of the most important factors that affects the performance on a single CPU is the data locality in a program. Programs that have a high spatial and temporal locality make extensive use of instructions and data that are near to other instructions and data, both in space and time. Several ways of improving data locality, such as changing the order of loops or unrolling them are frequently done automatically by the compiler (especially Intel's icc is good at this). Others such as changing the data layout must be done by the programmer.

> Obtaining high performance gain for data analysis using fine-grained parallelism is not a straightforward task and require to take into account many competing factors.

Data Layout. Data layout is an important issue that the programmer must address, especially for MPP systems. As a processor has access only to its local memory and disk, any data that is required by the processor, but is not in its local memory or disk, results in communication overhead. Therefore, it is important that a processor has most of the data it needs in its local memory. In addition, within each processor, the data should be arranged and the algorithm written, in such a way that cache misses and page faults are minimised. This benefits not only the single processor performance, but also reduces the demand for bus bandwidth in SMP systems.

Load Balancing. Another problem that usually degrades performance in both SMP and MPP systems is load imbalance. This occurs when one or more processors has less work to do than the other processors and is idle during part of the computation. In SMP systems, techniques such as guided self-scheduling can be used for load balancing. In MPP systems, initial load balancing is typically achieved by distributing the data equally across the processors. However, the load may become unbalanced during the computation, requiring a re-distribution of data, resulting in communication and synchronisation overhead.

Inter-processor Communication. In MPP systems, communication costs add to the overhead of the data analysis algorithm. They can adversely affect parallel performance to a great extent and considerable effort may be required to achieve scalability on MPP systems. Techniques to reduce the time spent in communication include merging several short messages into a long one, using an appropriate data layout, trading communication for redundant computation and overlapping communication with computation. Partitioning the problem so that the computation to communication ratio is high may require the complete rewrite of an implementation, or the development of new algorithms.

I/O. Both memory and disk bandwidth continue to grow more slowly than capacity. For example, while disk capacities have been increasing at about 60% per year, the I/O bandwidth has increased only 40% per year. As the bandwidth to capacity ratio shrinks, it takes longer to read or write all the memory or disk available to a system. The practical definition of *large* in large-scale data analysis tends to be set by the total size of memory and disk available to an application. This implies that the high cost of data movement, algorithm setup, data redistribution and data scanning will only worsen over time. At the very least, this will increase the pressure on developing algorithms that require fewer passes over the data. Designing algorithms that can make accurate decisions with less data will be important as well, in some cases.

7.8 The Parallel Interactive Analysis Facility (PIAF)

Around 1992 the PAW [4] (see Chap. 1) data analysis infrastructure, like the column-wise n-tuple data storage system and associated query mechanism, had matured. At the same time the amount of data collected by the Large Electron Positron (LEP) experiments at CERN was steadily increasing and analysis on a single machine was becoming quite time consuming. To bring the analysis time back to the interactive domain we decided to parallelise the PAW analysis engine so it could be run on a cluster of workstations. After some discussions in early 1993 we started the development of a prototype called PIAF [5], for Parallel Interactive Analysis Facility. We decided on a three tier architecture with the user PAW client session talking to a master and the master controlling a set of workers. The master would divide the set of events to be analysed in equal parts and submit to each worker a range of events to be processed. Due to the *embarrassingly parallel* nature of event processing (all events are independent), there was no need for communication between the workers and very little between master and workers, which allowed for a very high level of parallelisation and a good speedup. We designed PIAF such that it behaved exactly like a local PAW analysis session. To start PIAF, only one extra command was required and the user scripts did not have to be changed. This transparency considerably eased the adoption of the system once we released it to the users in the spring of 1994.

7.8.1 Push Architecture

PIAF was based on a so called *push architecture*, where the master pushes the workload to the workers (see Fig. 7.5). This was a simple but brute force approach to parallelism, which lacked the refinements needed to handle changing loads on the workers and did not take into account data locality, with as a result a possible uneven workload on the workers, where the slowest node could considerably delay the finishing of the query. However, with the typical analysis workload consisting of fairly short queries, in the range of a few minutes, this was not a serious problem. The PIAF communication layer was implemented by our own simple TCP/IP based message passing library written in C, while PAW was written in FORTRAN. We had evaluated several different existing communication libraries but they all provided much more than we needed, and by using our own little library we kept full control over the communication infrastructure.

> The Parallel Interactive Analysis Facility (PIAF) pioneered parallel analysis in HEP in 1993.

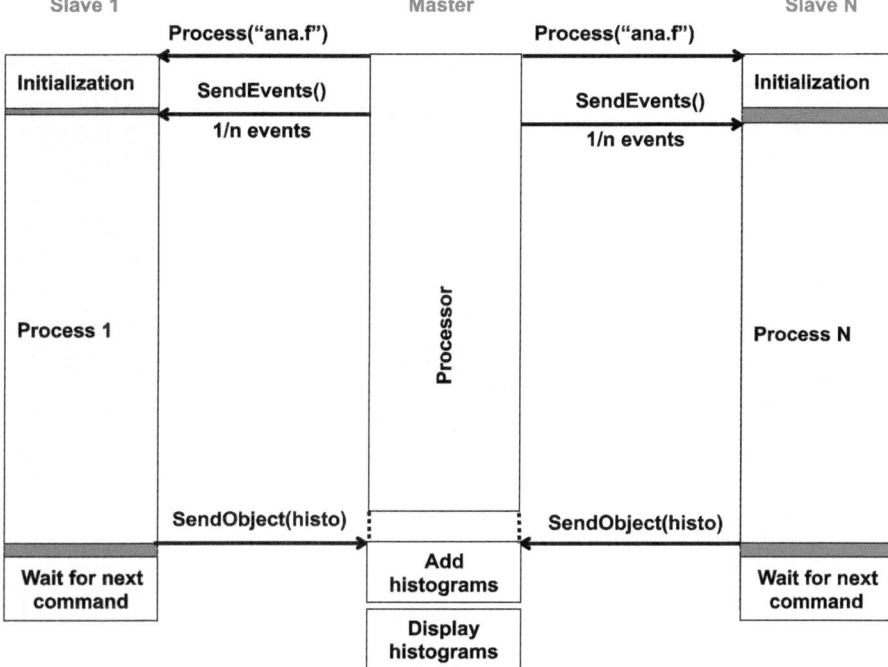

Fig. 7.5 The PIAF push architecture. After distributing the analysis macro *ana.f* to the workers, the master divides the total number of events by the number of workers and pushes to each worker a unique, equally sized, range of events to be analysed

7.8.2 A Joint Project with Hewlett-Packard

In the PAW development team we had a long history of joint projects with Hewlett-Packard (HP) and previously with Apollo before it was acquired by HP. The joint projects provided the PAW developers with state of the art HP workstations, with the new and very fast PA-RISC processors. In return HP received feedback on how their machines worked in the demanding HEP computing environment. When the idea of PIAF was mentioned during a joint-project meeting, the HP Bristol Lab manager saw the interest of this development and how it could be applied to large scale data analysis outside of HEP. He decided to send an HP engineer for 1.5 years to CERN, to fund a fellow already working on PAW and to fund a 5 node cluster of high-end PA-RISC workstations equipped with prototype FDDI networking cards. Initially the cluster was installed in one of the developers offices, where in the summer the temperature, in the air conditioned room easily exceeded 40°C. The HP hardware kept functioning perfectly under these extreme conditions, however the developer suffered under the heat and noise. After the summer the machines were moved to the computer centre. Having this private, powerful, cluster greatly helped the development and testing of the PIAF prototypes. Especially the FDDI networking, with a speed of about 100 Megabit/s, allowed efficient remote file access.

7.8.3 Deployment at CERN

By the end of 1993 the first version of PIAF was ready for deployment. We integrated our HP test cluster into the CORE [6]/SHIFT environment of the CERN CN (IT) department (see Chap. 3) so the users could easily copy their data files from the central tape and disk stores to the PIAF cluster. See Fig. 7.6 for the full hardware setup of the PIAF cluster. All the disks were cross-mounted on all nodes via NFS over FDDI allowing access to the data from all nodes.

Initially we worked with a small number of physicists from several LEP experiments to get their analysis started on PIAF. The advantages of using PIAF were quickly recognised and the user community grew to about 100 registered users.

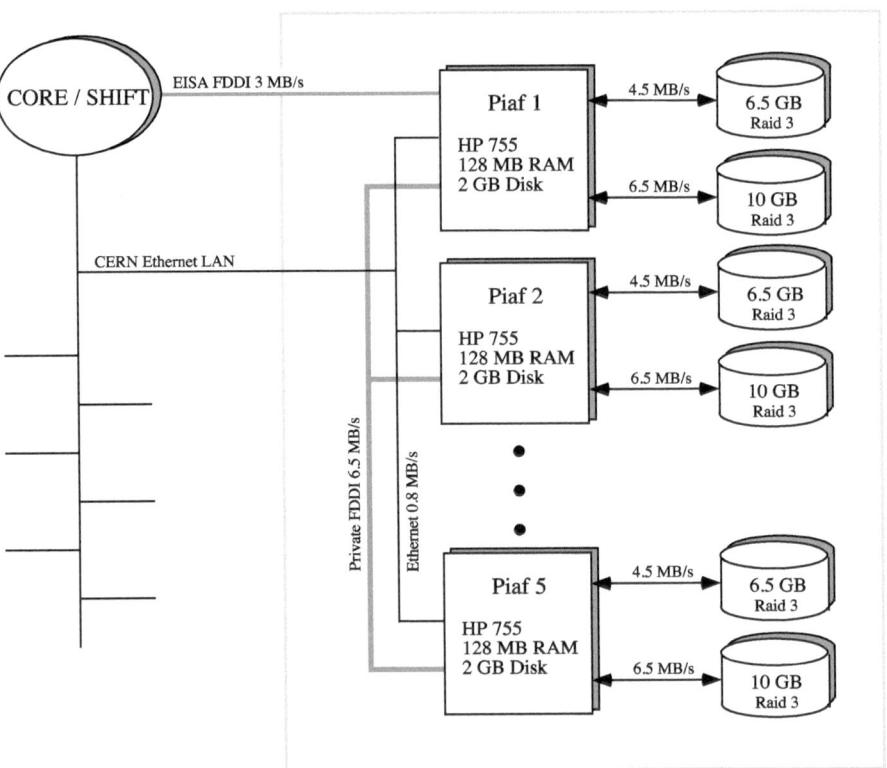

Fig. 7.6 The PIAF hardware as it was deployed by the end of 1993. The entire cluster packed less CPU, disk and network power than an average 2008 laptop, but was state-of-the-art in its time

7.8.4 PIAF Performance

To understand a complex distributed system good monitoring is a must. In PIAF we used the Unix *syslog* facility to monitor the PIAF master and workers performance. For each query the full query statement, the begin time, the real time, the total CPU time, the number of files, events and bytes processed, etc. were recorded. This detailed information allowed us to understand how the system performed (speedup, date processing rate, average query time) and what the users were doing and when they were doing it (all anonymously of course). From this data we found that:

- At the peak there were about 30 concurrent users.
- Queries typically lasted from several seconds to tens of minutes (with emphasis on tens of seconds).
- Most queries were only partially overlapping and hence the users often would get good speedup.
- Some queries enjoyed super linear speedup.

> Super-linear speedup was observed on the PIAF system.

According to Amdahl's law, the theoretical maximum speedup of using N processors would be N, namely linear speedup. However, it is not uncommon to observe in practise more than N speedup on a machine with N processors. This is called super-linear speedup. In the PIAF case this was due to memory aggregation. In parallel computers, not only does the number of processors change, but so does the size of accumulated memory from the different machines. With the larger accumulated memory size, more or even the entire data set can fit into memory, dramatically reducing disk access time and producing an additional speedup beyond that arising from pure computation.

7.8.5 PIAF Licensed by Hewlett-Packard

After HP had sponsored the development of PIAF with equipment and people it decided to license from CERN the final product. HP had a great interest in PIAF because it had customers facing the same type of data analysis problems as we did in HEP. One of their key customers was British Telecom which wanted to analyse call records for specific patterns so they could make their customers special offers depending on their calling habits. HP paid one million Swiss Francs for the PIAF license, which at the time, was one of the largest technology transfer contracts CERN had ever signed. However, due to a conflict between the CERN IT management and the PAW and PIAF development teams, this success was kept quiet and the PIAF team was silently dissolved by the end of 1994.

7.9 ROOT

After the dust had settled, the folding of the PAW and PIAF efforts turned out to be a boon which allowed us to make a clean break with the past. PIAF and PAW were coded in FORTRAN, but we felt this was not the appropriate language in which to write the next generation data analysis environment that should be able to handle the large LHC data volumes. We needed a language that allowed better abstraction and data encapsulation, which was efficient and which was widely available. The most logical choice at the time was to use C++. So in early 1995 René Brun and I started what would become ROOT [7] (see Chap. 1 for a complete account of the ROOT development). After two years of non-stop development on ROOT, we had histogramming, minimisation and basic math components, networking and messaging classes, an operating system abstraction layer, collection classes, a C++ interpreter, 2D/3D-visualisation tools, and, last but not least, a very efficient object data store with support for attribute-wise (column-wise) object storage. The object data store uses an automatic, dictionary based, object streaming technology which streams (flattens) an object into a data buffer before writing it to a file. Instead of writing this buffer to a file it can also be sent over a socket. This allowed us to easily implement the sending of complete objects between processes without having to resort to complex third party systems like CORBA [8].

> The development of ROOT allows a new depart for parallel analysis applications.

With all these components in place, albeit far from finished, we could start work on our second generation parallel analysis system.

7.10 Parallel ROOt Facility (PROOF)

Although ROOT was still quite young, a number of our early adopters had already produced large amounts of data in ROOT format. In particular the heavy ion experiments at Relativistic Heavy Ion Collider (RHIC) in Brookhaven were using ROOT and had created in the order of 10 TeraBytes of final event data. This amount of data was about 2 to 3 orders of magnitude more than what typically was analysed by the LEP experiments using PIAF and only about an order of magnitude less than what is expected at the LHC. A parallel version of ROOT was the next logical step. We set out developing PROOF [9] as an extension of the ROOT system enabling interactive analysis of distributed data sets on distributed clusters. PROOF stands for Parallel ROOT Facility.

7.10.1 PROOF Main Features

PROOF aims at processing the data as close as possible to where they are located, transferring back only the results of the analysis, which usually are much smaller in size than the data. In this way it minimises the overhead of data access, which may be significant for I/O bound analysis typically found in HEP.

The main design goals of PROOF are: *Transparency:* distributed system perceived as an extension of the local ROOT session (same syntax, same scripts); *Scalability:* efficient use of the available CPUs so that increasing their number results in a corresponding increase in performance and *Adaptability:* ability to adapt to heterogeneous resources (in CPU type and load).

> The PROOF system adopts the "pull" workload management for workers increasing the efficiency and extending the architecture to geographically separated clusters of machines.

PROOF has a flexible multi-tier architecture, the main components being the client session, the master server, and the worker servers. Additionally, PROOF supports the possibility to run in a cluster of geographically separated clusters (Grids) by allowing a hierarchy of masters, with a super-master serving as an entry point in the system, and the others coordinating the activities of the worker nodes on their respective clusters (see Fig. 7.7). The master is responsible for distributing the work. This is done using a *pull architecture* such that worker nodes ask for more work as soon as they are ready. In this way the faster worker nodes are assigned more data to process than the slower ones and load balancing is naturally achieved. The master is also in charge of merging the output it collects from the worker nodes, so that the client receives a single set of output objects, as would have been the case for local processing.

One of the main lessons learned from PIAF was that push architectures don't scale very well. In PIAF, the master pushed the work out to the workers. If one of these workers would experience a hiccup or slowdown due to some other task, it would delay the entire query from finishing. By employing a pull architecture in PROOF we made the system more resilient to variations on the worker nodes.

7.10.2 The Packetizer

The core of the system is the *packetizer* which runs in the master process. The packetizer is in charge of handing out work packets to the worker nodes. A packet consists of a file name containing the data and the range of events to be processed in this file. The packetizer takes into account the location of the files, the performance

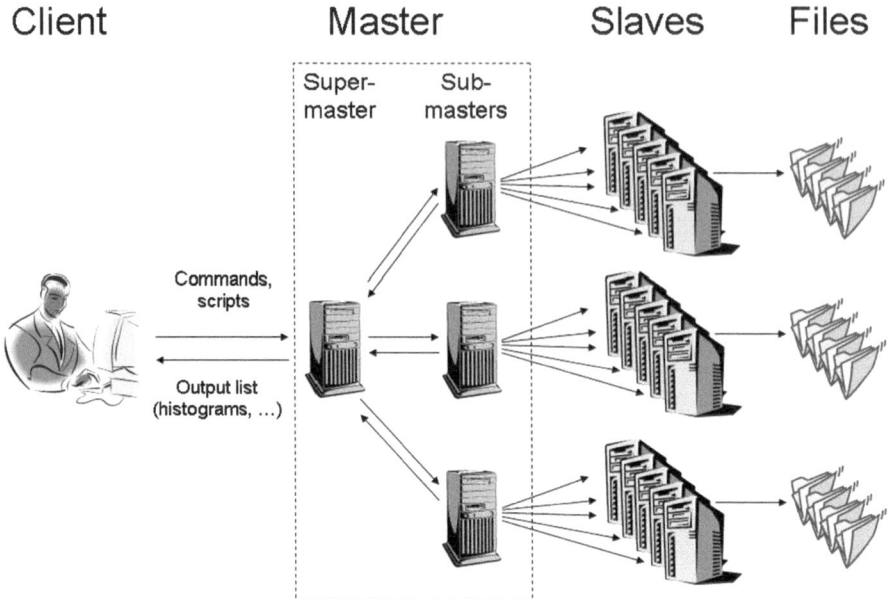

Fig. 7.7 PROOF multi-tier architecture

of the worker nodes, and the stage of the processing, reducing the packet size at the end of the query to avoid tail effects, where the slowest workers might delay the completion of the query. Different packetizers allow for different data access policies, e.g. a packetizer might be tuned to assign only local files to workers, avoiding network access.

7.10.3 Data Access Strategies

Performance measurements have shown good efficiency and scalability up to O(100) nodes. These tests have underlined the importance of low latency in accessing data from the worker nodes. For this reason PROOF always tries to assign data maximising locality: a worker node gets assigned first data located in its local pool. Data locality is, however, not always possible, hence the need to pursue techniques to minimise the latencies accessing remote data servers, both for opening the file and reading the data. In this respect, the usage of the high performance XROOTD [10] data server (see Chap. 10) is essential.

> PROOF aims at exploiting data locality wherever possible, bringing the processing to the data and not the contrary.

For larger clusters also the cluster topology will play an important factor, where data locality might be limited to the rack in which the node is placed. Typically nodes are interconnected via Gigabit Ethernet, and within a rack remote files can be accessed with nearly local disk access speeds (assuming a homogeneous spread of the files over the nodes). On the other hand, racks are often only interconnected via multiple 1 Gigabit or a single 10 Gigabit Ethernet link and hence the aggregate available bandwidth for extra-rack remote data access would become quickly a bottleneck.

7.10.4 Resource Scheduling

To face the needs of large, multi-user analysis environments expected in the LHC era, optimised sharing of resources among users is required. The resource scheduling improves the system utilisation, insures efficient operation with any number of users and realises the experiment's scheduling policy. To achieve that goal, two levels of resource scheduling are introduced. *At worker level*, a dedicated mechanism controls the fraction of resources used by each query, according to the user's priority and current load. *At master level*, a scheduler decides which number of workers can be used by a given query, based on the overall status of the cluster, the query requirements (data location, estimated time for completion, etc.), the client history, user priority, etc.

7.10.5 Interactive Batch

While PIAF was mainly geared toward short interactive processing, PROOF is expected and designed to support queries that may last for many hours. For these long queries it is not feasible nor practical to have the client session connected permanently to the remote master process. Ideally one wants to submit a query, monitor its successful start and disconnect. In the evening, at home, one would like to monitor the progress of the query by reconnecting to the desired master process, and in the morning, back in the office, one would reconnect again to retrieve the finished query results. To support this "interactive batch" feature we keep track of all the running master processes in the initial connection daemon.

7.10.6 The User Analysis Code

The user provides the analysis code by deriving from a generic ROOT analysis class called, TSelector. This class defines five main methods that the user must implement: Begin(), SlaveBegin(), Process(), SlaveTerminate()

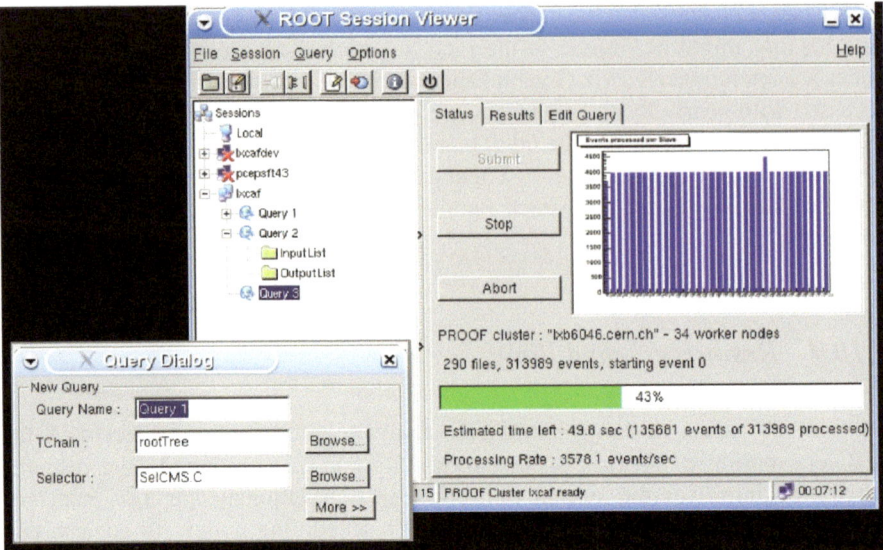

Fig. 7.8 PROOF session viewer

and `Terminate()`. The first and last methods are only executed on the client, while the other three are executed on all workers. The `Process()` method is called for each event. In this method the user puts all his selection and analysis code. Additional user code needed during the analysis can be made available to the workers by packing the sources or binaries of this code in, so called, PAR (PROOF Archive) files and uploading them to PROOF.

7.10.7 User Interface

PROOF provides a powerful API allowing the user to fully control the session environment. Methods are provided for management of the PAR files, of the query results, and to upload and distribute the data on the worker nodes. The full API can be controlled by a dedicated GUI (see Fig. 7.8).

7.10.8 Monitoring

To understand the behaviour of a complex system like PROOF good monitoring is of great importance. Detailed monitoring will help identifying bottlenecks or system malfunction. PROOF contains an internal monitoring system that gives the user feedback on how a query progresses and how the system performs. Global

7 Evolution of Parallel Computing in High Energy Physics

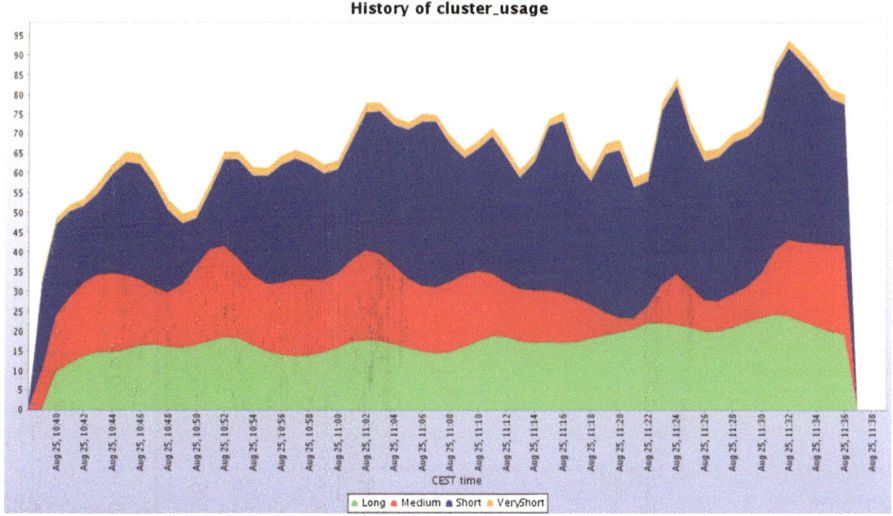

Fig. 7.9 MonALISA based monitoring showing the load on a cluster caused by different types of PROOF jobs

monitoring is implemented using the MonALISA system [11] which shows the status of the complete cluster and all PROOF sessions (see Fig. 7.9).

7.10.9 PROOF Lite

With the arrival of multi- and many-core machines PROOF becomes an ideal technology to harness the available cores. To avoid the overhead of the master process, TCP/IP sockets and connection daemons, we created a "PROOF Lite" version, where the client process directly creates the worker processes (typically one per available core), uses UNIX sockets for communication and tries to memory map the accessed files for efficiency. Using a multi-process application will add needed robustness compared to a multi-threaded solution. In addition, multi-threading a complex application like ROOT, that in addition runs user code, would be nearly impossible.

7.10.10 PROOF Performance

Extensive performance testing on large clusters with O(100) workers shows a near linear speedup with very high efficiency (see Fig. 7.10). As most analyses are I/O-bound a very good I/O infrastructure is needed to make the best use of the CPUs.

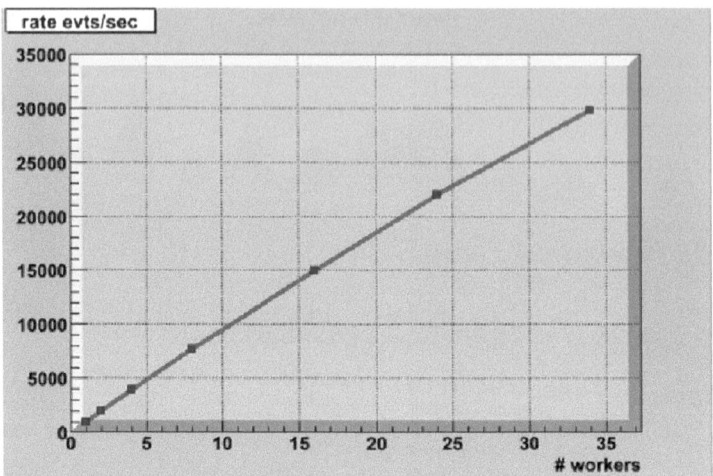

Fig. 7.10 PROOF shows excellent scalability with up to O(100) workers with basically a linear increase in processing rate with the number of workers

The more disks in parallel the better. The more evenly the data is distributed over the nodes the better. In many cases the same analysis is run over and over with slight adjustments in selection criteria. In these cases caching of the data becomes important and a lot of RAM will improve performance. This is especially the case on multi-core machines where one does not typically have one disk per core (see Fig. 7.10).

7.11 Challenges Ahead

Many challenges in parallel data analysis remain. Mainly in the area of file access and distributed file systems. Since most data analysis is I/O bound efficient data access is primordial. The recent Solid State Drive (SSD) technology may make a big difference as it could significantly speed up data access. The first 256 GigaByte drives are available but are still expensive. It is expected that SSD and HDD price/GigaByte "equivalence" will not be achieved before 2015, though it might be affordable enough to equip a dedicated analysis cluster with.

In addition to the increases in the number of cores in general-purpose CPUs there is the upcoming trend to off-load processing to GPUs (Graphics Processing Units). The latest version of GPUs can run many threads in parallel in a SIMD way and vendors start coming with C based programming environments to code algorithms for execution on the GPU, like the Nvidia CUDA [12] technology. Harnessing this technology for data analysis could bring big improvements in performance, but its usefulness in user coded Process() methods might be a challenge.

7.12 Conclusions and Lessons Learned

Basic parallelism is classically achieved in HEP by "manually" splitting a large job in many smaller ones and submitting them concurrently on a batch farm. With respect to this batch-oriented approach, PIAF and PROOF make optimal use of the resources for short and medium long running queries, for which the overhead from job preparation, submission and result merging may be substantial for batch systems. In addition PROOF provides continuous real-time monitoring of the running jobs, presentation of intermediate results and better control of the tails in the CPU-time distribution by allowing faster workers, once idle, to take over the packets assigned to slower workers. For very long running queries, PROOF will not be much faster than the batch approach but the previously mentioned advantages remain.

I expect a system like PROOF to play a major role in the analysis of the large quantities of Large Hadron Collider (LHC) data. At the same time it is an ideal platform to test and use new technologies in the areas of storage, CPU and GPU developments.

References

1. Amdahl, G.: Validity of the Single Processor Approach to Achieving Large-Scale Computing Capabilities. In: AFIPS Conference Proceedings, vol. 30, pp. 483–485 (1967)
2. http://www.mpi-forum.org
3. http://www.epm.ornl.gov/pvm
4. Brun, R., et al.: PAW – Physics Analysis Workstation. The Complete CERN Program Library, 1987, Version 1.07
5. Rademakers, F.: The Parallel interactive analysis facility. CERN-CN-94-9
6. See for instance Baud, J.-P., et al.: Mainframe Servicesfrom Gigabit- Networked Workstations Summer'92 USENIX – June 8-12, 1992 -SanAntonio, TX
7. http:/root.cern.ch
8. http://www.corba.org/
9. http://root.cern.ch/twiki/bin/view/ROOT/PROOF
10. http://www.slac.stanford.edu/xrootd
11. http://monalisa.cacr.caltech.edu/monalisa.htm
12. http://www.nvidia.com/object/cuda_home.html

Chapter 8
Aspects of Intellectual Property Law for HEP Software Developers

Lawrence S. Pinsky

In this day and age, when any significant software development project first gets going, every developer needs to consider the Intellectual Property impacts, whether it is to avoid doing anything that will infringe others rights, or simply that one wishes to either protect their own rights or preserve their ability to control the exploitation of their efforts. When one thinks of software development and Intellectual Property issues, the laws with respect to Copyrights and possibly Patents come to mind first. These areas of the law are clearly very important but the worlds of Trademarks and Trade Secrets, including employment agreements are many times just as crucial to consider. The advent of the Internet has also had an impact on the law in many ways that are not yet entirely settled. Finally, the European Union has embarked on a major experiment in Intellectual Property law with the creation of a wholly new paradigm in protecting databases. This chapter will present synopsis of the current state of these areas of the law with a slant towards their application to software development projects and the people involved in them both as developers and users.

8.1 Introduction

Any discussion of Intellectual Property must take into account the nature of property and the laws that grant and protect such rights in different societies across the world. Intellectual Property is intangible and the various forms that have evolved over the years have always been as a result of technological progress. Intellectual Property is also fundamentally of international interest because one of the principle purposes is to both foster the disclosure and availability of information on the widest scale, while at the same time protecting certain aspects of the efforts of the creators or inventors.

L.S. Pinsky (✉)
Physics Department, University of Houston, Houston, TX, USA
e-mail: pinsky@uh.edu

> Intellectual Property is intangible and the various forms that have evolved over the years have always been as a result of technological progress.

Patent law seeks to encourage the disclosure of inventions and to reward cleverness. It manifestly allows the owners to prevent others from using the inventions without permission. Copyright law is designed to foster creativity, and specifically enjoins copying but unlike patent laws, copyright protects independent creation whereas patents grant the rights to the first comer regardless of whether a subsequent inventor had any knowledge of the patentees efforts.

Trademark law is intended to protect consumers but is in practise used by producers against other producers of products. Trade Secrets laws protect businesses from surreptitious disclosure of information that is commercially damaging. Finally, the new Database Protection laws introduced recently by the European Union are specifically directed at protecting investment in databases.

The sections that follow in this chapter will discuss each of these issues in more detail with an emphasis on specific areas that are of relevance to software developers.

8.2 Property and the Law

Let's begin with the concept of property. Normally when we think of property, we think of tangible things like your laptop, or maybe the land your home is on. Property is linked to the concept of ownership. One can "own" property, of course, but just what does ownership mean? Actually, ownership is typically measured more by what you can prevent others from doing than by what you yourself are allowed to do with the property in question. Think about it for a moment. If you own land, the most precious rights are the ability to keep others off than any list of things that you are allowed to do yourself on the land. Generally, we describe the rights that an owner of some piece of property has being like a "bundle of sticks," each right being separate and distinct, and more importantly, where each right can be individually given to another person without the owner having to surrender any of the others. For example, if I own a piece of land, I may give to my neighbour the right to pass over a designated portion of it briefly each day to reach a road on the far side. Such a grant to another person to exercise one of my exclusive rights in property that I own or control is called a "license."

If another person exercises one of my exclusive rights in some piece of property without a license, that is called trespassing, or an infringement on my rights. A license is a complete defence to a charge of trespassing or infringement. This leads us to another point. Exactly where do these property rights come from? Who decides what rights exist? In the most primitive societies (and unfortunately among the most civilised nations) "might makes right," and the strongest individual determines what rights exist. Otherwise the answer is that the government does. That is, we determine by law what property rights the courts will enforce. If someone infringes one of your property rights, the law of each country determines whether or not the police will enforce your right, or if you try and enforce it yourself (that's called "self-help" in legal circles) whether they will consider that it is YOU who have broken the law in taking such an action. So, property rights are determined by laws, statutes that is, and in most cases just what items are considered to be "ownable" is codified within each countries legal infrastructure.

It is not surprising that there is some disagreement from one culture to another as to what things may be owned. For example, in some African tribal cultures the idea that an individual may own and bequeath to his or her heirs' real property (i.e. land) is totally foreign to them [1]. Rather, land is something that is owned by the tribe in common, and allocated temporarily to individuals during their lifetimes according to their needs. We will see that when it comes to Intellectual Property (IP), especially in the area of software, there is great disagreement even between advanced countries as to what rights should and should not exist.

When it comes to tangible things like real property and personal property (variously referred to as "personalty" or "chattel" like your iPod or your clothes), it is easy to view and understand the normal kind of exclusive ownership rights. For example, if someone steals my laptop, and I later discover it in some other place, if I can prove that it is my laptop (i.e that I never legally surrendered my ownership rights), generally I am entitled to recover it under the law without any need to establish who took it and irregardless of how the present possessor came into possession of it. That is because it is never possible for anyone to give "better title" to a piece of property to someone else than the possessor has. In other words, if I do not possess a particular one of the "sticks" in the bundle of ownership rights in that piece of property, then I cannot possibly give it to someone else. That is an essential attribute of property rights that distinguishes them from "contract" rights. I may sign a contract with another person, and thus be bound by the provisions of the contract, but in most cases those provisions are not binding on any third parties who are not directly "in privity" of contract. That is, the provisions of a contract generally bind only the parties to the contract, whereas a property right is good against the world-at-large.[1]

> How does it work when the property in question is intangible, like an idea? How does one own and idea?

How does all of this work when the property in question is intangible, like an idea? How does one own and idea? The answer is that it is just like tangible

---

[1] Note that there is a "tort" called tortuous interference with a contract which can reach an individual who knowingly tries to subvert the intention of a contract between other individuals.

property in terms of the way in which the rights work. Intellectual Property (IP) is intangible, but nonetheless can possess almost the same bundle of rights that tangible property can. Classically, IP has been divided into four generic categories: Patents, Copyrights, Trademarks and Trade Secrets. In recent years there has arisen an attempt to create still other rights, such as the new laws enacted in the E.U. creating property rights in the factual contents of databases.[2]

Virtually all countries have IP laws, and these laws are far more similar to each other than they are different. This is generally due to several major international treaties. The Paris Convention (see e.g. [2]) sets the standards for what are termed Industrial rights, which includes Patents and Trademarks. A treaty known as the Bern Convention (see e.g. [3]) covers copyrights. Both of these list minimum standards, which the signatories must grant under their laws. They also require what is termed "national treatment," which means that the rights of foreign nationals must at a minimum be the same as those afforded to the citizens of the granting country. Countries are generally permitted to give more than the minimum rights, which is where most of the disparities between different country's laws arise.

One overarching question to address before discussing the individuals types of IP in detail is why have IP at all? What are the possible motivations for a government to enforce such rights? Generally the rationales for IP fall into two categories, economic and moral. The U.S. and U.K. base their IP laws on the economic rationale, namely that by giving protection to IP creators, they will be provided with an incentive to create more. In the end the goal is to balance the limitation on other's rights to use the IP for some fixed period of time against the ultimate right of society to inherit the information. In the interim, the owner can choose to make the IP available for a fee (called a royalty), and the net effect is to provide a maximum of useful IP to society for its use, as well as to maximise the production of new IP. IP owners are never required to license their rights, except that governments usually retain an absolute right to license any IP for their own purposes for the payment of a "reasonable royalty."

The continental European governments come to the rationale for IP rights from the moral perspective, namely that inventors, authors and artists have the moral and ethical rights to the exclusive exploitation of their respective creations. This moral-rights approach is responsible for some important differences between the details of the respective laws crafted under these two rationales, a few of which will be pointed out along the way as we discuss the individual areas of IP law in more detail.

> Why do we need Intellectual Property at all?

One last point that is essential to keep in mind is that in general where there is more than one IP creator involved, there is joint ownership of the entire body,

---

[2] See The Directive 96/9/EC of the European Parliament and of the Council of 11 March 1996 on the legal protection of databases.

regardless of the relative quantity or merit of the individual contributions. This joint ownership is similar to a joint bank account, where one owner does not need the permission of the other owners to access any of the jointly owned resources. In some cases, however, an "accounting" of the use to the other owners may be required. The details of this accounting requirement differ from one jurisdiction to another.

This joint ownership characteristic of IP means that in general where collaborative projects are concerned, it is essential to consolidate the ownership rights into a single entity wherever possible. Typically this can be done by employment agreements such as those that are required to be signed by employees of many universities and national laboratories. However, may times, especially in the development of software for use in particle physics, a substantial number of contributors come from different countries with different laws governing the assignment of their rights, so particular attention must be paid to keeping clear records of the source and authorship of all contributions, and where possible to obtain an assignment of the rights in those contributions to some common entity. The detailed situation also differs from one type of IP to another as will be mentioned in the discussions that follow.

8.3 Patents

Patent law had its origins in Venice in 1474 with the world's first Patent Statue (see e.g. [4]) which was remarkably similar to the current patent infrastructure.

The motivation was economic in that it encouraged foreign inventors to come to Venice to teach the secrets of their inventions to the local artisans for which they would have the exclusive right to make, use and sell the inventions in Venice for a period of 20 years, after which the invention would become part of the public domain. The statute also provided to the patent holder a legal remedy against any infringers during the term of the patent. Modern patent law essentially provides the same provisions. The patentee is required to disclose all of the details of her invention in such a manner that it empowers anyone "skilled in the relevant art" to practise the invention. For that disclosure, the inventor gets a patent for the term of 20 years from the original filing date. The patent gives the patentee the exclusive right to deny others to make, use, sell, offer for sale or import the invention. It is very important to understand that the patentee does NOT get any specific rights to practise the invention herself.

Besides the detailed disclosure of the nature of the invention (called the "specification"), the patent application includes a set of very specific descriptions of unique details of the invention. Known as the patent claims, each claim effectively stands as a separate patent. Claims generally are composed of connected elements, schematically $A \implies B \implies C \implies D$. Patent applications must be novel. That means that it cannot be publicly known beforehand. There are specific lists of what kinds of existing information can be used as "prior art" to determine whether the invention is publicly known. Prior art can include publications or talks at professional meetings. Even a single discussion with another person without a covenant of confidentiality can be considered to have made the invention publicly known. Likewise, a single Ph.D. thesis if listed in the card catalogue in a single university library can also suffice.[3] In most countries, the invention must be novel at the moment the application is filed. This creates a problem because patents filed in other countries are considered a prior art, even if filed by the same inventor. Thus, a patentee would have to file simultaneous applications in every country that she wants to patent her invention in. Rather than inflict this burden on inventors, there is a Patent Cooperation Treaty (PCT) which holds that any application filed in a signatory country will be honoured as of that same date when later filed within a specified time in another signatory country by the same inventor. That time period can be extended for up to 30 months in total from the initial filing.[4]

> Patent law has its origin in Venice in 1474, and its first formulation was remarkably similar to the current patent infrastructure.

Novelty is determined by comparing the elements of the claims against the prior art. Any piece of prior art that a claim in the patent application "reads-on" (i.e. where the prior art has each element of the claim in question connected as described) is not novel. Thus a claim of $A \Longrightarrow B \Longrightarrow C \Longrightarrow D \Longrightarrow E$ would be patentable over a piece of prior art characterised by $A \Longrightarrow B \Longrightarrow C \Longrightarrow D$. However, even though the patent could issue, if there was a prior patent on a claim of $A \Longrightarrow B \Longrightarrow C \Longrightarrow D$, then the owner of the new patent for a claim of $A \Longrightarrow B \Longrightarrow C \Longrightarrow D \Longrightarrow E$ could be blocked from practising her patent by the owner of the prior patent. Likewise, the new patentee could block the old patentee from adding the new element "E" to his patent. This illustrates a case where a patent holder cannot practise her patent. IP rights tend to be of this sort, that is the right to deny others the opportunity to make use of the IP. Where multiple individuals collaborate on an invention, each is a joint co-owner of the patent (more on this later). To be a co-inventor, all one has to do is materially contribute to at least one element of one claim in the patent. In most countries, patent applications must be signed by all inventors.

For infringement one simply compares each claim of the patent to the infringed device, and if any claim reads-on the accused device (i.e. it has all of the elements connected as described in the claim, irregardless of whatever else it may have), then it infringes that claim of the patent. It does not matter that the accused infringer did not copy or even know about the existence of the patent but honestly re-invented it himself. If any claim reads on it, it is an infringement, period. Note that the infringing item can be a completely different type of entity from that contemplated by the patent, and intended for an entirely different purpose, it only matters that

---

[3] In re Hall, 781 F.2d 897, 228 USPQ 453 (Fed. Cir. 1986).
[4] See, e.g. http://www.wipo.int/pct/en/texts/articles/atoc.htm.

embedded somewhere within the infringing item one can find all of the elements connected as described by at least one claim of the patent.

In addition to being novel, patents must be "non-obvious" or as described in Europe, they must include a significant "inventive step" over the prior art. That is to say that patents will not be given for trivial or obvious improvements to existing publicly known inventions. Just what constitutes a significant inventive step is essentially the major time consumer of patent attorneys and patent examiners. Patents must also be useful, which seems like an innocent enough requirement until one considers patenting molecules like the proteins which various sequences of DNA code for when the patent applicant has no clue what use that protein might be put to. Note that while you do have to declare a valid use, that declaration is not a limitation, and the claims can create infringements in any use no matter how far removed from the original patentee's ideas, so long as the claims read-on the accused infringing entity. Patents can be given for physical inventions (e.g. a mousetrap), for compositions of matter (e.g. drug compounds), or for methods (e.g. the process to produce a drug). Method patents have the steps in the process as elements. This seems like a description that could fit a computer program, doesn't it?

> Patents can be given for physical inventions (e.g. a mousetrap), for compositions of matter (e.g. drug compounds), or for methods (e.g. the process to produce a drug).

So far, the descriptions given apply almost universally. However there are some terribly important differences. For example in the U.S. (and the Philippines), an invention does not have to be novel on the date of filing. Patentees have one year after they first publicly disclose their invention to file their application in the U.S. This leads to two other differences. Multiple inventors may file conflicting applications. In that case since there is no "first to file" rule, the U.S. uses a "first to invent" rule. That is the first person to fully conceive of the invention in its entirety in his mind and then who diligently moves to file the application without any undue delay will get the patent. Just how one provides such evidence is another matter. The other difference is that if an inventor in the U.S. publicly discloses her invention and then files the application in the U.S. within the 12-month "grace-period" she would still have lost the patent rights around the rest of the world where absolute novelty is required. The famous RSA encryption scheme fell victim to exactly this situation (see e.g. [5] for an excellent discussion of the history of the patent as well as a description of the actual mathematical process).

The authors presented their paper at an IEEE meeting, after which they were advised to patent it, which they did. The patent was valid in the U.S. for the full patent term, but unpatentable in the rest of the world. We will consider the conundrum of such inventions that manifest themselves on the web, later.

Perhaps the most important differences from one jurisdiction to another are concerned with what subject matter is and is not patentable. Generally, to be patentable the subject must have been invented and not discovered. As such, laws of nature or a naturally occurring thing cannot be patented. However, it is possible to include a physical law as part of an invention such as a process for curing rubber that uses a well-known equation from thermodynamics in determining some of the steps in the process.[5] Even more bizarre is U.S. patent law, contrary to that of many other nations, allows the patenting of living animals such as a strain of bacteria designed to clean up oil spills.[6]

The patenting of the human genome is also a matter of some contention as well. The rationale used so far is that when the genes are identified, specified exactly and in an "isolated and purified form," that this is somehow unnatural enough to avoid the discovery prohibition (see e.g. [6]).

In general, abstract algorithms cannot be patented, and in Europe this principle is employed to exclude patents on software per se. In the U.S. it is possible to patent software if you include the computer executing the code as part of the claimed device or method (e.g. a process for figuring out how much tax is owed comprising a computer running a specified tax preparation program). This has led to the more or less routine patentability of software in the U.S., including CPU microcode. The debate now raging in Europe over the wisdom of allowing software patents is also circulating in the U.S. as evidence mounts to support the suggestion that they have become far more of an obstruction than an incentive to the development of new software.

> Software can be patented in the U.S., under certain conditions and not in Europe, where however the debate is on.

The potential pitfall here is the fact that in developing new software, it may be possible for the programmer to purely accidentally include embedded in her code all of the elements of some claim from one of the many of the obscure software patents, which remain in effect for 20 years. Remember, that the original application intended for that code in the patent is irrelevant. The only thing that matters is whether or not the claimed elements are present in the same relative way in the accused infringing code.

Another recent area of patent subject matter evolution has been the gradual allowance of patents for "business models" (as method patents) in the U.S.[7] with the staunchly opposing view in the E.U. The celebrated case of Amazon and their patenting of the "one-click" method of doing business via a website is an example of this.

---

[5] See *Diamond v. Diehr*, 450 U.S. 175 (1981).
[6] See *Diamond v. Chakrabarty*, 447 U.S. 303 (1980).
[7] See *State Street Bank & Trust v. Signature Financial Services*, 149 F.3d 1368 (Fed. Cir. 1998).

Finally, there is a dark cloud in patent law that is causing a lot of sleepless nights, among patent attorneys at least. That cloud is called the Doctrine of Equivalents. The idea is that an infringer should not be able to avoid liability for infringement by making subtle changes in some element that avoids literal infringement. This means that the reach of the patent claims is slightly broader than their literal meaning. How much broader? Well, courts have used various phrases like "insubstantial differences" but the fact remains that a court must decide each case. So, even though you do not literally infringe any of the claims, if a court decides that your accused infringement is too close to the claims, you can still lose. The strongest argument to keep such a doctrine is "after-discovered-technology." That is, where a patentee could not have anticipated the advent of technology, which allows second-comers to avoid literal infringement in the latter years of the patents term. The leading example of this is a patent for stabilisation of orbiting satellites by calculating the timed firing of on-board thrusters. At the time the patent was filed, computers were far too big to be placed on the satellites themselves, so the patent called for radioing down to Earth the sensor information where the computers would do the calculation and radio the firing commands back to the satellite. With the advent of smaller computers, the radio step in the process could be eliminated avoiding literal infringement of the original patent. The Doctrine of Equivalents was used to preserve the patents scope to include the on-board computer as still being an infringement.[8]

The use of patents to protect software is relatively recent. Rather, from the earliest days of software development, Copyrights have been the IP of choice for software protection. Copyright is in some sense a complementary form of IP to patents. Unlike patents, copyright protects against explicit copying, and independent creation is totally allowed. Copyrights specifically do not protect ideas, nor do they protect any of the utilitarian features of a creation. Instead, copyright protects the creativity of the author or artist. The archetype of a copyright protected work is literary work (i.e. a book). When computer programs first came into being, the printouts of the code for a program looked like a literary work, so the natural response of the IP law community was to treat it like literary IP. So, let's explore briefly the extent of copyright protection.

8.4 Copyrights

Copyrights as legal rights came into being because of technology, specifically the advent of the printing press. When the tedious process of manually copying books by hand was required to reproduce books, there was no worry about the potential exploitation of the creative work of another author because the effort of the scribe was so considerable. However, with the printing press, came the ability for one publisher to easily and quickly profit significantly by copying the work of others

---

[8] See *Hughes Aircraft Co. v. United States*, 717 F2d 1351 (Fed. Cir. 1983).

without paying for it. The first copyright law was the Statute of Anne in England in 1710,[9] and it provided protection for the publishers (i.e. printers) of books rather than for the authors themselves. Ever since that first technologically motivated movement of the law into the provision of copyright protection, it has typically been the advent of new technology that has spurred the legal system to react and revise the reach and scope of copyright protection. Because, virtually all copyright laws begin by laying out the scope of the type of works that are covered, when a new technology evolves, as did photography in the mid-nineteenth century for example, or phonograph records in the early twentieth century, copyright law must decide whether to provide protection, and if so, what the scope of that protection will be. The advent of computers and its latest appendage, the Internet, have both provided a significant impetus to the modification and expansion of copyright law.

While authors and publishers do register their works and put ©notices on them, these are not a requirement for copyright protection to attach. The saying goes that "copyright attaches as the pen is lifted from the paper." Any copyrightable work is protected from the moment of its creation with no formalities required. Copyright applies to virtually all forms of authored creations from literary and artistic works to music and dance choreography. In some cases multiple forms of copyright are present in what seems to be single piece of work. Take a typical music CD for example. There is a separate copyright in the music by the composer, in the lyrics by the lyricist, in the performance by the singer, in the recording itself by the studio technician, and probably in the cover art by the graphic artist who designed it as well as by the author of any text printed on the CD or on the cover.

> The archetype of a copyright protected work is literary work (i.e. a book). Computer program printouts looked like a literary work, so the natural response of the IP law community was to treat them like literary IP.

The default owner of the copyright is the creator herself. However, typically in jurisdictions where the rationale for IP rights is based on the economic incentive, there is a doctrine called "works-made-for-hire." When an employer specifically employs a person for the purpose of creating a copyrighted work (i.e. hiring a person to write a user's manual for a piece of software), then in those jurisdictions, the copyright in the resulting work is owned from its creation by the employer without any need for an assignment by the employee. Generally in those jurisdictions any copyrightable work produced by an employee "within the course and scope" of his employment falls within the ambit of the works-made-for-hire doctrine. In moral rights jurisdictions there is no comparable doctrine, so employers have to be sure to get their employees sign a contract (an employment agreement) which specifically assigns to the employer the copyright in any works created by the employee within

---

[9] See e.g. http://www.copyrighthistory.com/anne.html.

the course and scope of their employment. In the case of patents, there is no comparable doctrine, and only the actual inventors may apply for the patents, so in all jurisdictions employers must obtain employment agreements requiring the assignment of all rights to any patentable inventions that they may make within the course and scope of their employment to their employers, along with their agreement to actually sign any relevant patent applications. The earlier remarks on collaborative works apply here.

So, once a work has been created, what rights does the copyright owner have? The basic right which gives its name to this general type of IP is the right to make copies, The copyright owner can prevent others from making copies of the work, or at least the protected portions of it. Recall that copyright protects only creativity, so it is only that portion of the work that is protected. Clearly if one copies the entirety of a work, then that would be an infringement because all of the embedded creativity would have been copied. However, if the copier only extracts unprotected portions, such as something that is utilitarian (i.e a menu structure in a user interface), or factual content, then that is permitted.

Since copyright protects only the creativity embodied in a work and not any of its utilitarian aspects, how does this affect replicating the functionality of a copyrighted program like say Microsoft Windows? Courts have wrestled with the problem of how to extract the creativity from the functionality in a computer program. Certainly, viewing the way a program functions in terms of its physical manifestations and then sitting down and writing from scratch an entirely new code to duplicate as closely as possible the functionality of original program, is generally permitted. Some aspects of the original program's features may be protected as creative, such as decorative colour choices, etc., but where such choices merge with functionality, the copyright scope of protection wanes. If one does something more than just observing the gross properties of the executing code, such as by reverse engineering the structure of the code down to the level of the subroutine structures and replicates that as well, then it becomes a closer question.[10]

Copying the actual code line-by-line is of course an infringement. However, if one separates the individuals who look at the original code and then abstract it into function elucidating "pseudo-code," which is turned over to "sterile" programmers (ones who have not actually ever seen the original programs detailed code), but who then write their own fully implemented version of the code, it is possible to avoid copyright infringement. Of course this is one of the reasons for the growing popularity of software patents.

Note that generally, the copyright owner cannot control the use that a purchaser of a bona fide copy of the work makes of it. If I buy a copy of a book that has been produced with the permission of the author, the author may not control the physical uses I make of that book, such as using it to hold up a short leg on a table or giving it to a friend when I am finished with it. I cannot make a copy of the book, of course,

---

[10]See e.g. *Computer Associates Int'l, Inc. v. Altai, Inc.*, 775 F.Supp.544, 549–55 [20 USPQ2d 1641] (E.D.N.Y. 1991).

and this is a crucial point for software. Courts have ruled that loading software from a hard-drive into RAM on a computer is copying the program.[11]

So, to use software as a practical matter the user must copy it, and to do that legally, you need a license from the copyright owner to exercise his exclusive right to make copies of the work. A license is a contract, and a contract can specify any reasonable terms. Because software can only be used with a license, the copyright owner has a vehicle unlike any other form of copyrightable subject matter to control the use the licensee makes of the software. Note that if the copyright owner sells the user the software, he loses that control, but if he merely grants the user a license with strings attached, then he can maintain the control. If you do not like the strings, then you should not accept the license, but if you do accept the license (usually by clicking the "I Accept" box at some point), then you have agreed to be bound by its terms. Microsoft never sells software, they only license it with conditions, many limiting conditions.

> To use software as a practical matter the user must copy it, and to do that legally, you need a license from the copyright owner to exercise his exclusive right to make copies of the work.

The GNU license is an interesting example of the use of copyright law to subvert it through licensing [7]. In the Open-Source movement, the idea is to produce software that is more or less free of the nominal copyright restrictions for subsequent users. To accomplish this, the software is actually licensed under copyright law where the ability to control uses is employed to keep that software from ever being incorporated as part of fully protected software in a derivative work, and to "infect" any work that does incorporate any part of it, requiring that the entire new work be subsumed within the GNU license. This movement is sometimes referred to as "Copyleft." Other restrictions such as having to incorporate the notices of the restrictions and the attribution of the source are also included. There is a continuum of possibilities, some examples of which are the less invasive restrictions of the BSD Unix license [8]. Lawrence Lessig, a professor at Stanford, has championed the Creative Commons approach to provide a range of license agreements that allow software developers (and other copyright owners) the ability to pick and choose what restrictions they want to impose on the subsequent users and developers [9].

There are many other rights associated with copyrights than just copying. Perhaps the most important for software developers is the right to make "derivative works." The usual concept of a derivative work is making a movie based on a book. In the case of software, it can be any obvious extension of the work. There is a limit to the reach of this protection in that if the new work is sufficiently "transformative": then

---

[11] See *Mai Systems Corp. v. Peak Computing, Inc.*, 991 F.2d 511 (9th Cir. 1993).

it may succeed in morphing into a wholly new work, free from the control of the copyright in the original. Remember that ideas are not protected by copyright.

Another exclusive right of the copyright owner is the right to publicly perform or display the work. If you buy a CD, you do not have the right to play it in a public place. There are "collective Societies" like ASCAP and BMI whose sole purpose is to license public performances of the copyrighted material produced by their subscribing authors and performers. Likewise, if you by a copy of a recently written play in book form, you do not have the right to perform the play in public. That would require a separate license from the author. Note that there is a popular misconception that if you personally do not profit from the infringement, then it is OK. This is absolutely wrong for all types of IP infringement. The issue is always, did the accused infringer exercise one or more of the exclusive rights of the IP owner without a license?

The Internet is a very interesting source of problems for this public performance or display right, along with the derivative work right. If I make a web page publicly available there is an implied license for visitors to my site to download and view my content on their compute screens (the license is required to copy the web page to their computer). However, there is no implied right to print out a hard copy of that page or any of its content. Doing so in the absence of an express license, would be an infringement. One subtle problem has to do with hyperlinks. Since I can create a web page that incorporates content from another site in the Internet, I can make something such as a graphic image from someone else's site appear on my page when it is loaded by another user on her own computer. Because I only give that person the address to go to where she can retrieve the content in question, and I never actually copy the content myself, I am not liable for direct copyright infringement. However, the web page that is viewed can be considered a derivative work, the making of which is an infringement, and if I induce another person to do that, then I can be liable as a contributory infringer.

> The Internet is a very interesting source of problems for this public performance or display right, along with the derivative work right.

Several rights that are recognised as part of copyright in Europe, but not in the U.S. because of their "moral rights" nature are the right of "attribution" and the right of "integrity." The former is perhaps the greatest sin that an academic professional can commit, while it is not even a protected right in the U.S. for all practical circumstances. It does exist in Europe. The right of Integrity is the right to prohibit the buyer of a legal copy of your work from defacing it or changing it in some manner. In a famous case from Germany, an artist was commissioned to paint a mural on wall bordering a formal staircase. The painting had some unclad female figures in it, which a subsequent home owner objected to, and so attempted to paint clothing over them. The artist sued and the court ruled that while the new owner had the full right to tear the house down, he did not have the right to deface the picture.[12]

As noted before, patents must be formally applied for and if granted, then their term extends from the date of issue to the 20th anniversary of the original application date. Fees are required for the application, at issue and periodically during the term. Patents are valid only in the issuing country, and must be applied for separately in each country where protection is desired. Copyrights persist for far longer, generally for the entire life of the author, and then for an additional 70 years after her death. In the case of works-made-for-hire, the term is 75 years, and as noted, copyrights exist automatically from the creation of the work, and there are no required fees. In the U.S., one can register copyrighted material for a modest fee, and for U.S. citizens, doing so allows them to qualify for some additional rights such as "statutory damages" should there be an infringement case. Non-U.S. citizens qualify without the registration requirement because of the provisions of the Bern Convention.

In cases of copyright infringement, the damages are usually in the form of a court order to cease and desist from continuing the infringing activity, as well as an award of any actual monetary damages that may have been suffered by the copyright owner due to the infringement. Statutory damages are available at the discretion of the court, which are intended to discourage small individual infringing activities that would otherwise be monetarily unprofitable to pursue. Currently in the U.S. the court can award up to $10,000 per infringement. The threat of this has been used recently in the U.S. by the music industry to fight on-line piracy by suing individuals who are caught downloading copyrighted materials illegally for the statutory damages, which could be as much as $10,000 per downloaded song. In contrast, patent infringement damages, besides court orders to stop the infringing activities, are usually actual monetary damages from lost sales or lost licensing fees. In both patent and copyright cases, the infringing materials as well as any items used to further the infringement are subject to confiscation and destruction. In cases of on-line music piracy, this typically means confiscation of the computer hardware used by the infringer.

Finally, in the case of copyright law, there is a special doctrine that legally grants people the right to infringe. It is called "Fair-Use." In special cases for things that the law considers worthy enough, individuals are allowed to exercise some of the exclusive rights of the copyright owner without permission. Should the copyright owner choose to sue for infringement, the user may offer Fair-Use as a defence in lieu of a license. Much of the Fair-Use doctrine is motivated by free speech or freedom of the press principles. For example, a book reviewer may quote small passages from the book being reviewed without a license from the author. Likewise, great deference is given to the use of materials in face-to-face education in the classroom. This does not mean that one can avoid paying for textbooks, but rather

---

[12]See Entscheidungen des Reichsgerichts in Zivilsachen 397, [1912] Droit D'Auteur 9.

that where getting a license is not practical due to time constraints, or the refusal of the author to grant a reasonable license, the Fair-Use defence would apply.

Perhaps the most famous case invoking Fair-Use was the lawsuit by the entertainment industry against Sony in the early days of home video tape players (VCRs).[13] The industry wanted to suppress the sales of VCRs or to collect a tax on their sales to offset the anticipated losses by the industry in the presumed massive infringing activities that would likely occur. The court ruled that while some infringing activities were possible, that there were also clearly many legal uses for the devices and that taping TV shows off of the air for later viewing (referred to as "time-shifting") was a Fair-Use. In retrospect, this was probably the most ill-advised action by the movie industry in its history, given the subsequent rise of the home movie rental business based on the wide availability of inexpensive VCRs to the public. The lesson in this is that one cannot always foresee the full impact of seemingly threatening new technology when it first emerges, and the political power of entities like the entertainment industry can be very counter-productive.

Like the Doctrine of Equivalents in patent law, one does not know beforehand whether the court will grant a Fair-Use exception in a particular case. Courts are given a set of factors to consider including: the nature of the use, the nature of the work, the amount of work taken, and the impact of the infringement on the copyright owner. So, while Fair-Use exists, it can still be uncomfortable to apply if the copyright owner persists in taking the matter to court. This uncertainty aspect of Fair-Use tends to reduce its usefulness considerably because in most cases the copyright owners are in a stronger financial position than the user, and the threat of legal action is often a sufficient enough discouragement.

> The lesson in this is that one cannot always foresee the full impact of seemingly threatening new technology when it first emerges, and the intervention of political power of industry can be very counter-productive.

Fair-Use has had a significant impact in the software world. In a case where the accused infringer wanted to produce and market games of their own original design for the copyright owner's game system, which employed special cartridges to provide the software for each the different games, the competitor needed to know the interface details. Naturally, the copyright owner, wishing to suppress competition and or collect royalty fees from other game designers, refused to disclose the needed information. So, the defendant in this case admitted to purchasing legal copies of the game cartridges, and reverse-engineering them to determine the interface specifications. This act was prohibited by the express license terms of the acquired cartridges, and in order to access the information on them, the defendant had to copy the code off of them, which was a direct infringement of the copyright. The

---

[13] *Sony Corp. of America v. Universal City Studios, Inc.*, 464 U.S. 417 (1984).

court ruled that where there was not other way to do so, and for the purpose of interoperability of different software systems, the needed acts were indeed a Fair-Use. The principles espoused in this case became widely incorporated in the Fair-Use doctrine around the world, and was used to defend Jon Johansen in the famous case in Norway where he broke the DeCSS encryption scheme used to protect the content of commercial movie DVD's, and published the first Linux-based DVD player, something that the movie industry had refused to license.

Subsequent to this setback, the entertainment industry in the U.S. had the Digital Millennium Copyright act passed which provides that any act intended to circumvent a copyright protection technology is per se illegal. Technically this is not about copyright infringement itself, because no infringing act is required to run afoul of this law. The mere act of circumventing the protection scheme, or for example publishing a written article merely describing how a scheme works can be actionable by itself in the total absence of any actual infringing activity. This act was used to suppress the distribution by websites in the U.S. of Linux-based DVD player programs making use of the DeCSS technology. Of course, the law does not reach outside of the U.S., so sites beyond the reach of U.S. law can and do still distribute such programs.

Another relevant application of the Fair Use Doctrine has been by the Internet search engines. When they display small samples of the text or thumbnails of pictures from the listed selections resulting from a search, these would in principle be copyright infringements. Because of the uncertainty as to just what is actually included within the scope of the Fair-Use Doctrine, the search engine owners typically do not press the issue, and when site owners object, they maintain that while they have the right to do it, they will "voluntarily" respect requests to exclude the objected content from their search engine displays. More about search engines later.

8.5 Database Protection

Historically, there have been cases where the courts have ruled that where someone invested a lot of time or money in producing something like a factual database, that the investment somehow made it protectable under copyright law where it would otherwise not be. This "sweat-of-the-brow" doctrine was resoundingly dismissed in a famous case over telephone books.[14] A publishing company wanted to produce a phone book that included the numbers of subscribers over a collection of small rural telephone companies, each of which published its own separate directory. The publisher contacted the companies and offered to pay for a list of the names and numbers. One of the telephone companies refused, so the publisher copied the names, addresses and phone numbers as published in that company's own directory.

---

[14] *Feist Publications Inc v Rural Telephone Service Co Inc.*, 113 L Ed 2d 358 (1991).

The phone company sued, and the case went to the U.S. Supreme Court, where the decision stated loudly and clearly that copyright protected creativity and not facts, no matter how much investment had been made in determining the facts. Facts are facts and not protectable under copyright law.

This decision sent a shock-wave through the database industry in this burgeoning era of on-line of electronic databases. The problem is the following. While I can use contract to protect my electronic database by insuring that everyone I grant access to has agreed not to further distribute it, this contract restriction does not extend to parties who are not directly bound by the contract. If someone who has agreed to the contract breaches that agreement and turns over to an innocent third party the extracted content of my database, I can sue the person that I have a contract with, who might not have sufficient resources to cover my losses, when the third party puts up his own database using the facts extracted from mine.

In Europe, the argument was made that this fear was suppressing the development of a database industry comparable to the one that existed in the U.S. So, the E.U. government enacted a sui generis (i.e. new) form of IP protection in database content.[15] The idea was to give a property right in the factual content of databases to the individuals who has invested sufficient resources in assembling the database. The point being that the database owners would "own" the facts in their database, and thus could reach any third parties, irrespective of any contractual relationships who had obtained the facts from the database. This is not an absolute right to the facts, as there is no exclusivity. Anyone is free to obtain the facts separately and then to disseminate such separately obtained facts without restriction. However, if the facts came from a protected database, then the database owner possesses the rights to control the use and distribution of those facts.

> Database IP rights aim at protecting the rapidly growing database industry, but it may hinder free and open access to factual information.

One of the very first cases to be decided under this new E.U. Database Protection law was the ownership of data about football matches in the U.K.[16]

A football league sued betting parlours for listing the dates and places of matches and the teams involved in their shops for their customers use in placing bets. The league claimed that this factual information was assembled by them at sufficient expense to qualify for protection under the new law, and that it was undisputed that the only source for such facts was the league itself, so it did not matter how the shops got the facts, the league owned them and could require a payment for their use by

---

[15]See The Directive 96/9/EC of the European Parliament and of the Council of 11 March 1996 on the legal protection of databases; or http://en.wikipedia.org/wiki/Directive_on_the_legal_protection_of_databases.

[16]Case C-46/02 *Fixtures Marketing Ltd v Oy Veikkaus Ab* [2005] ECDR 2, ECJ.

the shops. The league won the case initially, but more recently in a similar case on horse-racing the European Court of Justice ruled that the Database Protection Law did not apply in either situation.[17] So the situation is still evolving.

Similar legislation was introduced in the U.S., but at this time that initiative is essentially dormant. The potential problem for such legislation for the academic world is the need for the free and open access to factual information. While the economic IP rationale of providing an incentive to collect and distribute more facts with the protection than would be available without the protection has been suggested, there is little of no evidence of the net benefit of this legal infrastructure in the academic community, and the threat of the negative impact of such restrictions to access and use over the long run are more worrisome. Consider the following hypothetical situation.

Newton's gravitational constant, "G," is the presently the poorest known fundamental constant. There has been a proposal to improve our knowledge of G by an order of magnitude. The technique calls for building a special submarine apparatus and carefully measuring the acceleration of gravity as a function of depth as it descends in the middle of the Pacific Ocean. The cost of such an endeavour would clearly qualify the resulting information under the current E.U. law. Supposing this experiment is done, then the database in question would contain one decimal digit. Under the current law, the owners of that database could license the use of that information under contract to any interested parties. Note that as long as that is the only experiment that has been done, then anyone in the world who comes into possession of the information unquestionably got it from ultimately from the protected database, so it would remain under the control of the database owners, who could dictate the royalty terms for its use including in doing calculations in one's head. This hypothetical presents an interesting question in the case where the digit measured turns out to be zero!

8.6 Trademarks

Having discussed the two major areas of IP law, patents and copyrights, that impact software developers, let us not forget the relevant issues of Trademark Law. Trademarks have their origin in the guild-marks that craftsmen used to place on their products. Trademarks themselves have been around since trade itself began, and again, the earliest European "Trademark" laws were enacted in Venice around 1400 to protect "Guild-marks" [10].

The idea was to identify the source of a particular good so that the consumer would have confidence in the quality of the product. The fundamental underlying principle of trademark law is consumer protection, and not necessarily the protection

---

[17]*British Horseracing Board Ltd, Jockey Club, Weatherbys Group Ltd v William Hill Organization Ltd* (CA) 13.7.05.

of the trademark owner's rights. However, in practise it is the trademark owner who possesses the rights and who pursues infringers in the name of protecting the consumers. The basic issue in trademark law is "likelihood of confusion" on the part of the consumer. That is, does a confusingly similar mark serve to confuse the consumer as to the origin of the good in question?

Unlike the other areas of IP law we have discussed, trademark law is strictly about trade. That is it applies to goods and services for sale in the marketplace. Trademarks can only exist when there are goods and services to apply them too. In principle one cannot "warehouse" a trademark without having goods or services actually available in the market place. Recently, provisions for obtaining trademarks in advance of the actual marketing of the goods have been provided for, but only for very short periods of time. Trademarks are fundamentally specific to a type of good, and before the Internet at least, to a geographic region.

> Trademarks have their origin in the guild-marks that craftsmen used to place on their products, and again, the earliest European "Trademark" laws were enacted in Venice around 1400.

One recent modification to trademark law that defers to the trademark owners is the introduction of the concept of dilution. When a mark becomes famous (e.g. Coca Cola, Cadillac, etc.) then the mark can be protected against most other uses, even when there is no confusion and the products do not compete (e.g. Cadillac Cat Food would have been prevented from using the name Cadillac, even though consumers would not be likely to confuse it with a car manufactured by General Motors).

A trademark can be anything that is used to associate a product with a manufacturer. Logos and names are obvious, but they can also include colours, characteristic shapes (e.g the classic "Coke" bottle) and even the overall "trade-dress" appearance of a business like McDonalds or Starbucks. Like copyrights, trademarked characteristics cannot be used to prevent competitors from using functional features. Also, when a trademark becomes so successful that it is adopted by the public as the generic name for the general class of product itself, the trademark owner loses the exclusive right to the mark. "Aspirin," which was Bayer's original trademark name for acetylsalicylic acid became generic and is now usable by all drug makers. Xerox has had to battle to keep the use of their mark from becoming a generic term for making copies of documents.

> A trademark can be anything that is used to associate a product with a manufacturer and they can be even more powerful than copyright to prevent copying.

Trademarks do not expire arbitrarily. That is they do not have any specified term, and remain valid so long as they are in continuous use associated with a marketed product or service. Trademarks expire if they are not used (typically after 5-years of non-use), or as described above, if they become generic. In most jurisdictions you are required to register a trademark, and the application for registration is subject to a search for potentially confusing similarity with already registered trademarks. Unfortunately, like patents and unlike copyright, one has to register a trademark in every jurisdiction where you want protection.

Trademarks can be even more powerful then copyright to prevent copying. In the fashion industry, while clothing designs cannot be copyrighted (that is just a tradition which is arbitrary), manufacturers have learned to incorporate their logos into their fashion articles (e.g Louis Vuitton's LV logo, Chanel's interlocking C's, etc.) so that copiers will be infringing trademark law by making exact copies, whereas in the absence of the trademark protection, there would be no prohibition.

In the software development world, the trademark issues are modest at best, but can appear. While you might be able to copy Microsoft Windows functionality, you could not include the Windows trademark, and possibly even there default colour schemes. For web page design more issues can arise. While copyright protection may not cleanly protect against linking items on one web page to content from another, if the linked material includes a trademark, then protection under trademark law would kick in and make the linking an infringement. This was used to prevent the marketing of a web page that used frames to provide convenient links to major news sources while posting banner ads around them [11].

8.7 Internet Law

At one of the very first conferences on Internet Law in the mid-1990s, a famous U.S. Federal Judge and legal expert, Easterbrook, gave the keynote address [12]. He basically told the participants to go home, that there was not such thing as "Internet Law" any more than there had been a "Law of the Horse." This comment was a reference to a famous remark by Oliver Wendell Holmes, the Supreme Court Justice, referring to the then current idea that there was a separate "Law of the Railroad." Holmes actual remark was "There is no more a Law of the Railroad than there is a Law of the Horse..." What Holmes and Easterbrook meant is that the fundamental principles of the law are universal and independent of any application area like the railroad or the Internet. However, in recent years as the legal systems have had a chance to come to grips with some of the unique aspects of the Internet, it is not so clear that there aren't some new and different principles that needed to be incorporated into the law.

Some examples that have strained the application of pre-existing legal principles will suffice to illustrate the point. Search engines have created challenges to the law in many areas in addition to the interpretations of their employment of the Fair-Use doctrine in copyright law as described earlier. Sometimes site owners want to exclude them for anti-competitive reasons. In one case a search engine designed to provide comparisons of current bid-prices on competing auction sites like eBay, was sued because it was directing business away from the more popular sites. The auction sites won the case when the court decided to consider that the search engines were "trespassing" to the property of the auction-site owners. Consider the comparable situation in the case of physical stores like supermarkets where one storeowner has the right to enter the publicly accessible areas of his competitors store to ascertain the prices at which he is selling his goods. Clearly, that kind of activity is in the public interest and supports the fundamental principles of the market economy. Apparently, the courts have decided to treat the Internet differently.

> "There is no more a Law of the Railroad than there is a Law of the Horse...", but there may well be an Internet Law.

Another issue that has arisen is that when one posts something on a publicly accessible website, it is viewable all over the world, and as such it is subject to the restrictions and moral judgements as well as the IP laws of every jurisdiction simultaneously. An early but illustrative case actually predates the Internet. A couple in San Francisco, ran a dial-in site providing "indecent" material for download at a price, paid for by credit cards. Everything they had on the site was publicly available in San Francisco and in fact they had submitted it to the local prosecutors who ruled that it was "merely indecent" and not "pornographic." Legally, "pornography" is illegal where "merely indecent" material is not. The problem is in determining where to draw the line between the two. In the U.S., the law is based on "local standards." What is considered "merely indecent" can (and generally is) different in San Francisco than it is, for example in conservative rural Tennessee. In a sting operation, a postal inspector in Tennessee called into their site in the middle of the night and downloaded ten carefully chosen pictures. Once the downloads were completed, police, who were waiting outside of the couple's apartment arrested them. They were extradited to Tennessee and tried there for distributing pornography. They were convicted and served a jail sentence. In that case, because they could be extradited, they were vulnerable to prosecution in a distant venue for doing something that was entirely legal where they lived.[18] Imagine, the potential for international travellers to be arrested as they pass through distant lands for perceived IP infringements by things posted on their websites at home.

Another issue, which is not local to the Internet per se, is on-line identity verification. In conducting transactions one has to decide whether a digital signature will be accepted to bind the remote alleged party. A moment's thought will lead you to the conclusion that in all cases of trying to verify an individual's identity, one has

---

[18] *United States v. Thomas*, 74 F. 3D 701 (1996). The 6th U.S. Circuit Court of Appeals. See e.g. http://www.eff.org/pub/Legal/Cases/AABBS_Thomases_Memphis.

to find some trusted third party who is in a position to know the absolute identity of the person in question. This is as true in the real world as it is in the virtual one. In practise it is the government that is the guarantor of identities in the real world by issuing birth certificates and passports (although in the U.S. the very insecure use of driver's licenses is the most common form of identification). Ultimately too in the virtual world, governments will have to serve the same role. Private entities can do the bulk of the actual work, but ultimately they will have to be certified by the governments via secure links. The systems are evolving, and efforts are underway to make them operate as closely to the physical world as possible. The advent of "Certificate Authorities" is a manifestation of this philosophy.

While strictly speaking it is not Internet Law, one related issue is the recent advent of Preprint Servers in the world of Scientific Publications, and the rise of the electronic distribution of articles by the conventional print-based peer-reviewed journals. As most academics are aware, you are generally asked to surrender all copyrights in any article you agree to allow to be published by these journals. That assignment typically includes the rights to copies residing on the Preprint servers as well, which leads to the frustrating situation where once an article is formally published, you have to pay to download what was available for free a short time ago. Publishers are well aware, by the way, that old copyright assignments executed without specific inclusion for the right to distribute copies electronically, do not include that right by default. In a recent case, authors of submitted articles to the New York Times sued and won additional royalty compensation for the on-line distribution of their works by the Times as part of its normal publication of web-based archival copies of the newspaper.[19]

8.8 Conclusions and Lessons Learned

Well, we have covered a lot of ground, and the basic lesson is that whenever the product is intangible intellectual work product, the impact of the relevant IP laws must be considered.

The legal community, both internationally and within governing bodies such as the European Commission and the World Intellectual Property Organization in Geneva, are actively considering substantive changes in the law that have the potential to be of major impact to the software industry in general and whose corollary impact on HEP software developers may be far reaching. The Database initiative is but one example. The ownership of facts due to the expense of generating them is very problematic for science in general. While one may feel insulated from possible scenarios like charging for access to information, just consider the

---

[19]*New York Times Co. v. Tasini*, 533 U.S. 483 (2001) 206 F.3d 161, affirmed. Syllabus, Opinion [Ginsburg], Dissent [Stevens], Available at: http://www.law.cornell.edu/supct/html/00-201.ZS.html.

increasing tendency for Internet sites to charge for providing access to information that is not otherwise protected.

Finally, software developers need to keep clear records of the contributors to all projects and to maintain formal assignments of rights from all contributors in order to insure that the software they produce is not encumbered with liabilities that might prevent their use by the needed constituency.

References

1. Ault, D.E., Rutman, G.L.: The Development of Individual Rights in Property in Tribal Africa. J. Law Econ. **22**(1), 162–189 (1989)
2. http://www.wipo.int/treaties/en/ip/paris
3. http://www.wipo.int/treaties/en/ip/berne/trtdocs_wo001.html
4. http://www.experiencefestival.com/patent_-_history_of_patents
5. http://en.wikipedia.org/wiki/RSA
6. http://www.ornl.gov/sci/techresources/Human_Genome/elsi/patents.shtml
7. http://www.gnu.org/copyleft/gpl.html
8. http://opensource.org/licenses/bsd-license.php
9. http://creativecommons.org
10. Browne, W.H.: A Treatise on the aw of Trademarks, 1–14 (1885)
11. Kuester, J.R., Nieves, P.A.: Hyperlinks, Frames and Meta-Tags: An Intellectual Property Analysis. IDEA: J. L. & Tech. **38**(243) (1998)
12. Easterbrook, F.H.: Cyberspace and the Law of the Horse. U Chi Legal F 207 (1996). Available at: www.law.upenn.edu/law619/f2001/week15/easterbrook.pdf

Chapter 9
Databases in High Energy Physics: A Critical Review

Jamie Shiers

This chapter traces the history of databases in High Energy Physics (HEP) over the past quarter century. It does not attempt to describe in detail all database applications, focusing primarily on their use related to physics data processing. In particular, the use of databases in the accelerator sector, as well as for administrative applications–extensively used by today's large-scale collaborations–are only covered in passing.

A number of the key players in the field of databases–as can be seen from the author list of the various publications–have retired from the field or else this world. Given the fallibility of human memory, the need for a record of the use of databases for physics data processing is clearly needed before memories fade completely and the story is lost forever. It is necessarily somewhat CERN-centric, although effort has been made to cover important developments and events elsewhere. Frequent reference is made to the Computing in High Energy Physics (CHEP) conference series–the most accessible and consistent record of this field.

9.1 Introduction

The year 2000 is marked by a plethora of significant milestones in the history of High Energy Physics. Not only the true numerical end to the second millennium, this watershed year saw the final run of CERN's Large Electron-Positron collider (LEP)–the world-class machine that had been the focus of the lives of many of us for such a long time. It is also closely related to the subject of this chapter in many respects.

Classified as a nuclear installation, information on the LEP machine must be retained indefinitely. This represents a challenge to the database community that is

---
J. Shiers (✉)
CERN, Geneva, Switzerland
e-mail: Jamie.Shiers@cern.ch

almost beyond discussion–archiving of data for a relatively small number of years is indeed feasible, but retaining it for centuries, millennia or more is a very different issue. There are strong scientific arguments as to why the data from the LEP machine should be retained for a short period. However, the complexity of the data itself, the associated metadata and the programs that manipulate it make even this a huge challenge. The story of databases in HEP is closely linked to that of LEP itself: what were the basic requirements that were identified in the early years of LEP preparation? How well have these been satisfied? What are the remaining issues and key messages? Finally, the year 2000 also marked the entry of Grid architectures into the central stage of HEP computing. How has the Grid affected the requirements on databases or the manner in which they are deployed? Furthermore, as the LEP tunnel and even parts of the detectors that it housed were readied for re-use for the Large Hadron Collider (LHC), how have our requirements on databases evolved at this new scale of computing?

> The story of databases in HEP is closely linked to that of Large Electron Positron machine itself.

It is certainly appropriate at this point to mention the famous LEP Database Service – "LEP DB". Quoting from "LEP Data Base Information note number 1:"

"Oracle version 2 was installed at CERN in the summer of 1981, on a VAX system running VMS version 2. A pre-release of version 3 is presently under test and a production version is expected before the end of the year."

The LEP DB service led to the installation of the first VAX 11/780 into CERN's Computer Centre. This marked another significant change in HEP computing (at least at CERN!), as it marked an important change from batch-dominated computing: the strengths of VAX computing were its interactivity, its excellent (for the time) debugger and its well-integrated networking support. Although it was for the IBM VM/CMS system to introduce the concept of "service machines", the impact of these changes can still be seen today. Computing for the LEP and LHC experiments is largely based on services – experiment-specific or otherwise – of much higher level than the basic batch system and/or tape staging system – a trend that is strongly linked to a database-backend to maintain state, coupled to the rapid developments in computing power that allowed the necessary servers to be setup. A further significant event that occurred around the same time was the introduction of the first Unix system at CERN. Although reference has often been made to early highly conservative estimates of the growth of the Unix installed base, no one at that time predicated that it would soon dominate HEP computing – as it continues to do in its Linux guise today – and let alone on commodity PCs. Indeed, the reluctance to move to Unix – although relatively short-lived – gave a foretaste of the immense and lingering resistance to the demise of FORTRAN.

> Whilst size is only one measure of management complexity things clearly cannot scale indefinitely.

The rise of Linux on Intel-compatible platforms has also had a significant impact on database services. After the early popularity of VAX-based systems, Solaris was long the platform of choice (at least for Oracle – the DataBase Management System – DBMS – deployed at CERN). Solaris was displaced by Linux/Intel in recent years and has allowed database services to keep up with at least some of the demand. Not only has the number of database servers or clusters increased significantly, but also the volume of data thus managed. The great Jim Gray often referred to the "management limit" – somewhere in the low to medium multi-TeraByte region. Whilst only one measure of management complexity – and no one with Jim's great depth of insight would ever have meant otherwise – things clearly cannot scale indefinitely, even given the write-once, read-rarely nature of our bulk data. Early proposals (see below) called for solutions that required much less than one person per experiment for support. The required support level has clearly long passed this threshold, perhaps normal given the scale of HEP experiments in the LHC era. However, alarm bells should possibly be ringing. Are the proposed solutions compatible with the manpower resources that will be available to support them?

Finally, in addition to the core applications identified over 25 years ago, Grid computing has brought new requirements to the database arena – a large number of key Grid applications, such as the reliable File Transfer Service and storage services, are dependent on back-end databases. In reviewing the evolution of Databases in HEP during a quarter century of change, we try to establish the key discontinuities and to answer the many questions that have been raised.

9.2 ECFA Study

In the early 1980s, the European Committee for Future Accelerators (ECFA), launched a number of study groups into various aspects of HEP computing. One of these groups – subgroup 11 – reported [1] on "Databases and Book-keeping for HEP experiments". The goals of this working group were to (a) provide a guide to the database and bookkeeping packages used at present by HEP groups; (b) find out what future requirements (would) be; and (c) make recommendations as to how these (could) best be met. The working group defined a DataBase as: "A collection of stored operational data used by the application system of some particular enterprise.". It then went on to explain that "In the HEP context, the word 'database' is sometimes used to refer to the totality of data associated with a single experiment... We shall not use the word with that meaning... Instead, we shall use the word for more highly organised subsets of data such as:

- Catalogues of experimental data (with information such as run type, energy, date, trigger requirements, luminosity and detector status).

- Information on the status of the analysis (e.g. input and output tapes, cut values and pass rates).
- Calibration data.
- Summary information from the analysis (e.g. histograms and fitted parameters)."

Detail aside, such a definition would be instantly recognisable to a physicist of today.

The report also clarifies:

"It is further necessary to distinguish between: Database systems developed within the HEP community, sometimes for a single experiment, which are referred to as 'HEP databases' or 'simple databases' and Database management systems (DBMS), which may be classified as hierarchical, network or relational in structure."

> The ECFA Study Group on databases marks an important milestone in the evolution of Databases for HEP.

Finally, it records that, with very few exceptions, DBMS were not used by HEP experiments at that time.

The report continues with a long list of detailed requirements and surveys of packages in use at that time. We nevertheless include the summary of recommendations made by the working group:

There would be many advantages in using commercially available DBMSs in HEP to reduce the amount of work required to obtain a database or bookkeeping system tailored to the needs of a particular experiment. They will clearly have a place in HEP computing in the future and should be used for LEP experiments in place of complex user-written systems;

The requirements of flexibility and ease of use clearly point to the need for a relational DBMS;

Standardisation at the Standard Query Language (SQL [2]) interface level is suggested both for interactive terminal use and embedded in FORTRAN programs. This is an alternative to the implementation of a common DBMS at all centres of HEP computing;

Greater awareness is needed within the HEP community of what DBMSs offer. Pilot projects should be set up so that some experience can be obtained as soon as possible;

There is an immediate need for the major HEP computing centres, especially CERN, to make suitable relational DBMSs (e.g. SQL/DS or Oracle) available to users;

Simple HEP database packages will continue to be needed, especially in the short term. The KAPACK [3] system is recommended for this purpose. However, the basic KAPACK package should not be extended significantly. (If a much more sophisticated system is needed, then a DBMS should be used.);

A simple bookkeeping system could be written using KAPACK and supported in the same manner as KAPACK;

Users developing higher level software of a general nature on top of KAPACK or a DBMS should be urged to write as much as possible in the form of a standard add-on packages which can be used by other groups. Central support for such packages should be offered as an incentive to do this;

Before the development of very sophisticated or complicated packages is undertaken for a given experiment, careful consideration should be given as to whether the advantages to be obtained will justify the work involved. (Considerable effort has been expended in the past in providing facilities that would be standard with a DBMS.);

A greater degree of automation in the management of tape data would be desirable. If, as at DESY (the Deutsche Electron SYnchrotron laboratory in Hamburg Germany), users do not normally have to worry about tape serial numbers, the need for user tape handling packages is obviated and the problems of bookkeeping are considerable simplified.

The report also noted that DBMSs and data structure management packages were closely related – a fact borne out by many of the database-like developments for LEP, as we shall see later.

9.3 The Central Oracle Service at CERN

Following the recommendations of the ECFA report, and building on the experience gained with the Oracle service for the LEP construction project, a proposal to establish a central Oracle service on the CERNVM system was made in early 1984 – just a few months after the publication of the report.

Although, from today's point of view, the choice of Oracle appears almost automatic, things were much less obvious at that time. For example, the evaluation of replies to the 1982 LEP relational database enquiry – initially sent out to over 30 firms – resulted in only 6 replies that were considered to be relational systems. Of these, only two (SQL/DS and Oracle) were further considered, although SQL/DS had not yet been delivered to a customer. Furthermore, it only ran under DOS and would have required an additional system to support it. Oracle, on the other hand, was installed at over 70 sites, including 4 in Switzerland!

> Although obvious from today's point of view, the choice of Oracle was not so obvious back in 1984.

From such humble beginnings, the service has continued to grow with the years, with physics applications representing a relatively small fraction of the overall service, until the central cluster was logically separated into two in the early 2000s. At this time, a 2-node cluster running Solaris was established – using recycled Sun nodes and a small disk array – to host physics applications, being rapidly complemented by experiment-specific servers built on stovepipe systems, namely

"CERN disk-servers". The latter was never an optimal solution and following a lengthy study into Oracle's RAC architecture and its use on Intel systems with SAN storage, such a solution has now been adopted. Numerous additional database servers hosted applications related to the accelerator, experiment controls and AIS/CIS applications, but these are not the main thrust of this chapter.

9.4 Database Systems for HEP Experiments

In 1987, a review of database systems in HEP [4], primarily but not exclusively within the context of the L3 collaboration, evaluated a variety of database systems and described the L3 database system [5] (later DBL3), then under construction. The systems considered – Oracle, SQL/DS, Ingres, KAPACK and ZEBRA RZ [6] (see also Chap. 1) – were evaluated on the basis of the following criteria: Full features; Efficiency; FORTRAN access; Terminal access; Concurrent writes; Portability of FORTRAN; Portability of Data; Robustness; Security; Cheapness.

None of the systems excelled in all categories, although the commercial systems fared best in their feature set and clearly worst in terms of cost. Based not only on these criteria, but also performance measurements, the choice narrowed rapidly to Oracle, RZ or KAPACK – the latter two being part of the CERN Program Library. Given the more extensive feature set of RZ over KAPACK, this left only Oracle and RZ. However, at that time it was not considered realistic to require all institutes that were part of the L3 collaboration to acquire an Oracle license – an issue that has reappeared and been re-evaluated at regular intervals over the past two decades. Despite significant advances on this front, the requirement for all institutes in a HEP collaboration to acquire commercial licenses – and not just a strictly limited subset – is still as high a hurdle today as it was 20 years ago.

> The requirement for all institutes in a HEP collaboration to acquire commercial licenses is still as high a hurdle today as it was 20 years ago.

Thus, the DBL3 package was built using the ZEBRA RZ system – and ZEBRA FZ for the exchange of updates. A system with largely similar functionality – also built on ZEBRA RZ/FZ – was later developed by OPAL (the OPCAL system), whereas DELPHI had already developed a KAPACK-based solution. The ALEPH book-keeping and ADAMO systems are described in more detail below.

Whilst today's computing environment is clearly highly complex, it is worth emphasising that of LEP start-up was, for its time, equally challenging. The degree of heterogeneity – of compilers, operating systems and hardware platforms – was much greater. Networking was still primitive and affordable bandwidths only a trickle by today's standards. Just as today, every drop of ingenuity was required to squeeze out adequate resources and functionality – requirements that continue to maintain HEP computing ahead of the wave.

9.5 Computing at CERN in the 1990s

In July 1989, the so-called "green book" [7] on LEP computing was published. Amongst the many observations and recommendations made by this report – including spotting the clear trend to distributed computing and the potential use of workstations in this respect (a foretaste perhaps of the SHIFT project, see Chap. 3), it contained a chapter on Data Base systems. (Historically, the use of "database" as a single word was already common in the previous decade). The book was published simultaneously with the commissioning of the LEP machine and thus by definition covered most of the production systems deployed by the LEP experiments. By that time a central Oracle service – as opposed to the dedicated LEP DB service which continued to run on VAX hardware – had been setup on the central IBM systems. Moreover, two new packages had entered the scene which were set to influence LEP computing significantly. These were the ZEBRA (see Chap. 1) data structure management package – which can somewhat naively be thought of as combining the strengths of the HYDRA and ZBOOK packages before it – and the ALEPH Data Model (ADAMO) [8] system. The ADAMO system is particularly notable in that it brought the use of entity-relationship modelling to the mainstream in HEP computing.

The report presents a rather thorough analysis of the areas where database applications were in use, or where the use of such technology would make sense. The list included the following: Collaboration address lists; Electronic mail addresses; Experiment bookkeeping; Online databases; Detector geometry description databases; Calibration constants; Event data; Bookkeeping of program versions; Histograms and other physics results; Software documentation; Publication lists; Other applications.

Specific recommendations were made in a number of these areas, as described below:

Education and training:

"An effort should be made to make physicists in experiments more aware of the potentialities of commercial DBMS for their applications. This could be achieved by intensifying training in the area of data models (software engineering) and DBMS."

Design Support Team:

"Manpower should be made available to support centrally the experiments, starting with the design of the database and continuing during the whole life cycle, including the implementation of the application dependent code. This support team should also ensure the long term maintenance of the General Purpose applications described below."

Data Model Software:

"A package should be provided to design interactively a Data Model and to store the definition in the form of a dictionary in ZEBRA files. The Entity-Relationship Model and related software from ADAMO should be considered as a first step in this direction. This would allow to profit from the experience and possibly from existing tools, including commercial ones."

Portability of Database Information:

"A package should be provided to move data from Oracle to a ZEBRA (RZ) data structure. The reverse could also be implemented, providing a data model describes the structure of the data in the DB. A decent user interface should be written on top of these files to allow the users to enquire about the information contained in this structure and to update it. Tools provided with the ADAMO package could be used to learn from the existing experience and could possibly be used directly as part of the proposed package."

> The LEP Green Book contained a number of important recommendations that shaped the evolution of computing in HEP.

Experiment Administrative Databases:

"A data base should be set up covering all CERN (or HEP?) users and other people related to experiments. It should link with information and existing data bases. It should include the functionality required for experiment mailing lists and experiment specific data. Control of the data, i.e. entering and updating the information, should stay within the experiment concerned. We further recommend that a study be made on existing tools and their performance, in order to coordinate any future efforts, such as those that are being made around NADIR and EMDIR. The functionality should cover at least the one of the NADIR and AMSEND systems."

Bookkeeping Databases:

"A solution should be researched and developed urgently, in common between LEP experiments, in the area of tape bookkeeping, to avoid duplication of effort."

Documentation Databases:

"The redesign of existing documentation data bases (CERNDOC, HEPPI, ISIS etc.) into a common data base system (e.g. Oracle) should be envisaged."

Detector description/Calibration Constants Databases:

"This is probably the areas with the largest investment of manpower and the largest savings if a common solution could be found ..."

Interactive Data Analysis Databases:

"PAW data-sets are expected to play this role ..."

9.6 ALEPH Scanbook

The ALEPH bookkeeping system SCANBOOK [9] was developed starting in 1988. Originally based on CERNVM, it was re-written a number of times, most recently in 1999. It is now implemented using an Oracle database using a web interface written in Java. It is the basis of the Lack book-keeping system.

Quoting from the abstract of a presentation by ALEPH to LHCb in 2000:
"The Scanbook program has been used extensively over the last 10 years to access ALEPH data (Monte Carlo and real data). It enables the users to build a list of tapes suitable for input into the ALEPH analysis framework, based on parameters relevant for a given type of analysis. Selection criteria like year of data taking, detector condition, LEP energy etc ... can be combined and transformed into a set of "data cards". The latest version is based on an Oracle database, a set of stored procedures which perform the selections, and a user interface written in Java."

9.7 File and Tape Management (Experimental Needs)

Given all the discussion above, the situation was ripe for a more formal study into the needs of the LEP experiments for bookkeeping and data cataloguing and a possible common solution. Initiated by a discussion in the LEP computing coordination meeting, MEDDLE, a working group was setup in early 1989. This task force, which had the unfortunate acronym FATMEN (for File And Tape Management: Experimental Needs) had the following mandate:

"At the MEDDLE Meeting held on 6/12/88 it was decided that a small task force be performed to review with some urgency the needs of various experiments for a file and tape management system to be available from LEP start-up. The following is the proposed mandate of this task force.

1. The composition will be one representative from each major collider experiment (4 LEP, 2 UA), one representative for LEAR, one for the SPS fixed target programme, and 3–4 representatives from DD Division. You should feel free to seek advice and assistance from other experts as appropriate.
2. The task force will endeavour to specify the needs of the experimental program in the area of automatic tape and file management systems. It is suggested that three levels be specified:

   - Minimal and absolutely urgent requirements: solution needed by September 1989.
   - Minimal (less urgent) longer term requirements: solution needed by March 1990.
   - Optimal (perhaps too deluxe?) specifications of what we would really like, but which may not be available on a desirable timescale.
3. The task force will review the approach of the experiments to:

   - The generation of production jobs
   - The location of events at all stages of production
   - The location of magnetic cartridges, both at CERN and outside

   and make any recommendations that seem useful to avoid unnecessary duplication of effort. Transportability between operating systems, between sites, and between experiments should be considered.

4. In view of the short timescale before LEP data starts flowing, and the limited resources available, the task force is encouraged to look very seriously at basing the overall approached on a commercially available storage management package. If that proves unrealistic the task force should take all possible steps to encourage common development between the experiments.
5. Taking into account the probable diversity of tape management software that is likely to be installed at LEP processing centres (not all of which work exclusively for HEP), the task force should make recommendations for interfaces to be respected.
6. The committee is asked to reports its conclusions to MEDDLE at its meeting scheduled for 4th April 1989."

Two were most concrete outcomes of the report – dated April 6, 1989. *A Tape Management System (TMS)*, based on SQL/DS, was imported from the Rutherford Appleton Laboratory (RAL) in U.K. This system was also deployed at other HEP sites, such at the Centre de Calcul de l'Institut National Pour la Physique Nucléaire et des Particules (CCIN2P3) in France. The CERN version was later ported to Oracle – a non-trivial task, considering that parts of the RAL original were written in IBM 370 assembler with embedded SQL. Oracle had no plans for a pre-compiler for this language, nor was one ever produced. The TMS lived for many years, eventually being replaced by the volume manager component of today's CASTOR(2). *A File Catalogue*, based on ZEBRA RZ, and introducing the so-called generic name – equivalent to today's logical name – was written. This had both command-line and FORTRAN callable interfaces, hiding much of the underlying complexity and operating system specifics.

Despite such a late start to this project, a first pre-alpha release was made during the summer of 1989 for VM/CMS only. This allowed users to perform basic catalogue manipulations and access (i.e. stage-in) catalogued files. How was it possible to produce even an alpha version so rapidly? (An initial release, covering also VMS and Unix systems, was made in time for the MEDDLE meeting of October that year, although it was still several years before the full functionality was provided – partly due to the ever changing environment at that time, including the migration from mainframes to SHIFT). This was no doubt partly due to the mature and extensive CERN Program Library but also to the excellent and fertile working environment existing at that time. Young programmers could discuss on a peer basis with veritable giants of HEP computing and rapidly assimilate years of experience and knowledge by adopting a widely-used programming style, as well as debugging and testing techniques and an informal documentation process. This allowed a "jump-start" in proficiency and highlights the value of mixing experienced and less experienced developers in the same teams. A further concrete step in this case was an informal code review by a very experienced developer – Hans Grote – who highlighted key issues at an early stage. This practise would surely be equally valid in today's complex world of the Grid.

> The rapid development of an alpha version was fostered by the excellent and fertile working environment existing at that time.

Once again, the considerable heterogeneity of the early LEP computing environment has perhaps been forgotten. A simple program allowed a user to forget operating system and staging system details and access data, be it disk or tape resident, in a uniform many across a host of incompatible platforms. Three main platforms (VM/CMS, VMS and Unix, in all its many flavours, as well as also MVS) were supported, together with many times as many incompatible variants. The need for a standard and consistent interface to storage lives on today, albeit in a rather different guise.

As a file catalogue, the FATMEN package [10] of the CERN Program library was used by DELPHI, L3 and OPAL (ALEPH having their own SCANBOOK package), as well as many other experiments outside CERN (notably at the Deutsche Electron SYnchrotron laboratory – DESY – in Hamburg Germany and at the Fermi National Accelerator Laboratory – FNAL – in Chicago). The CERN based server was only closed down in April 2007, with read-only use continuing only from OPAL. At both DESY and FNAL there was strong collaboration between the CERN and local teams – integrating with DESY's FPACK system and D0/CDF's computing environments respectively. The latter involved multi-laboratory collaboration, with the STK robot control software for VMS systems coming from Stanford Linear ACcelerator – SLAC – Laboratory.

Originally, FATMEN supported both Oracle and RZ back-ends, although the Oracle version was later dropped, for reasons discussed in Sect. 9.13 below.

The way that users were able to update the FATMEN catalogue and the techniques used for distributing updates between sites was extremely similar to that adopted by other packages, such as DBL3 (and hence HEPDB [11]), and OPCAL, and is discussed in more detail below.

Some two million entries from all catalogues at CERN were used relatively to stress test the European DataGrid "Replica Location Service" catalogue.

Given the Oracle backend, the package attracted quite some interest from Oracle corporation, which led in turn to regular visits to their headquarters to argue for product enhancements – such as those delivered with Oracle 10g – for the HEP community. One of the first such proposals was for a distributed lock manager – now a key feature behind Oracle's Real Application Cluster architecture.

> Some key features of Oracle's Real Application Cluster Architecture came from HEP requirements.

The FATMEN report also recommended that mass storage systems built according to the IEEE Computer Society's reference model be studied. Indeed, several such systems are used today in production – notably HPSS at Brookhaven National Laboratory (BNL), IN2P3 and SLAC and OSM at DESY and Thomas Jefferson lab. The CASTOR system is also based on this model.

Originally designed to handle disk or tape resident files – the latter by invoking the appropriate staging system or requesting direct access to a mounted volume – the package was extended to support "exotic opens", whereby the underlying system – such as those mentioned above – hid the gory details of file recall or equivalent operations. This was done using a syntax eerily similar to today's storage URL (SURL) – namely *protocol:path*.

The system proved extremely stable over many years, although younger "administrators" preferred the technique of dumping the entire catalogue and manipulating it with their preferred scripting language – by far from the most efficient mechanism but one that is echoed today with the LHC Computing Grid (LCG) File Catalogue (LFC), as is described later. The final "change" to the system was to relink one of the utility programs (which made a backup of RZ catalogues) that had been omitted from regular rebuilds as part of CERNLIB and was hence not Y2K safe.

9.8 CHEOPS

The computer centre at CERN boasts a large satellite dish on the roof, marking one of several attempts to distribute scientific data by such means. Requests to transfer files – aka today's FTS – could be made through the FATMEN Application Programming Interface (API) or Command Language Interface (CLI) to the CHEOPS [12] system – a batch data dissemination system based on the OLYMPUS satellite [13].

CHEOPS was a collaboration between CERN, LIP and INESC in Portugal, SEFT in Helsinki and four Greek institutes – the up-link station being in Athens. The CHEOPS earth stations had access to the Olympus satellite on an overnight schedule, each site having a local Unix management server.

It entered operation early in 1992, but was destined to be somewhat short-lived. Unfortunately, after an earlier incident due to operator error was recovered, the satellite was silenced forever in a freak meteorite shower.

9.9 Data Structures for Particle Physics Experiments

A workshop held in Erice, Sicily in November 1990 – the 14th Eloisatron project workshop – covered many of the data structure/data base managers in HEP at that time. It included not only position papers from the authors of the various systems, but also experience papers from the user community. In addition, future directions and the potential impact of new programming languages were hotly debated. Quoting from the proceedings [14]:

9 Databases in High Energy Physics: A Critical Review

"The primary purpose of the Workshop was to compare practical experience with different data management solutions in the area of:

- Simulation of Interactions and their Detection;
- Data Acquisition, On Line Management;
- Description of Detector and Other Equipment;
- Experiment and Data Processing Bookkeeping;
- Reconstruction Algorithms;
- Event Display and Statistical Data Analysis."

One paper at this workshop described "A ZEBRA Bank Documentation and Display System", known as DZDOC [15]. This was an initiative of Otto Schaile, then of the OPAL collaboration, and consisted of

"[...] a program package which allows to document and display ZEBRA bank structures. The documentation is made available in various printed and graphical formats and is directly accessible in interactive sessions on workstations. FORTRAN code may be produced from the documentation which helps to keep documentation and code consistent."

Another idea that (re-)arose during this workshop was that of a common "HEPDDL". Some discussions – particularly between ZEBRA and CHEETAH [16] – took place, but the great tsunami of object-oriented programming and design was soon to engulf us.

9.10 ADAMO

The following description of the ADAMO system is copied verbatim from the abstract of the corresponding paper presented in Erice by Paolo Palazzi:

"The ADAMO (ALEPH DAta MOdel) system was started in the early eighties in the ALEPH experiment as an attempt to apply state of the art concepts of data modelling and data base management systems to algorithmic FORTRAN programs, especially particle physics data reduction and analysis chains for large experiments.

The ADAMO data model marks the introduction of the Entity-Relationship data model in High Energy Physics.

The traditional FORTRAN + memory manager style of programming had several drawbacks that limited programmer's productivity and made projects difficult to manage: obscure reference to data objects by offsets in a large vector, arbitrary use of pointers and no automatic correspondence between data structures and their documentation.

ADAMO adopted the principles of database systems, separating the internal representation of the data from the external view, by reference to a unique formal description of the data: the Entity-Relationship model..."

9.11 The 1991 Computing in High Energy Physics Conference (CHEP'91)

At CHEP'91 two important papers were presented summarising the status of databases in HEP. One of these papers – Database Management and Distributed Data in HEP: Present and Future [17], by Luciano Barone, described the current state of deployment of database applications and raised the issue of "event databases" – somewhat akin to today's event tag databases but with a very reduced amount of information per event, as a key challenge for future work. The other – Database Computing in HEP [18], by Drew Baden and Bob Grossman – introduced the idea of "an extensible, object-oriented database designed to analyse data in HEP at the Superconducting SuperCollider (SSC) Laboratory".

This was clearly not "business as usual" and was subject of much – often heated – debate during the rest of that decade. To skip ahead, the end result – seen from the highest level – was that both viewpoints could be considered correct, but for different domains. For the applications identified at the time of the ECFA study group, the "classical approach" is still largely valid. However, for event data, we have – according to the prediction of Jim Gray "ended up building our own database management system". Will these two domains ever converge, such that a single solution can be used across both? Is this even desirable, given the markedly different requirements – e.g. in terms of concurrency control and other database-like features?

> At CHEP'91 the discussion about DataBases was not "business as usual" and it opened a new era for DataBases in HEP.

Barone's paper summarised the key characteristics of databases in HEP, as well as describing the experience of the four LEP experiments. The similarities between the global approaches of DELPHI, L3 and OPAL were stressed, as well as the close resemblance in many ways of the L3 and OPAL solutions. ALEPH was different in that the initial (see also the discussion on this point in [9]) size of the database was significantly smaller – some 5 MegaByte as compared to 60 MegaByte for OPAL and 400 MegaByte for L3. He also high-lighted ALEPH's use of ADAMO and its Data Definition Language (DDL) for building their system.

Finally, he summarised the work on event directories/tags, as well as event servers. This activity was relatively young at the time, but set to become an important component of future analyses. Event directories were typically very concise – a given file of run/event numbers – together with their offsets in a file – corresponded to a specific selection. Today's tags are significantly larger and correspond to the input to the selection, rather than the result set.

His definition of databases is interesting in that it had already expanded somewhat from that of the ECFA report. This is primarily in his final (4th) criterion, namely:

"A HEP database is accessible and used on different computers and different sites. This is inherent to the nature of present collaborations, geographically very distributed, and with relevant computing resources at home institutes."

9.12 HEPDB

Following on from the discussions in Erice and at CHEP, a small group was setup to study the possibility of a common solution to the experiments' needs in terms of calibration databases – much as proposed by the "green book". As had already been revealed, there was a high degree of commonality not only in the requirements but also in specifics of the various implementations – some 20 packages were reviewed at that time. It was fairly quickly – although not unanimously – agreed to build a package based on either OPCAL or DBL3, re-using as much code as possible. In the end, the DBL3 base was preferred, due to its additional functionality, such as client-side caching, and both OPCAL and DBL3 compatibility interfaces were produced. Sadly, neither of these experiments ever migrated to the new code base. However, possibly 20 experiments worldwide went on to the use the system – with continued use by NA48 for its 2007 data-taking. The central server is no longer maintained by IT, with an AFS-based copy of the previous RZ database available for both read only and update access – the latter under control of NA48 experts.

The main "added-value" of the central service was to run a centrally monitored (console operators) service, with operations procedures; provide regular backups and data integrity checks of the Database files and perform recovery if required.

Due to unfortunate bugs in the area of record re-allocation, the latter primarily plagued FATMEN – it being a mantra of DBL3 and hence HEPDB [19] "never delete". FATMEN – on the other hand – by default updated the catalogue on each file access with the last use date and use count. Whether this was ever more than academic interest is far from clear, but it certainly helped to debug the record allocation routines!

HEPDB was supported on VM/CMS, Unix and VMS systems, the latter being plagued by a host of TCP/IP implementations, some of which were not available at CERN and hence could not be fully tested. In terms of a common development, it represents an interesting example of a package almost entirely developed within an experiment that is subsequently taken over centrally. In this respect, as well as the benefit that it gave to smaller experiments, unable to devote the manpower to (unnecessarily) develop their own solution, it can be considered a success.

> HEPDB was an interesting example of a package almost entirely developed within an experiment that is subsequently taken over centrally.

As suggested above, the update mechanism for all of these packages was via the exchange of ZEBRA "FZ files" between client and server. On VM/CMS systems, these files were sent to the virtual card reader of the corresponding service machine, prompting the server to leap into action. On VMS and Unix systems they were written into a special directory which was polled at regular (configurable) intervals. The updates could be replayed if required and similar queues – i.e. directories – were established to exchange updates with remote servers, typically configured in a similar arrangement to that later proposed by the MONARC project (see Chap. 3) and adopted by World LHC Computing Grid [20] (see again Chap. 3) in its Tier 0/Tier 1/Tier 2 hierarchy. DBL3 and hence HEPDB had a concept of a "master" server – which assigned a unique key and timestamp – and hence updates made at remote sites were first transferred – using the above mentioned routing – to the master site before redistribution. In the case of FATMEN, all servers were equal and updates were processed directly and then dispatched to remote sites. This update mechanism also allowed for recovery – a not uncommon operation in the early days was the excision of a complete directory or directory tree that was then recreated by replaying the corresponding update or "journal" files. To reduce overhead, the journal files could be batched as required. However, although essentially any manipulation was possible through the API and CLI, global changes were performed much more efficiently by writing a special program that worked directly on the catalogue/database. Such a change would typically come from the change of name of a host or to perform bulk deletions or other operations – a requirement that still exists today. The results could be dramatic – one listing operation that took many hours when using the standard (necessarily general) API took only seconds using a program optimised for that sole purpose.

9.13 Computing in High Energy Physics 1992 (CHEP'92)

A panel [21] on Databases for High Energy Physics held at CHEP '92 in Annecy, France attempted to address two key questions, namely whether we should buy or build database systems for our calibration and book-keeping needs and whether database technology will advance sufficiently in the next 8 to 10 years to be able to provide byte-level access to PetaBytes of SSC/LHC data?

In attempting to answer the first questions, two additional issues were raised, namely: whether it is technically possible to use a commercial system whether it would be manageable administratively and financially?

> The question of the usage of the usage of a database for HEP data, commercial or home-grown, was raised at the CHEP'92 conference.

At the time of the panel, namely in September 1992, it was pointed out that the first question had already been addressed during the period of LEP planning: what was felt to have a technical possibility in 1984 had become at least a probability by 1992, although the issues related to licensing and support were certainly still significant.

We follow below the evolution of the use of Databases in High Energy Physics between two CHEPs – in Annecy and Mumbai – and then revisit these questions in the pre-LHC era.

9.14 Calibration and Book-Keeping

At the time of this panel and as described above, two common projects that attempted to address general purpose detector calibrations ("conditions") and book-keeping and file catalogue needs were the two CERN Program Library packages *HEPDB* and *FATMEN*. At a high-level, these packages had a fair degree of commonality: both were built on top of the ZEBRA RZ system, whilst using ZEBRA FZ for exchanging updates between client and server (and indeed between servers). Both implemented a Unix file-system like interface – and indeed shared a reasonable amount of code.

Indeed, one of the arguments at the time was that the amount of code – some tens of thousands of lines – would be more or less the same even if an underlying database management system was used. Furthermore, it was argued that the amount of expert manpower required at sites to manage a service based on a was higher – and more specialised – than that required for in-house developed solutions.

The ZEBRA RZ package had a number of restrictions: firstly, the file format used was platform dependent and hence could not easily be shared between different systems (e.g. using NFS) nor transferred using standard ftp. This restriction was removed by implementing "exchange file format", in analogy with the ZEBRA FZ package (Burhardt Holl, OPAL). In addition and in what turned out to be a disturbingly recurrent theme, it also used 16-bit fields for some pointers, thereby limiting the scalability of the package. ZEBRA RZ was improved to use 32-bit fields (Sunanda Banerjee, TIFR and L3), allowing for much large file catalogues and calibration files, as successfully used in production, for example by the FNAL D0 experiment.

9.15 CHEP'92 and the Birth of the Object-Oriented Projects

For many people, CHEP'92 marks the turning point away from home-grown solutions, which certainly served us extremely well for many years, towards "industry standards" and Object Orientation. In the case of programming languages, this meant away from "HEP FORTRAN" together with powerful extensions provided by ZEBRA and other memory and data management packages, to C++, Java and others. This has certainly not been a smooth change – many "truths" had to be unlearnt, sometimes to be re-learnt, and a significant amount of retraining was also required.

> For many people, CHEP'92 marks the turning point away from home-grown solutions towards "industry standards".

Notably, CERN launched the Research and Development project N. 41 (RD41) nicknamed "MOOSE" [22], to evaluate the suitability of Object Orientation for common offline tasks associated with HEP computing, RD44, to re-engineer the widely-used GEANT detector simulation package [23], RD45 to study the feasibility of object-oriented Databases (ODBMS) for handling physics data (and not just conditions, file catalogue and event metadata) [24], LHC++ (a CERNLIB functional replacement in C++) [25] and of course ROOT [26].

With the perfect 20–20 vision that hindsight affords us, one cannot help but notice the change in fortunes these various projects have experienced. At least in part, in the author's view, there are lessons here to be learnt for the future, and which are covered in the summary.

9.16 The Rise and Fall of Object Databases

This is well documented in the annals of HEP computing – namely the proceedings of the various CHEP conferences over the past decade or so. Object Databases were studied as part of the PASS project [27], focusing on the Superconducting SuperCollider (SSC [28]) experiments. The CERN RD45 project, approved in 1995, carried on this work, focusing primarily on the LHC experiments, but also pre-LHC experiments with similar scale and needs.

> At the time of writing, the use of Object Databases in HEP for physics data is now history, although some small applications for conditions data still remain.

At the time of writing their use in HEP for physics data is now history, although some small applications – such as the BaBar [29] conditions DataBase – still remain. To some extent their legacy lives on: the POOL [30] project builds not only on the success of ROOT, but also on the experience gained through the production deployment of Object Databases at the PetaBytes scale – successes and shortcomings – as well as the risk analysis proof-of-concept prototype "Espresso", described in more detail below.

9.17 RD45: The Background

Of the various object-oriented projects kicked off in the mid-90's, the RD45 project was tasked with understanding how large-scale persistency could be achieved in the brave new world. At that time, important international bodies to be considered were the Object Management Group (OMG), as well as the similarly named Object Data(base) Management Group. The latter was a consortium of Object Database vendors with a small number of technical experts and end-users – including CERN. Whilst attempting to achieve application-level compatibility between the various ODBMS implementations – i.e. an application that worked against an ODMG compliant database could be ported to another by a simple re-compile – it had some less formal, but possibly more useful (had they been fully achieved) goal, namely that the Object Query Language (OQL) be compliant with the SQL3 Data Manipulation Language (DML) and that no language extensions (thinking of C++ in particular) would be required for DDL.

ODMG-compliant implementations were provided by a number of vendors. However, as was the case also with relational databases, there are many other issues involved in migrating real-world applications from one system to another than that of the API.

9.18 RD45: Milestones

There is a danger when reviewing a past project to rewrite – or at least re-interpret – history. To avoid this, the various milestones of the RD45 project and the comments received from the referees at the time are listed below.

The project should be approved for an initial period of one year. The following milestones should be reached by the end of the 1st year.

1. A requirements specification for the management of persistent objects typical of HEP data together with criteria for evaluating potential implementations. [Later dropped – experiments far from ready]
2. An evaluation of the suitability of ODMG's Object Definition Language for specifying an object model describing HEP event data.
3. Starting from such a model, the development of a prototype using commercial ODBMSes that conform to the ODMG standard. The functionality and performance of the ODBMSes should be evaluated.

It should be noted that the milestones concentrate on event data. Studies or prototypes based on other HEP data should not be excluded, especially if they are valuable to gain experience in the initial months.

The initial steps taken by the project were to contact the main Object Database vendors of the time – O2, ObjectStore, Objectivity, Versant, Poet – and schedule presentations (in the case of O2 and Objectivity also training). This lead to an initial selection of the two latter products for prototyping, which rapidly led to the decision to continue only with Objectivity – the architecture of O2 being insufficiently scalable for our needs. Later in the project, Versant was identified as a potential fallback solution to Objectivity, having similar scalability – both products using a 64 bit Object Identifier (OID). Here again we ran into a familiar problem – Objectivity's 64 bit OID was divided into four 16 bit fields, giving similar scalability problems to those encountered a generation earlier with ZEBRA RZ. Although an extended OID was requested, it was never delivered in a production release – which certainly contributed to the demise of this potential solution.

The milestones for the second year of the project were as follows:

1. *Identify and analyse the impact of using an ODBMS for event data on the Object Model, the physical organisation of the data, coding guidelines and the use of third party class libraries.*
2. *Investigate and report on ways that Objectivity/DB features for replication, schema evolution and object versions can be used to solve data management problems typical of the HEP environment.*
3. *Make an evaluation of the effectiveness of an ODBMS and MSS as the query and access method for physics analysis. The evaluation should include performance comparisons with PAW and n-tuples.*

These were followed, for the third year, with the following:

1. *Demonstrate, by the end of 1997, the proof of principle that an ODBMS can satisfy the key requirements of typical production scenarios (e.g. event simulation and reconstruction), for data volumes up to 1 TeraByte. The key requirements will be defined, in conjunction with the LHC experiments, as part of this work.*
2. *Demonstrate the feasibility of using an ODBMS + MSS for Central Data Recording, at data rates sufficient to support ATLAS and CMS test-beam activities during 1997 and NA45 during their 1998 run.*
3. *Investigate and report on the impact of using an ODBMS for event data on end-users, including issues related to private and semi-private schema and collections, in typical scenarios including simulation, (re-)reconstruction and analysis.*

Finally, the milestones for 1998 were:

1. *Provide, together with the IT/PDP group, production data management services based on Objectivity/DB and HPSS with sufficient capacity to solve the requirements of ATLAS and CMS test beam and simulation needs, COMPASS and NA45 tests for their '99 data taking runs.*
2. *Develop and provide appropriate database administration tools, (meta-)data browsers and data import/export facilities, as required for (1).*
3. *Develop and provide production versions of the HepODBMS class libraries, including reference and end-user guides.*
4. *Continue R&D, based on input and use cases from the LHC collaborations to produce results in time for the next versions of the collaborations' Computing Technical Proposals (end 1999).*

9.19 Why Event Data?

The footnote to the first milestone given to the RD45 collaboration deserves some explanation. At the time, it was not felt realistic to use a single solution for the full problem space – from simple objects, such as histograms, to the event data of LHC-era experiments. The initial ideas – as borne out by paper-only records from that time – were to use a common interface, with a backend tailored to the particular domain. There was strong interest in the ODMG 93 standard at that time and this was rapidly proposed as such an interface. It was upon discovering more than one database with an architecture that scaled on paper – borne out by initial functionality and scaling tests – that the focus on a single solution appeared.

> At the beginning of the project it was not felt realistic to use a single solution. Only with the discovery of scalable DataBase architectures the focus on a single solution appeared.

CERN joined the vendor-dominated ODMG standards body with "reviewer" status. Meetings were held quarterly, with CERN representation at least twice per year. One such meeting was held in *Providenciales* – an island in the Caribbean, named after a ship that had wrecked off its coast. The group of islands is so remote that a former flag of the currently British colony lying between the Bahamas and Cuba – which was intended to depict a pile of salt (the islands then main source of income) – was retouched to represent an igloo. Even in as remote a location as this – far from any hadron collider – HEPDB support questions were to be found on the sparsely populated beach.

9.20 RD45: Risk Analysis

The CMS Computing Technical Proposal, Sect. 3.2, page 22), contains the following statement:

"If the ODBMS industry flourishes it is very likely that by 2005 CMS will be able to obtain products, embodying thousands of man-years of work, that are well matched to its worldwide data management and access needs. The cost of such products to CMS will be equivalent to at most a few man-years. We believe that the ODBMS industry and the corresponding market are likely to flourish. However, if this is not the case, a decision will have to be made in approximately the year 2000 to devote some tens of man-years of effort to the development of a less satisfactory data management system for the LHC experiments."

As by now is well known, the industry did not flourish, so alternative solutions had to be studied. One of these was the Espresso proof-of-concept prototype, built to answer the following questions from RD45's Risk Analysis:

- Could we build an alternative to Objectivity/DB?
- How much manpower would be required?
- Can we overcome limitations of Objectivity's current architecture?
- To test/validate important architectural choices.

The Espresso proof-of-concept prototype was delivered, implementing an ODMG compliant C++ binding. Various components of the LHC++ suite were ported to this prototype and an estimate of the manpower needed to build a fully functional system made.

The conclusions of an IT Programme of work retreat on the results of this exercise were as follows:

Large volume event data storage and retrieval is a complex problem that the particle physics community has had to face for decades.

The LHC data presents a particularly acute problem in the cataloguing and sparse retrieval domains, as the number of recorded events is very large and the signal to background ratios are very small. All currently proposed solutions involve the use of a database in one way or another. A satisfactory solution has been developed over the last years based on a modular interface complying with the ODMG standard, including C++ binding, and the Objectivity/DB object database product.

The pure object database market has not had strong growth and the user and provider communities have expressed concerns. The "Espresso" software design and partial implementation, performed by the RD45 collaboration, has provided an estimate of 15 person-years of qualified software engineers for development of an adequate solution using the same modular interface. This activity has completed, resulting in the recent snapshot release of the Espresso proof-of-concept prototype. No further development or support of this prototype is foreseen by DB group. Major relational database vendors have announced support for Object-Relational databases, including C++ bindings.

Potentially this could fulfil the requirements for physics data persistency using a mainstream product from an established company.

CERN already runs a large Oracle relational database service.

> The Espresso project provided the proof-of-concept that an Object-relational DataBase could be build with the Objectivity modular interface, albeit at a rather high manpower cost.

This was accompanied by the following recommendation:

The conclusion of the Espresso project, that a HEP-developed object database solution for the storage of event data would require more resources than available, should be announced to the user community.

The possibility of a joint project between Oracle and CERN should be explored to allow participation in the Oracle 9i beta test with the goals of evaluating this product as a potential fallback solution and providing timely feedback on physics-style requirements. Non-staff human resources should be identified such that there is no impact on current production services for Oracle and Objectivity.

9.21 The 1997 Very Large Databases Conference (VLDB'97)

A paper [31] presented at this conference on "Critical Database Technologies for High Energy Physics" by David Malon and Ed May addressed the following issues:
"A number of large-scale High Energy Physics experiments loom on the horizon, several of which will generate many PetaBytes of scientific data annually. A variety of exploratory projects are underway within the physics computing community to investigate approaches to managing this data. There are conflicting views of this massive data problem:

- There is far too much data to manage effectively within a genuine database.
- There is far too much data to manage effectively without a genuine database.

and many people hold both views."

The paper covered a variety of projects working in this area, including RD45, the Computing for Analysis project (CAP) at FNAL, the PASS project and a recent Department of Energy "Grand Challenge" project that had recently been launched.

The paper included a wish-list of DBMS systems, which included:

- *Address at least tens – eventually, hundreds – of PetaBytes of data.*
- *Support collections of 10^9 or more elements efficiently.*
- *Support hundreds of simultaneous queries, some requiring seconds, some requiring months to complete.*
- *Support addition of 10 Terabytes of data per day without making the system unavailable to queriers.*
- *Return partial results of queries in progress, and provide interactive query refinement.*

as well as a number of requirements related to mass storage systems, either as back-ends or else integrated into the DBMS.

This confirmed that there was some commonality in the approaches of the different projects but that there were still many issues that remained unresolved – the stated goal of the paper being

" ...to begin a dialogue between the computational physics and very large database communities on such problems, and to stimulate research in directions that will be of benefit to both groups."

In passing, it is interesting to note the relatively modest ATLAS event sizes foreseen at that time, with 100 kiloBytes/event at the event summary data (ESD) level, compared with 500 kiloBytes/event at the time of writing.

9.22 LHC Computing Review Board Workshops

During this period a series of workshops focusing on LHC computing was organised by the LHC Computing review Board (LCB). These took place in Padua in 1996, in Barcelona in 1998 and in Marseille in 1999. For a short period, it looked as though the combination of Objectivity/DB together with HPSS might even become a semi-standard across HEP laboratories, with experiments from many sites investigating these as potential solutions. However, with time, opinions began to diverge, fuelled in part by the slowness in delivery of important features – such as a non-blocking interface to mass storage (designed by SLAC), the full Linux port, support for the required compilers and so forth. The mass storage interface – which would probably never have been delivered had it not been for SLAC's design and indeed proximity to Objectivity's headquarters in Mountain View, allowed the system to be deployed in production. This interface was both powerful and flexible and allowed CERN to later move the backend to CASTOR in a largely transparent way.

9.23 Computing in High Energy Physics 2000

"All is not well in ODBMS-land". This quote from Paris Sphicas in his summary talk [32] at CHEP 2000 effectively acted as a death knell for object databases in HEP.

One of the key presentations at this conference was BaBar's experience in scaling to full production level. Many adjustments had to be made to achieve the required degree of performance and scalability, leading to the conference quote "either you have been there or you have not" – and at the time of writing, there are still number of important aspects of the LHC experiments' computing models – not just limited to database services – that have not yet been fully demonstrated at production load, for all experiments at all relevant sites concurrently.

> The CHEP'2000 conference effectively acted as the death knell of object-oriented Databases in HEP.

Also during this CHEP, not only were the various aspects of the RD45 risk analysis presented, but also a number of experiments presented their experience with hybrid or non-ODBMS solutions. Questions were clearly raised as to whether an ODBMS solution was the only path ahead or even a useful one. Although the formal decision to change the baseline persistency solution was still some distance away, the community in general had lost confidence in this approach and by this stage it was simply a question of time. As more and more effort was devoted to investigate alternative solutions, a swing back in favour of a commercial ODBMS became increasingly unlikely. The only remaining issues being how to rapidly identify and if necessary provide such an alternative and what to do with existing data.

9.24 LCG Requirement and Technical Assessment Group N.1

The newly formed LHC Computing Grid project setup its first Requirements and Technical Assessment Group (RTAG 1) in February 2002 with the following mandate:

"Write the product specification for the Persistency Framework for Physics Applications at LHC:

- Construct a component breakdown for the management of all types of LHC data.
- Identify the responsibilities of Experiment Frameworks, existing products (such as ROOT), and as yet to be developed products.
- Develop requirements/use cases to specify (at least) the meta-data/navigation components.
- Estimate resources (manpower) needed to prototype missing components.

RTAG may decide to address all types of data, or may decide to postpone some topics for other RTAGs, once the components have been identified. The RTAG should develop a detailed description at least for the event data management. Issues of schema evolution, dictionary construction/storage, object and data models should be addressed."

Based on the final report of this RTAG and the recommendations of the LCG, the POOL project was established, which is now the baseline persistency solution for ATLAS, CMS and LHCb – ALICE using native ROOT for this purpose.

9.25 The Triple Migration

Following the decision to move away from Objectivity/DB at CERN, the data of the experiments that had used this system had to be migrated to a supported alternative. The needs of the pre-LHC – i.e. running – experiments were somewhat more urgent and could not wait for a production release of the POOL software. Hence, the following strategies were proposed. The (simulated) data of the LHC experiments would not be migrated but maintained until rendered obsolete by a sufficient quantity of newly simulated data in the agreed LHC persistency format. The data of the pre-LHC experiments would be migrated to a combination of Oracle (for the event headers/tags/metadata) and DATE (ALICE raw data format).

More than 300 TeraByte of data was migrated in all – a triple migration [33] as it involved migration from one persistency format to another, migration from one storage medium to another and finally migration of the associated production and analysis codes.

It also required a degree of R&D on the target solution – not only Oracle as a database system but also Linux/Intel as a hosting platform. This work is described in more detail below.

> The triple data migration from ODBMS required a significant amount of human effort and computing resources.

This triple migration required a significant amount of human effort and computer resources. However, as we shall see later regarding maintaining long-term scientific archives, such migrations need to be foreseen if data is to be preserved even in the medium term – it is far from guaranteed that the media chosen at the beginning of LHC will be readable by the end, and a migration of tape format is a convenient time to perform other pending migrations.

"...we describe the migration of event data collected by the COMPASS and HARP experiments at CERN. Together these experiments have over 300 TeraBytes of physics data stored in Objectivity/DB that had to be transferred to a new data management system by the end of Q1 2003 and Q2 2003 respectively. To achieve this, data needed to be processed with a rate close to 100 MegaBytes/s, employing 14 tape drives and a cluster of 30 Linux servers. The new persistency solution to accommodate the data is built upon relational databases for metadata storage and standard 'flat' files for the event data. The databases contain collections of 10^9 events and allow generic queries or direct navigational access to the data, preserving the original C++ user API. The central data repository at CERN is implemented using several Oracle9i servers on Linux and the CERN Mass Storage System CASTOR [34]."

9.26 Security Issues

> It has been agreed that response to severe security threats must receive top priority – even if it meant stopping the accelerator.

A well known security incident in recent years drew attention to the amount of responsibility a site such as CERN can have for database servers deployed at external sites. The clear answer is *none*. Although there are a number of well documented practises that can significantly reduce exposure to typical security exploits – and the consistent use of *bind variables* is one of them – the responsibility for site-local services must run with the site concerned. Nevertheless, in the aftermath of this event it was agreed that response to severe security threats must receive top priority – even if it meant stopping the accelerator. This was the first time that such agreement was reached but can be expected to have similar consequences to other Grid-related services and beyond.

9.27 Lessons Learnt in Managing a PetaByte

BaBar's experience in managing a PetaByte database using Objectivity/DB and HPSS, the enhancements that they found it necessary to introduce and their subsequent migration to a 2nd generation solution provide an extremely valuable case study in this story [35]. Of particular note:

"The commercial ODBMS provided a powerful database engine including catalogue, schema management, data consistency and recovery, but it was not deployable into a system of BaBar's scale without extra effort. Half a million lines of complex C++ code were required to customise it and to implement needed features that did not come with the product."

The paper describes in detail the enhancements that were required to run a production service and – of particular relevance to the Grid community – how to deal with planned and unplanned outages. Less than three full time DataBase administrators (DBA)s were required to manage the system – although this in itself would have raised scalability concerns for the LHC, where each experiment is generating roughly this amount of data per year. Hiring an additional three DBAs per experiment per year would clearly not be affordable.

Again, the lessons learnt from the 2nd generation re-factoring can clearly influenced the LHC programme, particularly as BaBar "led" by "following" the LHC decision.

"Planning for change makes inevitable migrations practical."

The paper concludes (penultimate sentence) with:
"Planning for change makes inevitable migrations practical."
A lesson we would clearly be advised to follow for the LHC.

9.28 VLDB 2000 Predictions

The 26th Very Large Database (VLDB) conference, held in Cairo in September 2000, included a panel on predictions for the year 2020. One of these was that YotaByte (10^{24} B) databases would exist by that time. Now a YotaByte is a lot of bytes. By 2020, the LHC might have generated around 1 ExaByte – 10^{18} B of data. 1 YotaByte is 10^6 times larger – and would require not only significant advances in storage but also in processing capacity to handle effectively. In particular, we cite Jim Gray's work on the need for balanced systems. Finally, 2020 is perhaps three – maybe four – product cycles away. Today's largest databases are perhaps scraping a PetaByte. What will be the driving forces behind the need for such massive data volumes?

9.29 ODBMS in Retrospect

It would be easy to dismiss Object Databases as a simple mistake. However, their usage was relatively widespread for close to a decade (CERN and SLAC in particular). Was there something wrong in the basic technology? If not, why did they not "take off", as so enthusiastically predicted?

> It would be easy to dismiss Object Databases as a simple mistake. However, their usage was relatively widespread for close to a decade.

Both of the two laboratories cited above stored around 1 PetaByte of physics data in an ODBMS, which by any standards has to be considered a success. There were certainly limitations – which is something to be expected. The fact that the current persistency solutions for all LHC experiments (which differ in some important respects in detail) have much in common with the ODBMS dream – and less with those of the LEP era deserves some reflection.

There was certainly some naïvety concerning transient and persistent data models – the purist ODBMS view was that they were one and the same. As a re-learnt lesson, RD45 pointed out very early that this was often not viable. More importantly, the fact that the market did not take off meant that there was no serious ODBMS vendor – together with a range of contenders – with which to entrust LHC data.

9.30 ORACLE for Physics Data

Following the recommendations above at the end of the Espresso study, and based on Oracle's 9i and later 10G release, the feasibility of using Oracle to handle LHC-era physics data was studied. This included the overall scalability of the system – where once again 16 bit fields raised their ugly heads (since fixed) – as well as the functionality and performance of Oracle's C++ binding "OCCI". As a consequence of this work, the COMPASS event data was migrated out of Objectivity into flat files for the bulk data together with Oracle for the event headers – of potential relevance to LHC as this demonstrated the feasibility of multi-TeraByte databases – similar to what would be required to handle event tags for LHC data.

However, the strategy for all LHC experiments is now to stream their data into ROOT files, with POOL adopted as an additional layer by all except ALICE.

In parallel, the database services for detector related and book-keeping applications – later also Grid middleware and storage management services – were re-engineered so as to cope with the requirements of LHC computing. A significant change in this respect was the move away from Solaris for database servers to Linux on PC hardware. Initial experience with the various PC-based systems at CERN showed that the tight coupling between storage and CPU power inherent in a single box solution was inappropriate and a move to SAN-based solutions, which allow storage and/or processing power to be added as required, has since been undertaken.

At the time of writing, the CERN physics database services consists of "Over 100 database server nodes are deployed today in some 15 [TeraByte sized] clusters serving almost two million database sessions per week. [36]"

9.31 OpenLab and Oracle Enhancements

Although the explosion in Oracle database applications had yet to happen, a concerted effort was made to ensure that any necessary enhancements were delivered in production well ahead of LHC data taking. The main areas targeted were:

- Support for native IEEE float and double data types.
- Removal of any scalability limitations, such as 16 bit fields etc..
- Support for Linux and commodity hardware.
- Improvements in the area of transportable tablespaces – foreseen not only for bulk data exchange between sites, but also for building a potential interface to mass storage systems.
- Reduction in administrative overheads.

Work on these issues was initially started as a continuation of the longstanding relationship between the company at CERN and then continued more formally as part of CERN's OpenLab [37] – designed to foster exactly such industrial partnership in Grid-related areas. As part of the OpenLab work, a variety of high-availability and related techniques were evaluated and prototyped, with the clear goal of production deployment (where appropriate) in the short to medium term. Areas studied included the use of commodity Linux systems to host database clusters, Oracle's DataGuardDataGuard for high availability and to help perform transparent upgrades, as well as Oracle Streams for data distribution. All of these solutions are now routinely used as part of the production services deployed at CERN and elsewhere.

> Although the explosion in Oracle database applications had yet to happen, a concerted effort was made to ensure that any necessary enhancements were delivered in production well ahead of LHC data taking.

Indeed, at the time of the Oracle 10g launch in San Francisco, CERN was publicly acknowledged for its contribution in driving the database area forward.

9.32 Clusters

Clusters have played an important role in database deployment at CERN throughout this quarter century. From the first VAXCluster in the mid-eighties, which hosted the LEP DB and other services, through the Oracle Parallel Server some ten years later, to today's Real Application Clusters (RAC). These systems are linked by more than name: the clusterware of VMS was later made available on Digital Unix systems, and is now used on Linux systems in RAC environments. Architecturally, a RAC and VAXCluster have a number of similar features – not only the distributed lock manager but also a dedicated interconnect for cluster communication. Indeed, many of the centres of excellence for VAXClusters – such as Valbonne in southern France and Reading in the U.K. – are now centres of excellence for RAC systems. The LEP DB service also implemented disk-resident backup – again close to two decades before its time.

The use of clusters has a number of advantages – not only a high(-er) availability solution, they also allow more flexible CPU and storage allocation than in a single server solution, such as a conventional disk server. However, not all applications scale well in a cluster environment: conventional wisdom being that those that perform well on an SMP will adapt well to a cluster.

9.33 Enter the Grid

The LHC Computing Grid (LCG) has a simple hierarchical model where each Tier has specific responsibilities (see Chaps. 3 and 4). There is a single Tier 0 – CERN, the host laboratory, with O(10) Tier 1 sites and O(100) Tier 2s. To first approximation, the sum of resources at each level is roughly constant. The different Tiers have separated roles.

Tier 0: safe keeping of RAW data (first copy); first pass reconstruction, *distribution of RAW data and reconstruction output to Tier 1*; reprocessing of data during LHC down-times.

Tier 1s: safe keeping of a proportional share of RAW and reconstructed data; large scale *reprocessing* and safe keeping of corresponding output; *distribution of data products to Tier 2s* and *safe keeping* of a share of simulated data produced at these Tier 2s.

Tier 2s: Handling *analysis* requirements and proportional share of *simulated event* production and reconstruction.

> One does not have to dig very deep to find that Databases are behind virtually all services in the Grid.

Whilst databases are not explicitly mentioned in this high level view, one does not have to dig very deep to find that they are behind virtually all services in the Grid. Many, as we shall see, had their counter-part in the LEP era. Some – in particular in the case of workload management and the handling of Grid certificates – are new and – at least when all relevant components handle roles and groups correctly – can be considered defining elements of the Grid.

9.34 European Data Grid RLS Deployment

One of the first Grid services to be deployed that required an Oracle database (in fact also the Oracle Application Server) was the European Data Grid (EDG, see Chap. 3) [38] Replica Location Service (RLS). This was a critical service, which, if unavailable, meant that running jobs could not access existing data and scheduling of jobs at sites where the needed data was located was not possible.

The Grid – if not down – was at least seriously impaired. As a result this was taken into account when designing the service deployment strategy and procedures. In addition to trying to define a service that was highly available and for which all possible recovery scenarios were tested and documented, an attempt was made to package the software – together with the underlying Oracle components – in a manner that made them trivial to install, both on CERN instances and at Tier 1 sites outside. This proved to be an extremely difficult exercise – in part as many of the sites involved had at that time little or no experience with the technologies involved. Furthermore, despite repeated attempts at producing some sort of "appliance" that simply ran unattended, such a self-managing, self-healing database system still seems to be as far off today as when first suggested more than ten years ago. The only possible alternative to in-house expertise is for "hosted applications", as has been done successfully at CERN for the Oracle*HR service. Could this ever be extended to Grid middleware services?

9.35 Jim Gray's Visit

Having followed the progress in HEP on using databases for physics applications for many years, he visited CERN in 2001 and attempted to convince us to: "Put everything online, in a database".

One concrete proposal that he made at the time was for a *geoplex* – namely where data is stored (online) in two or more places (as is largely done in the LHC Computing Grid) and to "scrub it continuously for errors" (as is not).

He continued: "On failure, use other copy until repaired – refresh lost copy from safe one(s)."

As a further potential advantage, the copies may be organised differently, e.g. optimised for different access patterns. As we are now witnessing "silent corruption" at a level that is bound to impact the large volumes of data already collected – let alone those that are being produced by LHC – this wisdom now seems particularly pertinent.

He also argued: "In reality, its build versus buy. If you use a file system you will eventually build a database system with metadata, query, parallel ops, security, ... in order to safely and efficiently reorganise, recovery, distribute, replicate ..."

Finally, his top reasons for using a database were:

1. Someone else writes the million lines of code
2. Captures data and *Metadata*
3. Standard interfaces give *tools* and *quick learning*
4. Allows *Schema Evolution* without breaking old apps
5. *Index* and *Pivot* on multiple attributes space-time-attribute-version...
6. *Parallel TeraByte searches* in seconds or minutes
7. Moves *processing & search close to the disk arm* (moves fewer bytes (qestons return datons))
8. *Chunking* is easier (can aggregate chunks at server)
9. Automatic *geo-replication*
10. *Online update and reorganisation*
11. *Security*
12. If you pick the right vendor, ten years from now, there will be software that can read the data

Jim is well known for his work on databases in astrophysics, where he demonstrated that quite complex queries can indeed be expressed in SQL. Some examples include:

Q1: Find all galaxies without unsaturated pixels within 1' of a given point of ra $= 75.327$, dec $= 21.023$.

Q2: Find all galaxies with blue surface brightness between and 23 and 25 mag per square arc-seconds, and $-10 <$ super galactic latitude (sgb) < 10, and declination less than zero.

Q3: Find all galaxies brighter than magnitude 22, where the local extinction is > 0.75.

Q4: Find galaxies with an isophotal surface brightness (SB) larger than 24 in the red band, with an ellipticity > 0.5, and with the major axis of the ellipse having a declination of between 30" and 60"arc seconds.

Q5: Find all galaxies with a deVaucouleours profile (r sup $\frac{1}{4}$ falloff of intensity on disk) and the photometric colours consistent with an elliptical galaxy.

9.36 Database Applications in the LHC Era

Whilst the database applications for the LHC experiments can be broadly categorised as was done for LEP in the "Green Book" [39], there are a number of distinguishing characteristics that require additional attention. Namely, those applications that are critical to the experiments production processing and data distribution and those that require some sort of distributed database solution. Some may fall in both categories.

In this section we focus on the latter, as the techniques for handling the former are largely the same as for production Grid services and are hence discussed below.

To date, the only application in this category is that of detector calibrations/conditions (for LHCb, a replicated file catalogue [40] is also made available using the same technologies that we shall describe, once again echoing the situation in the LEP era).

ALICE have chosen to base their conditions data on ROOT files [41], distributed in the same way as for event data, together with the AliEn [42] file catalogue.

CMS have implemented their own conditions application on top of Oracle, which uses caching techniques to make conditions data available to Tier 1 sites and thence out to Tier 2s. Based on experience at FNAL, the overall system consists of an Oracle database together with a FroNTier [43] server at the Tier 0 and Squid web caches at the Tier 1 and Tier 2 sites. Data is also exchanged between online and offline systems, using the same Oracle Streams [44] technology that is used in a wider sense by ATLAS and LHCb.

ATLAS and LHCb have adopted a common solution based on the COOL package [45]. The data maintained in the backend databases is replicated using Oracle Streams to Tier 1 sites, with data flows also to/from the online systems. ATLAS has the largest number (10) of Tier 1 sites and also has three special "muon calibration centres" that are not Tier 1s but play a specific role in this exercise, with calibration data flowing back to CERN and then out again.

"To enable LHC data to flow through this distributed infrastructure, Oracle Streams, an asynchronous replication tool, is used to form a database backbone between online and offline and between Tier-0 and Tier-1 sites. New or updated data from the online or offline database systems are detected from database logs and then queued for transmission to all configured destination databases. Only once data has been successfully applied at all destination databases is it removed from message queues at the source."

The distributed solutions for all experiments except ALICE are coordinated by the LCG 3D project [46]:

"describes the LCG 3D service architecture based on database clusters and data replication and caching techniques, which is now implemented at CERN and ten LCG Tier-1 sites. The experience gained with this infrastructure throughout several experiment conditions data challenges and the LCG dress rehearsal is summarised and an overview of the remaining steps to prepare for full LHC production will be given."

Whilst extensive testing of these solutions continues, full scale LHC production experience is now ironing out any remaining issues.

Whilst extensive testing of these solutions has been done, full scale LHC production experience is now ironing out any remaining issues.

Since the adoption of Objectivity/DB by the BaBar experiment at SLAC, a whole host of conditions database implementations have been produced. The first such implementation, by Igor Gaponenko [47], was introduced at CERN and eventually migrated to Oracle. A new implementation – COOL [48] – was subsequently made at CERN, this being the baseline choice of ATLAS and LHCb. The COOL system itself is based on CORAL [49] – the The COmmon Relational Abstraction Layer:

"the LCG Conditions Database Project …COOL, a new software product for the handling of the conditions data of the LHC experiments. The COOL software merges and extends the functionalities of the two previous software packages developed in the context of the LCG common project, which were based on Oracle and MySQL. COOL is designed to minimise the duplication of effort whenever possible by developing a single implementation to support persistency for several relational technologies (Oracle, MySQL and SQLite), based on the LCG Common Relational Abstraction Layer (CORAL) and on the SEAL libraries."

9.37 Event Tags Revisited

For many analysis application only one part of the events is needed. The criteria with which a sub-sample of the event is selected are an important part of the analysis strategy and are optimised via several iteration on the data. As analysis is often an I/O-bound operation, it is important to be able to select events without necessarily reading them all. For this reason, events are "tagged" with a set of features, and these "tags" are collected in special files that allow to select events without reading them all in memory. Once a given collection of events is created based on the "tags", only these events are read and analysed. The point here is that the tag files are much smaller than the event files and usually easier to search.

> It is at the moment still unclear whether the currently used database synchronisation mechanism would be able to handle the volumes and rates involved in a nominal LHC year of data taking.

At the time of writing, ATLAS is the only LHC experiment potentially interested in storing event tags in an Oracle database. The experiment with the most experience in this respect is COMPASS, who currently store some 6 TeraByte in Oracle, following their migration from Objectivity/DB. However, the COMPASS tag database is maintained centrally at CERN, with a small subset of the data copied to Trieste. (BaBar also maintain a bookkeeping database that is replicated to some ten sites and even some laptops, but it is at a much higher level and only contains a few GigaByte of data.) Until recently, ATLAS foresaw maintaining tag databases at least at all of their ten Tier 1 sites. It is unclear whether the currently used database synchronisation mechanism would be able to handle the volumes (6 TeraByte of data in a nominal year of LHC running) and rates involved, and other techniques – such as transportable tablespaces – are also being considered. Recently, this model changed and the latest proposal is to store the tags at those Tier 1 sites that volunteer to host them.

9.38 Database Developers' Workshops

Given the very large number of database applications – and indeed database developers – foreseen for the LHC, a workshop focusing on LHC online and offline developers was organised for early 2005. Around 100 developers signed up for this week-long session, consisting of both lectures and hands-on exercises. Although previous and subsequent training events have taken place, this workshop was unique in focusing on the needs of the physics community.

All attendees at the workshop were given a copy of Tom Kyte's excellent book – "Effective Oracle by Design". Shortly after the workshop, Tom himself visited CERN and gave a series of tutorials, including one on "The Top ten Things Done Wrong Over & Over Again". Such events are essential given such a large and geographically distributed community and are to be encouraged if they do indeed reduce the support load on the DBA teams, as well as producing applications that are both more robust and performant. It certainly goes in the direction of the ECFA recommendation, although it is unlikely that a DB developer community of more than 100 was imagined at that time. Given that the type of application is largely as predicted, can the growth in number of applications – and hence developers – be purely explained by the magnitude of today's detectors? It is surely also related to the fact that databases are a well understood and widely taught technology, whereas the number of true experts in the dark arts of ZEBRA were closer in number to those in the early days of relativity.

9.39 Grid Middleware and Storage Solutions

A number of the Grid middleware and storage solutions that are deployed in the LCG rely on a database backend. However, there is no unique solution: IBM's HPSS now used DB2 internally. Sites running dCache typically use PostgreSQL, whereas those deploying CERN Data Pool Manager (DPM) use MySQL (Oracle is also supported). CASTOR2 sites run Oracle. The gLite FTS is only supported on Oracle, whereas the LHC File Catalogue (LFC) can use either Oracle or MySQL back-ends – the former being preferred for larger sites, i.e. Tier 1s and the Tier 0. The use of databases in these applications is described in [50].

The Virtual Organisation Membership (Relational) System (VOM(R)S) applications were recently ported to Oracle, whereas some Grid components – in the particular the Resource Broker – still only support MySQL.

Given the impressive degree of standardisation elsewhere, why is there so much diversity at this level? In the case of IBM's storage solutions, the choice of DB2 is mandated by the vendor. For dCache, PostgreSQL is preferred for licensing reasons. For the other data management middleware, MySQL makes more sense for smaller sites, whereas the additional features of Oracle are required for larger scale production services.

Despite this seeming diversity, there appears to be a set of problems that affect many of the implementations and this is largely related to database housekeeping. Unless maintained – preferably by the application – some tables grow indefinitely until queries first become inefficient and later grind to a halt. Whilst not explicitly covered by the ECFA recommendations, there is clearly a list of "best practises" that it would be useful to establish to guide not only existing sites but also those yet to deploy the above storage and data management solutions.

9.40 Those Questions Revisited

After more than a decade it seems that the questions posed at CHEP'92 still have some relevance. Today, it is common practise that applications in the area of storage management, experiment book-keeping and detector construction/calibration use a database backend. However, the emergence of open-source solutions and indeed much experience has changed the equation. Nowadays, it is common practise to use a database backend (where the distinction between object/object-relational/pure-relational is very much blurred). However, the licensing, support and deployment issues are still real.

So in summary, should we buy or build database systems for our calibration and book-keeping needs? It now seems to be accepted that we *build* our calibration and book-keeping systems *on top* of a database system. Both commercial and open-source databases are supported by this approach.

The other important question is: will database technology advance sufficiently in the next 8 to 10 years to be able to provide byte-level access to PetaBytes of LHC data? We (HEP) have run production database services up to the PetaByte level.

> After more than a decade it seems that the questions posed at CHEP'92 still have some relevance..

The issues related to licensing, and – perhaps more importantly – *support*, to cover the full range of institutes participating in an LHC experiment, remain. Risk analysis suggests a more cautious – and conservative – approach, such as that currently adopted, such as finding out who are *today* the concrete alternatives to the market leader.

As regards lessons for the future, some consideration of the evolution of the various object-oriented projects – RD45, LHC++ and ROOT – is deserved. One of the notable differentiators of these projects is that the former were subject to strict and frequent review. Given that the whole field was very new to the entire HEP community, some additional flexibility and freedom to adjust to the evolving needs – and indeed our understanding of a new technology – would have been valuable.

As we now deploy yet another new technology for LHC production purposes, there is at least the possibility of falling into the same trap.

Food for thought for CHEP 2030 or thereabouts?

9.41 The ECFA Report Revisited

It would be hard to argue that there was a concerted effort to systematically address the recommendations of the ECFA report (apart from in the initial years – leading to the first central Oracle services and some specific enhancements to KAPACK). Nevertheless, there has been significant progress on all of the issues raised. As described above, databases are now an integral part of current experiments on- and off-line environments and an essential component of the overall production and analysis chain.

Perhaps two of the ECFA recommendations deserve further attention. Further effort in training for database developers could reduce the amount of effort required to solve key implementation mistakes, such as the infamous lack of use of "bind variables". Also, the cost of administering the databases for the experiments is significant. Anything that can be done to reduce this effort – over and above the reduction in support load that would come from better design and implementation – would be welcome.

9.42 Issues on Long-Term Archives

Very long term data archives are far from a solved problem – maintaining scientific or other data in a way that it is still usable hundreds or thousands of years hence is still not understood. However, there is recent experience in maintaining scientific data with the specific goal of a reanalysis in the light of new theories and/or experimental results. Such a reanalysis was performed on data from the JADE collaboration at the PETRA accelerator at DESY was made in the mid-1990s.

> Maintaining scientific or other data in a way that it is still usable hundreds or thousands of years hence is still not understood

Apart from the rather obvious issues of maintaining the data (the tapes in question were found abandoned in the corner of an office), there are issues related to programming languages, which may be obsolete after even a few years – as happened in this case – or more likely the program execution environment. However, the biggest problem as seen in the JADE case and rediscovered in the various attempts at a LEP data archive, has been in the area of metadata – maintaining enough information about the detector and the experiments' bookkeeping so that the bits – even if they can be read – can be meaningfully used. This is a big challenge for the database area, in that the necessary care to identify and preserve all of the necessary metadata must be made well in advance. Waiting until the necessary experts have retired or moved on is simply too late. There are many arguments that scientific data – such as from LEP or the LHC – should be maintained for posterity. However, if we are unable to analyse it even a few years hence, there is little chance of achieving such a notable goal. Arguably, however, this is tantamount to destroying our scientific legacy and is an area that should be addressed with priority.

9.43 The State of the Grid

As experienced by BaBar and indeed many other experiments beforehand, operating reliable distributed services is a challenge. In the case of a number of the middleware services, redundancy is provided by load-balanced servers, deployed in such a way as to avoid single points of failure, such as power, network switches and so forth. Whilst high availability database technology is well understood in theory, it can be both costly and complex to implement. Indeed, unnecessary complexity – such as cross-site services – may do little to enhance actual availability and may even make it worse. A further element in the equation is that Grid users typically care about much higher level applications than the core Grid services. Often an experiment-level service may be built on a combination of a number of experiment-specific services – some of which may have a database component – as well as Grid services likewise.

On the positive side, the Grid is basically a batch environment and so resilience to shortish-term glitches is acceptable and even "transparent". However, it is not sufficient to list the basic technologies involved – an in-depth study of the key services and their criticality, followed by a specific implementation consisting of hardware, middleware, procedures and application are required to achieve this goal. At the time of writing this work is clearly "in progress", but it is well understood that the benefits to both service providers and service users is significant and well worth the effort. The first step has been to perform an analysis of the services required by CMS and – once those were deployed at an acceptable level – the equivalent analysis has been performed with the other LHC VOs. Clearly, the experience of previous experiments, together with high-availability database techniques, will be essential components of this strategy. The target was to have the key services deployed in this manner early enough to be reported on at CHEP 2009 (March 2009 in Prague). Due to problems with the LHC machine the first CHEP conference that reported on data taking from collisions between accelerated beams was CHEP 2010 in Taipei.

9.44 CHEP 2006

A review of earlier technology predictions highlighted:
"Object databases may change the way that we view storage".
It is hard to guess exactly what was behind this remark. If it was that we would be using commercial object databases to manage all LHC data, then the story is told above. If, however, it was intended to mean that we would finally treat disk storage as random-access, and not just "fast tapes", then indeed the prediction can be considered correct. Furthermore, based on the definition from the early ECFA report, and indeed Jim Gray's analysis of our work, it also correct that we are using databases (commercial or open source) to manage book-keeping and other non-event data, whereas we have built a powerful – albeit not fully featured – object-oriented database system in which the full event data – from raw to tags – is maintained. Indeed, in many aspects this is very similar to the work reported on at CHEP '91 – "Database Computing in HEP".

9.45 CHEP 2010

By the time of CHEP 2010 we had gathered almost six months of experience with data taking at the LHC [52]. This confirmed the readiness of the services which provided to be sufficiently robust as to handle the needs of production data taking, processing and analysis, even though the rates and volumes where in some notable cases higher than those that had been foreseen and tested. The procedures that had been put in place to handle problems – such as the judicious use of alarms for critical services – proved to be sufficient, although a tail of problems that were not resolved within the target timeframe (see paper for full details) still remained.

The challenge for 2011 and beyond will be to maintain the current level of service with increased demands and steadily improve the service so that these tails no longer remain.

9.46 Conclusions and Lessons Learned

There is no doubt that the era described above was at times turbulent – both the move to distributed computing and from "FORTRAN to object-oriented" resulted in heated debates and often diametrically opposed opinions. However, the ECFA report turned out to be remarkably prescient – apart from relatively minor details, such as the use of ZEBRA RZ in most cases for home-grown solutions, rather than KAPACK. A number of official or semi-official joint projects were established addressing the areas raised by the report – it being in many cases the smaller experiments that benefited most from this work. At the same time, the emergence of commodity computing and a convergence of technologies have made a new era of computing possible, namely that of Grid computing.

We have not yet gained sufficient experience in this environment for a fully objective analysis – this must wait another few years, including the onslaught of full LHC data taking and analysis.

The full story of databases in HEP is worthy of a much longer treatise and an event modelled on the "SQL 25 year reunion" held in Palo Alto in the mid-1990s is clearly called for.

Acknowledgements Numerous people have contributed to the story of Databases in HEP, including the many who worked on various aspects of the CERN Program Library and the ZEBRA, FATMEN and HEPDB packages. Members of the PASS and RD45 projects, together with the ROOT and POOL and related projects as well as all those who have contributed to database deployment at the various HEP sites throughout the years also played key roles in this story. Particular thanks are due to Harry Renshall for his careful reading of this and many other documents during the past 25 years, together with frequent guidance and mentoring. Finally, the "Grid Guys & Gals" (3G), working on the various middleware components and associated services, for deploying a host of database applications and in anticipation of the many years of LHC data taking to come. *LHC data archive anyone?*

References

1. Databases and bookkeeping for HEP experiments, ECFA Working Group 11, ECFA/83/78 (Sept 83)
2. See http://en.wikipedia.org/wiki/SQL
3. KAPACK - Random Access I/O Using Keywords, CERN Program Library Entry Z303
4. Mount, R.P.: Database systems for HEP Experiments. Comput. Phys. Comm. **45**, 299–310 (1987)
5. The L3 database system, Nuclear Instruments and Methods in Physics Research A309, 318–330 (1991)
6. ZEBRA Data Structure Management Package, CERN Program Library Entry Q100
7. Computing at CERN in the 1990s, July 1989
8. The ALEPH DAta MOdel - ADAMO. In: The proceedings of [14]. http://adamo.web.cern.ch
9. ALEPH bookkeeping and SCANBOOK, Jacques
10. FATMEN - File and Tape Management Package, CERN Program Library Entry Q123
11. HEPDB - Database Management Package, CERN Program Library Entry Q180
12. Isnard, C., et al.: CHEOPS: A batch data dissemination system based on the OLYMPUS satellite. In: Proceeding of the International Conference on the Results of the OLYMPUS Utilisation Programme, Sevilla, 20-22 April, 1993
13. see for instance http://ieeexplore.ieee.org/stamp/stamp.jsp?arnumber=00182285 and also http://www.esa.int
14. Data Structures for Particle Physics Experiments: Evolution or Revolution? ISBN 981-02-0641-0

15. Goossens, M., Schaile, O.: DZDOC – Bank Documentation Tool. In: ZEBRA, CERN Program Library Long Writeups Q100/Q101
16. Kunz, P., Word, G.: The Cheetah Data Management System, SLAC-PUB-5450 March 1991 (E/I)
17. Barone, L.M.: Database Management and Distributed Data in High Energy Physics: Present and Future. In: The Proceedings of CHEP "91. ISBN 4-946443-09-6
18. Database Computing in High Energy Physics. In: The Proceedings of CHEP91, ISBN 4-946443-09-6
19. HEPDB Package, CERN Program Library Entry Q180
20. http://lcg.web.cern.ch/LCG
21. Linder, B., Mount, R., Shiers, J.: Databases for High Energy Physics. In: The proceedings of CHEP92. ISBN 0007-8328
22. Bos, K.: The Moose project. Comput. Phys. Comm. **110**(1-3), 160–163 (1998)
23. http://geant4.web.cern.ch/geant4
24. See http://wwwasd.web.cern.ch/wwwasd/cernlib/rd45/
25. See http://wwwasd.web.cern.ch/wwwasd/lhc++/indexold.html
26. ROOT - an Object-Oriented Data Analysis Framework - see http://root.cern.ch/
27. May, E., et al.: The PASS Project: Database Computing for the SSC. Proposal to the DOE HPCC Initiative, Argonne National Lab, 1991, unpublished
28. See http://www.hep.net/ssc/
29. http://www.slac.stanford.edu/BF/
30. POOL - Persistency Framework - see http://pool.cern.ch/
31. Malon, D., May, E.: Critical Database Technologies for High Energy Physics. In: The Proceedings of VLDB97, available from vldb.org
32. Sphicas, P.: Conference summary talk, CHEP 2000, Padua, Italy
33. Objectivity Data Migration. In: The Proceedings of CHEP 2003, La Jolla, California
34. Nowak, M., Nienartowicz, K., Valassi, A., Lbeck, M., Geppert, D.: Objectivity Data Migration, Computing in High Energy and Nuclear Physics, La Jolla, California. March 2003 24–28
35. Becla, J., Wang, D.: Lessons learnt from Managing a Petabyte, SLAC. In: The Proceedings of the 2006 CIDR Conference
36. Girone, M.: Database services for Physics at CERN. In: The Proceedings of CHEP07
37. http://proj-openlab-datagrid-public.web.cern.ch/proj-openlab-datagrid-public/
38. http://eu-datagrid.web.cern.ch
39. Computing at CERN in the LEP era, CERN, May 1983
40. Martelli, B., et al.: LHCb experience with LFC database replication. In: The Proceedings of CHEP07
41. The ALICE Collaboration, Technical Design Report of the Computing, CERN-LHCC-2005-01, ALICE TDR 012, 15 June 2005
42. See http://alien2.cern.ch
43. Lueking, L., et al.: FroNtier: High Performance Database Access Using Standard Web Components in a Scalable Multi-Tier Architecture. In: Proceedings of CHEP04, Interlaken (September 2004)
44. Oracle streams - see http://www.oracle.com/
45. See http://lcgapp.cern.ch/project/CondDB/
46. Düllmann, D., et al.: Production experience with Distributed Deployment of Databases for the LHC Computing Grid. In the Proceedings of CHEP07
47. Gaponenko, I., et al.: Using Multiple Persistent Technologies in the Conditions Database of BaBar. In: Proceedings of CHEP06, February 2006
48. Valassi, A., et al.: COOL Development and Deployment: Status and Plans. In: The Proceedings of the CHEP 2006 Conference
49. Papadopoulos, I., et al.: CORAL, A Software System for Vendor-neutral Access To Relation Databases. In: The Proceedings of the CHEP 2006 conference
50. Abadie, L., et al.: Grid-Enabled Standards-based Data Management, submitted to the 24th IEEE Conference on Mass Storage Systems and Technologies. Boucrot, in "The ALEPH Experience", p 128

51. In Proceedings of Symposia On Advanced Computer Programs for Digital Computers, sponsored by ONR. Republished in Annals of the History of Computing, Oct. 1983, pp. 350–361. Reprinted at ICSE'87, Monterey, California, USA, March 30-April 2, 1987, June 1956
52. Girone, M., Shiers, J.: WLCG Operations and the First Prolonged LHC Run to appear in the proceedings of CHEP 2010

Chapter 10
Towards a Globalised Data Access

Fabrizio Furano and Aandy Hanushevsky

In this chapter we address various solutions to initiate a smooth transition from a high performance storage model composed by several nodes in different distant sites to a model where the nodes cooperate to give a unique file system-like coherent view of their content.

This task, historically considered very problematic, has to be considered from several points of view, functional and not, in order to be able to deploy a production-quality system, where the functionalities are effectively usable, and the data access performance is as close as possible to the one reachable by the used hardware. This can be considered as the major challenge in such systems, even because the general expectations about performance and robustness are very high, and very often they are compared to distributed file systems, whose evolution over time has been quite promising, but not to the point of being able to fully satisfy the requirements of large scale computing. To reach this objective, among other items, a design must face the difficulties dealing with the network latency, which varies of more than two orders of magnitude between local network access and Wide Area Network access.

Moreover, the high latency of the typical wide area network connections and the intrinsically higher probability of temporary disconnections among clients and servers residing in different sites have historically been considered as very hard problems. From this perspective, this work addresses in a generic way the problems related to design, functionalities, performance and robustness in such systems, with a strong focus in enabling a generic data-intensive application to compute its inputs and save its outputs in a distributed storage environment, where the most stringent non functional requirements are concerned with the data access speed and the processing failure rate.

F. Furano (✉)
CERN, Geneva, Switzerland

A. Hanushevsky
SLAC, CA, USA
e-mail: Fabrizio.Furano@cern.ch

This also forces us to address the issues related with the scalability of the whole design, and the fact that the focus on performance is a major requirement, since these aspects can make, in practise, the difference between being able to run data analysis applications and being not. The case study for these methods is the ALICE experiment at CERN, where a system based on these principles is currently being incrementally deployed in an already-running production environment among many sites spread all over the world, accessed concurrently by tens of thousands of data analysis jobs. However, given the generality of the discussed topics, the introduced methods can be extended to any other field with similar requirements.

We will address this complex problem describing the way in which the Scalla system [14, 15] is trying to solve it. Although we will go into some depth explaining the architecture of this system, the analysis of the problem we provide and the general architecture of the system we propose are quite general.

10.1 Introduction

A common historical strategy, much adopted in High Energy Physics (HEP), to handle very large amounts of data (on the order of the dozens of PetaBytes per year) is to actually partition the data among different computing centres, and assigning a section of the data to each of them, eventually having some sites which are more important than others.

Among others, one of the consequences of this kind of architectures is that, for example, if a processing job needs to access two data files, each one residing on a different site, the typically adopted simple-minded solution is to make sure that a single site contains both of them, and the processing is started there. Unfortunately, the situation can become more complicated than this, once one starts considering some more characteristics of the High Energy Physics software analysis systems, for example the fact that each application may access hundreds of files, and when those files have to be replicated, very often the software designers introduce some form of replica catalogue, which is needed to keep track of the (eventually automatic) data movements. Moreover, if the overall software systems is based on the concept of "local replicas", the latency in scheduling the massive replication of big data files can be a serious problem, together with the need of transferring the data to the right place. With respect to this, we would like to highlight the high inefficiency of this approach if the analysis applications do not read entirely their input data files.

It is very straightforward to see some simple consequences that come from the described scenario. In fact, an hypothetical data analysis job must wait for all of its data to be present at a site, before starting, and the time to wait before starting can grow if some file transfers fail and have to be rescheduled. In its more complex form, the task of assigning files to sites (and replicating them) can easily degrade into a complex n-to-m matching problem, to be solved online, as new data comes and the old data is replicated in different places. An optimal, or even sub-optimal but satisfying solution can be very difficult, if not impossible, to achieve.

One could object that the complexity of an n-to-m online matching problem can be avoided in some cases by assigning to each site something like a "closed" subset of the data, i.e. a set of files whose processing does not need the presence of files residing elsewhere. This, depending on the problem, can be true, but generally implies a reduced number of choices for a job scheduler, since generally there are very few sites eligible for a given kind of processing. This situation can theoretically evolve to one where the sites containing the so-called "trendy datasets" are overloaded, while others can be idle.

This problem is very well known in the field of High Energy Physics, where the users submit a great number of data analysis jobs a few weeks before the conferences where they expect to present their results. Another direction towards which this approach can evolve is the so-called "manual data management", which we did not consider in this work. Moreover, we could also object that making data processing jobs wait too much just because the distribution of files is not yet optimal for a given analysis could become a major source of inefficiency for the whole system, here referring to the users and their results as parts of it.

Even if these issues are well known, however, it is in our opinion that the initiatives related to what we can call "transparent storage globalisation" are not apparently progressing fast enough to allow a system designer to actually design a major distributed storage deployment using them. What happens in many cases is that those designs include some component whose purpose is to take decisions about jobs or file placements, based on the knowledge of the supposed current status of the storage system, i.e. the current (and/or desired) association of data to sites. Invariably, this kind of component takes the form of some kind of a simple-minded "file catalogue", and its deployment generally represents a serious performance bottleneck in the whole architecture and also can be a single point of failure, or a source of inconsistencies.

We must clarify here the subtle difference that lies between a file catalogue and a meta-data catalogue. A file catalogue is a list of files, associated to other information, e.g. the sites which host a particular file or the URLs to access it. When part of a design, it typically is the only way an application has to locate a file in a heavily distributed environment, hence it must be contacted for each file any application needs to access. A meta-data catalogue, instead, contains information about the content of files, but not about where or how they can be accessed. Hence it can contain meta-data about a non-existing file, or a file can have in principle no meta-data entries. This typically causes no problems to carefully designed applications, and, instead, can be highly desirable for the intrinsic robustness of the concept. A file catalogue instead must deal with the fact that the content of the repository might change in an uncontrollable way (e.g. a broken disk or other issues), opening new failure opportunities, e.g due to mis-alignments between its content and the actual content of the repository.

To give a simpler example, we consider a completely different domain: the tags for music files. A very good example of a comprehensive meta-data catalogue for them is e.g. the *www.musicbrainz.org* website. We could also think about creating a file catalogue, indexing all the music files owned by a state's population. The subtle difference among these two is that everybody is allowed not to own (or to loose or to damage) a CD listed in this meta-data catalogue. This would not be considered a fault, and has no consequence on the information about that particular CD. In other words, this is just information about music releases, which somebody could own, and somebody not. Hence, an application willing to process a bunch of music files will just process them, since they already contain their meta-data, or they are accessible. On the other hand, a file catalogue wants to accurately reflect the reality, adding some layers of complexity to the task, last but not least that it is computationally very difficult (if not impossible) just to verify that all the entries are correct and that all the CDs owned (and sold/exchanged) by the persons are listed. Our point, when dealing with this kind of problems is that, from a general perspective, our proposal is not to promote a meta-data catalogue to being a file catalogue, since the worldwide location of the requested resources should be a task belonging to the storage system, not to the one which handles the information bookkeeping. This might be considered a minor, or even philosophical aspect, but it makes a very big difference from the architectural point of view.

10.2 Objectives

Given all these considerations, in our work we considered multiple overall objectives, first of all how to simplify the simple minded fully replica-based approach. In our view, to avoid this kind of dependency, there is the need of a data storage subsystem able to cope with storage globalisation. A very important aspect which would support such a choice, is to determine at which extent this kind of solutions can be effective. Can they support for instance some interactive form of data analysis?

We also wanted to find a way to insert in a working data analysis computing model (which represents our use case) some key features that enhance it, while heading for more ambitious scenarios, where the components are better integrated.

While the given ones are to be considered just as simple examples, in this work we address the various solutions we considered and adopted in order to pass from a high performance storage model composed by several different distant sites coordinated by some software (or human-based actions) component to a model where the remote storage nodes cooperate in some way to give a unique file system-like coherent view of their content.

Doing this way, we could argue that a hypothetical job scheduler could choose for instance to schedule a job in a site even if the storage local to the site is missing some of the needed data files. Of course, the design ought to comprehend how this case should be handled, since there could be several options, for instance immediately pulling the requested data files into the local storage, or letting the analysis software access data that is in another site, by means of a direct connection.

Even if simple to understand, replica-based data management hides critical difficulties. Directly accessing data worldwide could reduce the need for replicas and enhance the performance of the system and its stability.

10.3 An Unique View Versus an Unique Thing

The requirement of efficiently accessing a huge data repository distributed through different sites can be linked to the size of the data repository itself or to the policies dealing with the data replication. For example, it could look easier or more efficient to assign to a particular site a part of a big data storage, especially if there is some other reason to keep that data in that site. For example, in the High Energy Physics world, the case could be that of a site which hosts a group of experts in some particular topic. This kind of site generally shows great interest in asking to physically host such a data partition. But in this case we do not need to focus on the requirements of the High Energy Physics computing, which represent a particular case. Another fair example could be that every town stores the photographs of all its citizens, and this storage site collaborates with the others in order to give a uniform, catalogue-free nation-wide repository of the citizen's photographs, able to efficiently feed several computing farms.

From this perspective, hence, it is somehow immaterial if the final purpose of the system is to handle High Energy Physics data or a very high number of pictures. The only important thing in our vision is that the characteristics of the chosen storage model must match the characteristics of the computing model in which it is inserted.

In other words, one could wonder if, in the example dealing with the pictures, an application could process a big part of this repository, without forcing the creation of replicas of the pictures' files just because there might be the need to process them. The overhead of such an approach can be simply much greater than the task of just processing them. With this statement we mean that the decision of creating replicas should be based on other reasons, with respect to the pure functional requirement of being able to access the data.

The answer is not necessarily obvious, and largely depends on the efficiency of the communication paths between the analysis processes, but also in the efficiency of the mechanisms that locate the data in the network of the storage sites. We believe that, by exploiting these two points, one can build a file-system-like view of a very large storage distributed through many sites, suitable for a wider range of applications willing to access them in real time.

There are caveats, however. For instance, one could analyse how the AliEn [1, 6, 9] user interface presents a unique view of all the data repository of the ALICE experiment, and argue that this goal has already been achieved. Despite the very good success of this kind of idea in the AliEn framework, unfortunately this is not completely true. In fact, what AliEn does is to "glue" the pieces of information regarding the content of many sites into a somehow coherent interface, showing a common POSIX-like namespace. The questionable point of this is that it does it by using a centralised database, which grows very large and needs constant maintenance, although it is stable up to now. One question could come at this point: since distributed file systems (like e.g.. AFS [16] or LUSTRE [17]) are common, why cannot one build a worldwide file system using those? Doing this we could avoid all the problems dealing with keeping a database aligned with the reality, and probably avoid having a unique bottleneck/point of failure. The answer is very difficult, and justifying the fact that it is negative would need a technological insight which we will try to keep as short as possible in this work, preferring to give a greater emphasis to the choices made and which could be made in the future.

Hence, the case study which spurred the development of more advanced techniques for handling very large data repositories was spurred to a great extent by the technical evolution of the data storage of the ALICE experiment (proton-proton and lead nuclei collisions at the LHC), but great care was put in order to evolve the system as a generic set of packaged tools, together with new generic methods and ideas. Historically speaking, the other very important starting point has been the work done in order to refurbish the BaBar [21] computing model, after the awareness of the fact that using a central Database Management System containing all the data was not a very productive approach.

We also believe that the purpose of this kind of statement lies also in the bigger freedom that there is in making a generic system evolve through time, with respect to one which contains too many parts which are too tightly linked to the particular requirements of the moment. We also preferred to introduce the system by trying to informally describe the various issues of this kind of systems, also in order to emphasise the fact that such a big deployment has not to be considered as a static instance, but instead as a place where to constantly evolve the used technologies together with the evolution of the expectations.

10.3.1 The Storage Reference Model

The purpose of building data access systems in order to publish data is common to several application domains, and various kinds of approaches are possible to fulfil this objective. Such approaches are not equivalent, since they refer to very different requirements and collocations. But they share and give different proposals to the common problems of coherency in data replication, performance in data access, scalability of a complex system and robustness of a complex system in the case of faults.

Most of the work presented in literature concerning fault tolerant and fast data access only marginally deals with both communication robustness and highly available systems. An interesting effort in order to standardise the way storage and data access systems can be designed is represented by IEEE 1244 [22]. However, our point is that the sophisticated way in which they describe the components of a storage system might lead to sparse systems where it is mandatory to use some form of database technology just in order to glue together all the pieces of information distributed through the components. This could be considered a very controversial choice, and being absolutely forced, by design (and not by the true requirements), to adopt it, was not considered as a fair one.

One of the approaches that are closer to the one we consider is the one of the distributed file systems. Generally, the distributed file system paradigm is tied to policies dealing with distributed caching and coherency, path and filename semantics. The algorithms that handle such policies are known to cause network and CPU overhead when dealing with the operations on the files they manage [12, 13, 25]. Hence, there is a general consensus that the network overhead due to the synchronisation of the internal data structures in distributed file systems becomes a serious issue when dealing with PetaBytes of data continuously accessed by thousands of clients. This is common to organisations that rely on massive data sharing, and, generally, is independent from the software framework or field of study.

Other interesting works can be found in the area of design of peer-to-peer file sharing networks [4, 5, 24]. There is general consensus that the peer-to-peer paradigm is a good proposal for the solution of problems dealing with scalability and fault tolerance [2, 19]. Moreover, the illusion of having a unique huge repository obtained by aggregating a very large number of small nodes is very appealing from the point of view of the features that this approach provides.

10.4 XRootd and the Scalla Software Suite

The basic system component of the Scalla software suite is a high-performance data server akin to a file server called *xrootd*. However, unlike NFS, the xrootd protocol includes a number of features like:

- Communication optimisations, which allow clients to accelerate data transfers (e.g., overlapping requests, TCP/IP multistreaming, etc.).
- An extensive fault recovery protocol that allows data transfers to be restarted at an alternate server.
- A comprehensive authentication/authorisation framework.
- Peer-to-peer elements that allow xrootd's to cluster together while still providing a uniform name space.

xrootd's clustering is accomplished by the *cmsd* component of the system. This is a specialised server that can co-ordinate xrootd's activities and direct clients to appropriate servers in real-time. In essence, the system consists of a logical *data* network (i.e., the xrootd servers) coupled with a logical *control* network (i.e., the cmsd servers).

The control network is used to cluster servers while the data network is used to deliver actual data. We define a *node* as a server pairing of an xrootd with a cmsd.

A cmsd can assume multiple roles, depending on the nature of the task. In a manager role, the cmsd discovers the best server for a client file request and coordinates the organisation of a cluster. In a server role, the cmsd provides sufficient information to its manager cmsd so that it can properly select a data server for a client request.

Hence, a server cmsd is essentially an agent running on a data server node. In a supervisor role, the cmsd assumes the duties of both manager and server. As a manager, it allows server cmsd's to cluster around it, it aggregates the information provided by the server cmsd's and forwards the information to its manager cmsd. When a request is made by its manager, the supervisor cmsd determines which server cmsd will be used to satisfy the request. This role parcelling allows the formation of data server cells that cluster around a local supervisor, which, in turn, clusters around a manager cmsd.

We can now expand the definition of a node to encompass the role. A data server node consists of an xrootd coupled with a server cmsd, a supervisor node consists of an xrootd and a supervisor cmsd, and a manager node consists of an xrootd coupled with a manager cmsd. The term node is logical, since supervisors can execute on the same hardware used by data servers.

10.4.1 Cell-Based Organisation

To limit the amount of message traffic in the system, a cell consists of 1-to-64 server nodes. Cells then cluster, in groups of up to 64. Clusters can, in turn, form superclusters, as needed. As shown in Fig. 10.1, the system is organised as a B-64 tree, with a manager cmsd sitting at the root of the tree. Since supervisors also function as managers, the term manager should subsequently be assumed to include supervisors.

A hierarchical organisation provides a predictable message traffic pattern and is extremely well suited for conducting directed searches for file resources requested by a client. Additionally, its performance, induced by the number of collaborating nodes, scale very quickly with only a small increase in messaging overhead. For

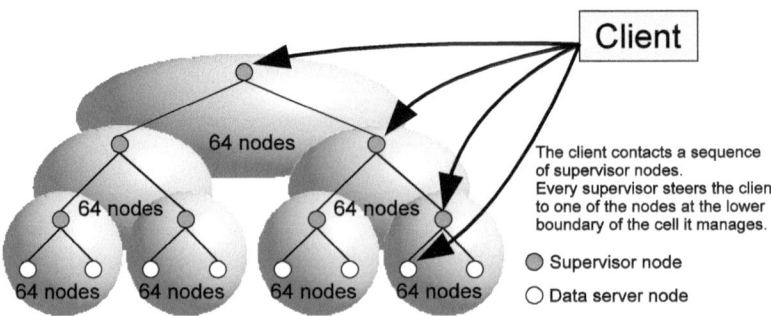

Fig. 10.1 Cell-like organisation of an xrootd cluster

instance, a two-level tree is able to cluster up to 262,144 data servers, with no data server being more than two hops away from the root node.

In order to provide enhanced fault-tolerance at the server side, manager nodes can be replicated. When nodes in the B-64 tree are replicated, the organisation becomes a directed acyclic graph and maintains predictable message traffic. Thus, when the system is configured with n managers, there are at least n control paths from a manager to a data server node, allowing for a significant number of node failures before any part of the system becomes unreachable. Replicated manager nodes can be used as fail-over nodes or be asked to load balance the control traffic. Data server nodes are never replicated. Instead, file resources may be replicated across multiple data servers to provide the desired level of fault-tolerance as well as overall data access performance.

The main reason why we do not use the term "File system" when referring to this kind of system is the fact that some requirements of file systems have been relaxed, in order to make it easier to head for extreme robustness, performance and scalability. The general idea was that, for example, it is much easier to avoid relying in the transactions' atomicity than being unable to make the system grow if its performance/size is not sufficient. For example, right now there was no need to dedicate a serious effort in order to allow for atomic transactions or distributed file locks. These functionalities, when related to extreme performance and scalability are still an open research problem, even more critical when dealing with high latency networks interconnecting nodes, clients and servers.

These kinds of relaxations were also related to the particular peer-to-peer like behaviour of the xrootd mechanism behind the file location functionalities. In the xrootd case, moreover, we can claim for instance that the fact that there is no database is not entirely true. In fact, the database used to locate files is constituted by the message-passing-based aggregation of several hierarchical databases, which are the file systems of all the aggregated disk partitions. Of course, these are extremely well optimised for their purpose and are up to date by definition; hence there is no need to replicate their content into a typically slow external database, just to locate files in sites or single servers.

10.4.2 Fault Tolerance

Some other aspects, which we consider as very important, are the ones related to robustness and fault tolerance. From one side, a system designer has to be free to decide what the system should do if, for instance, a disk breaks, and its content is lost. Given the relative frequency of this kind of hardware problems, the manual recovery approach should be reduced to a minimum or not needed at all. The typical solutions exploited in wiser HEP data management designs are typically two, i.e.:

- The disk pool is just a disk-based cache of an external tape system, hence, if a disk breaks, its content will be (typically automatically) re-staged from the external units. Of course, the system must be able to work correctly even during this process.

- The files are replicated somewhere else. Hence, some system could schedule the creation of a new replica of the lost ones. The point is that finding out which files were lost can be problematic.

The other aspect of fault tolerance is related to the client-server communication, typically based on TCP/IP connections between clients and servers, and, in the xrootd case, also between server nodes. The informal principles that we applied rigorously were that a client must never return an error if a connection breaks for any reason (network glitches, etc), unless it retried for a certain number of times and/or tried to find an alternate location where to continue. Moreover, the server can explicitly signal every potentially long operation, so that the client goes into a pause state from which it can be woken up later. This avoids having to deal too much with timeouts. To also reduce the overhead of creating an excessive number of TCP/IP connections, every inter-server connection is kept alive and eventually re-established if it seems to be broken.

10.4.3 Facing a Wide Area Network

The previous considerations about robustness acquire a stronger meaning when the TCP/IP connections are established through a Wide Area Network, which gives a higher probability of disconnections over long times, together with more difficulties in using efficiently the available bandwidth. From this perspective, we can say that our experience with the heavy usage of a "retry somewhere else and continue" mechanism has been extremely positive. A very welcome side effect of this is that typically no action is required to alert users, or pause any processing if one or more servers have to be upgraded or restarted.

Given the stability of the communication mechanism, one could argue that it could finally let an application access efficiently the data it needs even if the repository is very far, or, eventually through a fast ADSL connection from home. Unfortunately, for data intensive application, this is in general very difficult. The easy thing could be that, given enough bandwidth, WANs seem capable of delivering enough data to a running process. However, the achievable performance has historically been very low for applications in the High Energy Physics domain. This is generally due to the characteristics of WAN data streams, often called the *long, fat pipe problem* (high latency paid once per request, low achievable bandwidth per TCP/IP connection). This aspect also become more important when it comes to considering the usual structure of HEP data analysis applications, which, in their deeper simplification, consist in a loop of *get chunk – compute chunk* instructions.

In [10] the characteristics of WAN networks and of some HEP requirements are discussed, in order to introduce some methods to enhance their data throughput. Such techniques, schematised in Figs. 10.3 and 10.4 have been implemented in the Scalla/xrootd system during the various development cycles, and reached now a good maturity.

Typically, the data centres offering computing services for HEP experiments include computing facilities and storage facilities. In this environment, a simple application that has to analyse the content of some files typically will just open the files it has to access, then cycle through the stored data structures, performing calculations and internally updating the results, and output in some way the final results.

In this context, we assume that these data access phases are executed sequentially, as it is in most data analysis applications. These considerations refer to the concept of a file-based data store, which is a frequently used paradigm in HEP computing; however, the same performance issues can affect other data access methods. One of the key issues we tried to address is the fact that the computing phase of a HEP application is typically composed by a huge number of interactions with the data store. Hence, a computing application must deal with the fact that even a very short mean latency (e.g. 0.1 ms) would be multiplied by the large number of interactions (e.g. 10^8). This argument has historically been considered a serious issue which makes it impossible for data analysis applications to access remote repositories with a high degree of efficiency.

However, very often, this is not true. A trivial example of such a situation is when the application does not need to read the entire content of the files it opens [10]. A more complicated scenario involves complex data analysis applications which can predict in some way which data chunks they are going to access. If supported by adequate data access technologies, these applications can enhance their performance by up to two orders of magnitude, reaching levels comparable to those achievable through local access.

> If there is a way to predict (precisely or statistically) the data chunks needed, the latency problem can be circumvented efficiently.

In practise, this would mean that it could be possible, for example, to run a data analysis on a desktop or laptop PC, without having to gather all of its inputs before starting it. Let's see how this can be accomplished.

10.4.4 Read Requests and Latency

Figure 10.2 shows that, in a sequence of read requests from a client to a server, the transmission latency affects the requests twice, and is located before and after the server side computation. If we assume that the latency is due to the network, and that the repository is not local to the computing client, the value of the latency can be even greater than 70–80 ms. With a latency of 80 ms, an application in Padova, Italy, requesting data from a server at SLAC, located in California, USA, for example, will have to wait about 160 s just to issue 1,000 data requests requesting 1 Byte each.

Fig. 10.2 How the communication latency impacts data access

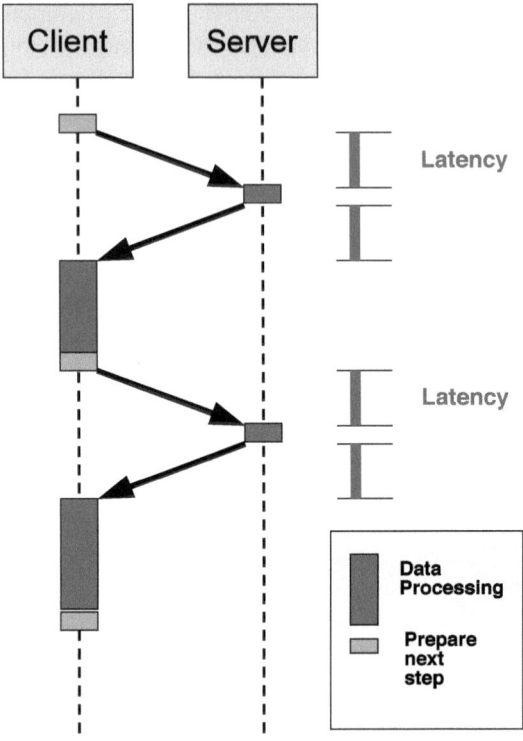

Fig. 10.3 Prefetching and outstanding data requests

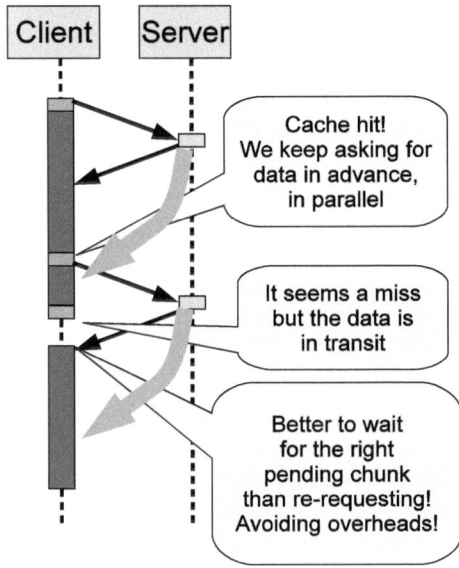

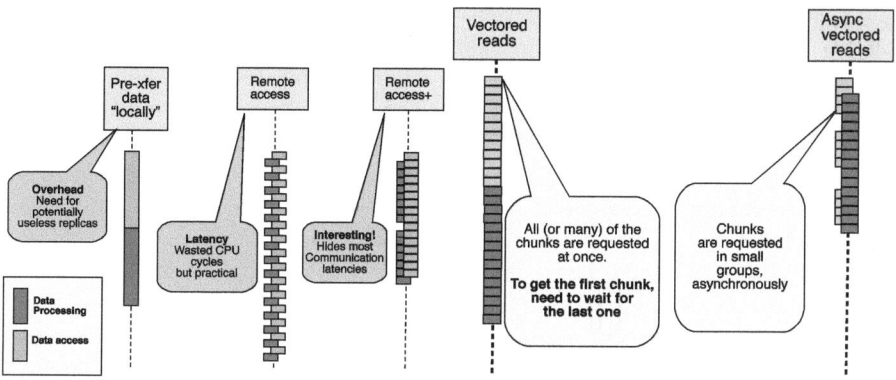

Fig. 10.4 Simple and advanced scenarios for remote data access technologies

However, the work done at INFN Padova, SLAC and CERN addresses the problem in a more sophisticated way. If we can get some knowledge about the pattern (or the full sequence) of the data request issued by the application, the client side communication library can, in general, request in advance the data for the future requests, and store portions of it in memory as it arrives, while other portions of it are still "travelling" in the network. Moreover, issuing the data requests in parallel, with respect to the present ones, has the advantage of having a very high probability of outstanding data containing responses of future data reads.

10.4.5 Read Requests and Throughput

The other parameter that is very important when dealing with data access is the data rate at which the computing process requests new data chunks to process, and its comparison with the data throughput that the data network can sustain. One of the characteristics which make WANs difficult to deal with in the case of data access is the fact that, even if the throughput of the network could be sufficient to feed the data processing, a single TCP/IP stream through a WAN is typically able to transfer data only at a fraction of the available bandwidth. Moreover, whenever the TCP/IP stack supports advanced features like the "windows scaling", this typically translates into a so-called "ramp-up phase", during which the achievable bandwidth is not very high because the communicating machines are adjusting the internal TCP/IP parameters. This can be related to the characteristics of the commonly used TCP/IP stacks, and also to the common configurations of the network devices, and led to the development of high performance file copy tools, such as BBCP, GRIDFTP, FDT and XRDCP, able to transfer data files at high speed by using a large number of parallel TCP/IP connections.

The approach of the Scalla/xrootd system with respect to this aspect is to make it possible to apply a similar mechanism to the transfer of a sequence of analysis application-requested data chunks. In the default case, the regular TCP/IP window scaling (if present in the operating system) will be used. The main difference from a copy-like application is that a data analysis application generates a stream of requests for data chunks whose sizes are orders of magnitude smaller (some kiloBytes versus up to several MegaBytes) than the chunk sizes used for copy-like applications, which typically are all equal and quite large. This is supposed to have some impact on the achievable data rate in the case of a data analysis application. Moreover, the fact that the data blocks are in general small and have a very variable size, adds a layer of complexity to the problem of correctly optimising their flow.

In general, a desirable feature would be the ability of the communication library to split a request for a big data chunk into smaller chunks, if it has to be distributed through multiple parallel TCP/IP streams, and then correctly handle a stream of multiple overlapping requests for small chunks, in order to achieve a higher efficiency in the WAN data transfer, even through a single TCP/IP stream. Even more important, all this complexity should be completely transparent to the application.

10.4.6 Current Status

To lower the total impact of latency on the data access, in the client side communication library (XrdClient or TXNetFile) [7, 8, 10, 11, 23, 26], the following techniques are exploited:

- A memory cache of the read data blocks.
- The client is able to issue asynchronous requests for data blocks, which will be carried out in parallel, and with no need for their responses to be serialised.
- The client exposes an interface that allows the application to inform it about its future data requests.
- The client keeps track of the outstanding requests, in order not to request twice the data chunks which are in transit and to make the application wait only if it tries to access data which is currently outstanding.

Figure 10.3 shows a simple example of how such a mechanism works. The client side communication library can speculate about the future data needs, and prefetch some data in advance at any moment. Alternatively, the client API can be used by external mechanisms (like *TTreeCache* in the ROOT framework [3, 20]) to inform it about the sequence of the future data chunks to request.

This kind of informed prefetching mechanism drastically optimises the flow of the requests which have to be forwarded to the server, thus *hiding* the overall data transfer latency, and is used as a buffer to coordinate the decisions about the outstanding data chunks. Moreover, for the intrinsically asynchronous architecture involved, the client side communication library is able to receive the data while the application code performs its processing on the previous chunks, thus reducing the impact of the data access on the data processing. For more information about caching and prefetching, the reader is encouraged to see [2, 18].

Figure 10.4 shows some typical scenarios for remote data access, sorted in the way they have (or will be) implemented and evaluated. The most obvious one is the first one, where the needed files are transferred locally before starting the computation. This solution, much used for historical reasons, forces the application to wait for the successful completion of the data transfer before being able to start the computation, even if the accessed data volume is a fraction of the transferred one.

The second scenario refers to the application paying for the network latency for every data request. This makes the computation extremely inefficient, as discussed.

The third one, instead, shows that, if the data caching/prefetching is able to keep the cache miss ratio at a reasonably low level (or the application is able to inform the communication library about the future data requests), the achievable result is to keep the data transfer going in parallel with the computation, but a little in advance with respect to the actual data needs. This method is more efficient than the ones discussed previously, and in principle could represent a desirable solution to the problem. However, if the data demanding application generates requests for very small data chunks, the overhead of transferring and keeping track of a large number of small outstanding chunks could have a measurable impact. The challenge is to keep this impact as low as possible.

If the client side concern moves to the fact that individually transferring small chunks (although in a parallel fashion) can degrade the data transfer efficiency, one obvious solution is to collect a number of data requests and issue a single request containing all of them. This kind of policy, known as vectored reads, takes into account the fact that a unique chunk containing all the requested ones constitutes the response itself. This kind of solution can result in an advantage over the policy of transferring individual small chunks. On the other hand, it has the disadvantage of requesting both client, network and server to serialise the composite data request, in order to manage its response as a unique data block to transfer. This means that the application has to wait for the last sub-chunk to arrive before being able to process the first one. Moreover, the server does not send the first sub-chunk until it has finished reading them all. These aspect, in practise, make this solution less appealing and performant. An interesting idea, still to evaluate in practise, is to try to merge the advantages of individual asynchronous chunk transfers with the vectored data transfers, leading to the last scenario visible in Fig. 10.4.

The basic idea shown is to request data transfers through vectored reads which contain enough sub-chunks to be able to reduce the protocol overhead linked to a client/server request, generate responses whose data chunk is sufficiently large to be efficiently transferred through the network pipe, and are small enough to avoid an excessive serialisation of the requests sequence.

The result would be that the client application reads the data chunks it needs in small bursts, spread through the time line of the application, thus giving a more uniform load profile on the involved data server, and reducing the negative impact of the chunks' serialisation in the communication pipe.

10.5 The Virtual Mass Storage System

The set of Wide Area Network related features of the Scalla/Xrootd system, as described, plays a major role in giving ways to make a storage system evolve in the direction of being able to exploit Wide Area Networks in a productive way. The major design challenge is how to exploit them in order to augment the overall robustness and usefulness of a large distributed deployment.

Nevertheless, coming back to our real-life example, in this perspective we must consider the fact that the ALICE computing model [1, 6] refers to a production environment, which is used heavily and continuously. Any radical change, hence, was not to be considered a good idea (given also the high complexity of the whole ALICE software infrastructure). Moreover, the idea of accessing remote data from data analysis jobs was not new in the ALICE computing framework. For instance, this is the default way that the analysis jobs have to access the so-called "conditions data", which are stored in a Scalla/Xrootd-based storage cluster at CERN. Each analysis job accesses 100–150 conditions data files (and $\geq 20,000$ concurrent jobs are considered a normal activity): historically the WAN choice always proved itself a very good one, with a surprisingly good performance for that task, and a negligible failure rate.

On the other hand, the AliEn software [1], supported with its central services, already gives some sort of unique view of the entire repository, faking the namespace coherence across different sites by mapping the file names using a central relational database. The way it is used also confirms the fact that it would be desirable to move towards a model where the storage coherence is given by the storage system itself and does not need computationally very heavy namespace translations in a centralised catalogue. One more very important consideration comes also from the fact that in the ALICE computing model any storage cluster can be populated only by writing directly to it from an external machine, where a software takes responsibility of the transfer or writing to the backend store (e.g. a tape system behind the storage).

For instance, in the ALICE computing model these tasks are accomplished by a daemon called FTD (File Transfer Daemon), which runs in every site. Although functional and relatively good performance, there are some issues which are inherent in this kind of design, and which we can shortly summarise as follows:

- All the data transfer load hits the same machine where FTD runs for the site, since what it does is, essentially, to start processes copying data from one site to another.
- Several concurrent transfers can consume a high amount of system resources.
- All the external traffic of the site goes through one machine, which by construction is in the site which is the destination of the data move.
- If a file is missing or lost for any reason (e.g. broken disks, errors in the central catalogue, etc...):

    All the processing jobs scheduled to run on that file instance will inevitably fail, and the trouble can be solved only with a careful human intervention.

With an order of 10^7 online files per site, finding the source of such a file miss can be extremely problematic.

An additional consideration is that, from a functional point of view, we considered as a poor solution the fact that, once a missing file is detected, nothing is done by the storage system to fetch it. This is especially controversial when, for instance, the file that is missing in a site is available in a well-connected one, and pulling it immediately (and *fast*) could be done without interrupting the requesting job.

A careful consideration of all these aspects led to thinking that there was a way to modify the ALICE Scalla/Xrootd-based distributed storage in a way which automatically tries to fix this kind of problems, and at the same time does not disrupt the existing design, allowing for an incremental evolution of the storage design, distributed across many almost independent sites contributing to the project.

The generic solution to the problem has been to apply the following statements. First, all of the xrootd-based storage elements are aggregated into a unique worldwide meta-cluster, which exposes a unique and namespace-coherent view of their content. Second, each site that does not have tapes or similar mass storage systems considers the meta-cluster (except itself) as its mass storage system. If an authorised request for a missing file comes to a storage cluster, then it tries to fetch it from one of the neighbour ones, as soon and as fast as possible.

The host which manages the meta-cluster (which in the Scalla/Xrootd terminology is called *meta-manager*) has been called *ALICE Global redirector*, and the idea for which a site considers the global namespace (to which it participates with its content) as its mass storage has been called *Virtual Mass Storage System* (VMSS).

In this context, we consider as very positive statements the facts that in the xrootd architecture, locating a file is a very fast task (if the location is unknown to a manager server, finding it takes, in average, only a network round-tripthe path of a message from a sender to a receiver and then back to the sender over the network time plus, eventually, one local file system lookup). Also, no central repositories of data or meta-data are needed for the mechanism to work, everything is performed through real-time message exchanges which at the end, can trigger direct actions in the involved file systems. Moreover, all of the features of the Scalla/Xrootd suite are preserved and enhanced (e.g. scalability, performance, etc...). Also, this design is completely backward compatible, e.g. in the ALICE case no changes were required to the complex software which constitutes the computing infrastructure.

Hence, the system, as designed, represents a way to deploy and exercise a global real-time view of a unique multi-PetaBytes distributed repository in an incremental way. This means that it is able to give a better service as more and more sites upgrade their storage clusters and join in.

Figure 10.5 shows a generic small schema of the discussed architecture. In the lower part of the figure we can see three cubes, representing sites with a Scalla/Xrootd storage cluster. If a client instantiated by an analysis job (on the leftmost site) does not find the file it needs to open, then the GSI cluster asks the global redirector to get the file, just requesting to copy it from there. This copy request (which is just another normal xrootd client) is then redirected to the site

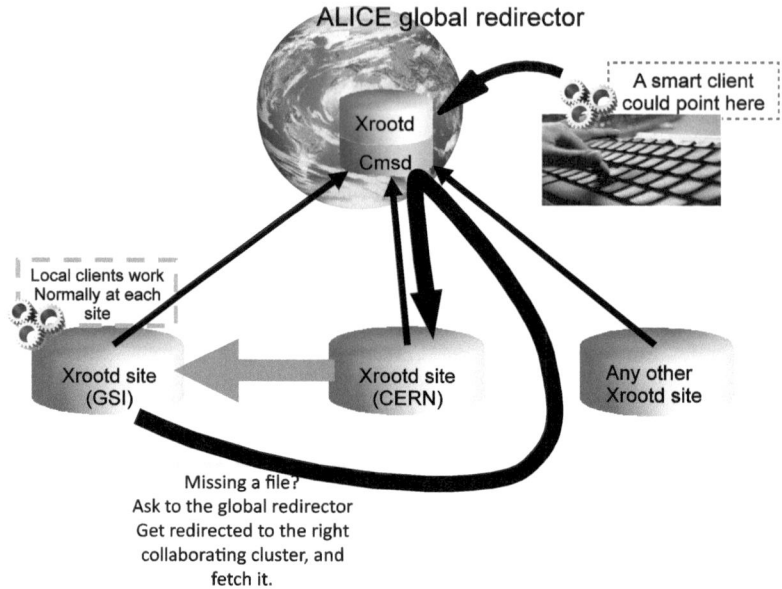

Fig. 10.5 An exemplification of the virtual mass storage system

that has the needed file, which is copied immediately to the leftmost site. The previously paused job can then continue its processing like if nothing particular had happened.

Hence, we could say that the general idea about the Virtual Mass Storage is that several distant sites provide a unique high performance file system-like data access structure, thus constituting some sort of cloud of sites which collaborate. The outcome of this collaboration is that some recovery actions are made automatically and that we are able to give a performant way to move data between sites if useful for the overall design.

10.6 Conclusions and Lessons Learned

The most obvious conclusion which we would like to emphasise is that there is not yet one, in the sense that this was (and is) an ongoing attempt to evolve some technologies in what we believe is a desirable direction. In the meantime, while the requirements and the technologies become more clear, we can claim that the reached milestones are very satisfying. The fact itself that such technologies are being exercised in a non trivial production environment is a strong confirmation of the goodness of several ideas, and a starting point to make them better, together with the tools based on them.

Some more aspects which are very important and quite controversial, however, deal with software quality and with where it is used. This means that it is our commitment to strive for the best quality of the tools we provide, however, we completely realise that a sophisticated storage model for data access can be better exploited by a data management strategy based on compatible principles. The relative ease of integration with the ALICE computing framework, for example, was also due to the fact that similar ideas were at least considered at the time of its design, even if there was no clarity about how to get to those results. Hence, from our perspective, the process is an ongoing one.

From the technological side, one of the upcoming initiatives will be a performance evaluation of the maturity of the described techniques in a controlled Wide Area Network environment. The point with this is that, in any benchmark involving public WANs, it is very difficult to argue if a given load pattern is due to a network performance variation or to the software which could behave better. After countless tests between different sites, the hope is to understand if there is space for more performance improvements at the client side. Nevertheless, a side effect of this would be to be able to personally and informally evaluate at which extent these technologies can be used to support interactive forms of data analysis in the years to come.

> Performance evaluations are very difficult to do in Wide Area Networks. These future efforts will help us in optimising the system.

One more field suitable for improvements could lie in the applications or the analysis framework which ties them to the data (which for High Energy Physics is very often ROOT). In this side we could even now wonder better algorithms to feed the latency hiding algorithms of the xrootd client. Definitely, this would increase the data access efficiency.

> It seems possible to solve the problem to choose the best site to fetch a replica from.

Other aspects worth facing are those related to the so-called "storage globalisation". For instance, a working implementation of some form of latency-aware load balancing would be a very interesting feature for data analysis applications in the need to access remote files efficiently. At the same time, also, it could become possible to design a file copy tool able to read from multiple sources the chunks it needs, in order to maximise the throughput and, even more important, dynamically adapt to variations in the available bandwidth in the connections with the remote sites it fetches data from.

References

1. ALICE: Technical Design Report of the Computing. June 2005. ISBN 92-9083-247-9. http://aliceinfo.cern.ch/Collaboration/Documents/TDR/Computing.html
2. Anderson, T.E., Dahlin, M., Neefe, J.M., Patterson, D.A., Rosselli, D.S., Wang, R.Y.: Serverless Network File Systems. ACM Trans. Comput. Syst. **1**, (1996)
3. Ballintijn, M., Brun, R., Rademakers, F., Roland, G.: Distributed Parallel Analysis Framework with PROOF. http://root.cern.ch/twiki/bin/view/ROOT/PROOF
4. Bar-Noy, A., Ho, C.-T.: Broadcasting Multiple Messages in the Multiport Model. IEEE Trans. Parallel Distr. Syst., Sept, 1988
5. Bar-Noy, A., Freund, A., (Seffi) Naor, J.: On-line Load Balancing in a Hierarchical Server Topology. In: Proceedings of the 7th Annual European Symposium on Algorithms, 1999
6. Betev, L., Carminati, F., Furano, F., Grigoras, C., Saiz, P.: The ALICE computing model: an overview. Third International Conference "Distributed Computing and Grid-technologies in Science and Education", GRID2008, http://grid2008.jinr.ru/
7. Dorigo, A., Elmer, P., Furano, F., Hanushevsky, A.: Xrootd - A highly scalable architecture for data access. WSEAS Trans. Comput., Apr. (2005)
8. Dorigo, A., Elmer, P., Furano, F., Hanushevsky, A.: XROOTD/TXNetFile: a highly scalable architecture for data access in the ROOT environment. Proceedings of the 4th WSEAS international Conference on Telecommunications and informatics (Prague, Czech Republic, March 13 - 15, 2005). Husak, M., Mastorakis, N. (eds.) World Scientific and Engineering Academy and Society (WSEAS), Stevens Point, Wisconsin, 1-6
9. Feichtinger, D., Peters, A.J.: Authorization of Data Access in Distributed Storage Systems. The 6th IEEE/ACM International Workshop on Grid Computing, 2005. http://ieeexplore.ieee.org/iel5/10354/32950/01542739.pdf?arnumber=1542739
10. Furano, F., Hanushevsky, A.: Data access performance through parallelization and vectored access. Some results. CHEP07: Computing for High Energy Physics. J. Phys. Conf. 119 **119** (2008) 072016 (9pp)
11. Furano, F., Hanushevsky, A.: Managing commitments in a Multi Agent System using Passive Bids. iat, 2005 IEEE/WIC/ACM International Conference on Intelligent Agent Technology (IAT'05), pp. 698-701, 2005
12. Garcia, F., Calderon, A., Carretero, J., Perez, J.M., Fernandez, J.: A Parallel and Fault Tolerant File System based on NFS Servers" Euro-PDP'03, IEEE Computer Society, 2003
13. Ghemawat, S., Gobioff, H., Leung, S.-T.: The Google file system. In: Proceedings of the nineteenth ACM symposium on Operating systems principles, Oct, ACM press, 2003
14. Hanushevsky, A., Weeks, B.: Designing high performance data access systems: invited talk abstract. In: Proceedings of the 5th international Workshop on Software and Performance (Palma, Illes Balears, Spain, July 12 - 14, 2005). WOSP '05. ACM, New York, NY, 267-267. DOI= http://doi.acm.org/10.1145/1071021.1071053
15. Hanushevsky, A.: Are SE architectures ready for LHC? In: Proceedings of ACAT 2008: XII International Workshop on Advanced Computing and Analysis Techniques in Physics Research. http://acat2008.cern.ch/
16. Howard, J.H. et al.: Scale and performance in a distributed file system. ACM Trans. Comput. Syst. **6** (1988)
17. Lustre: a scalable, secure, robust, highly-available cluster file system. It is designed, developed and maintained by Sun Microsystems, Inc. http://www.lustre.org/
18. Patterson, R.H., Gibson, G.A., Ginting, E., Stodolsky, D., Zelenka, J.: Informed prefetching and caching. In: Proceedings of the 15th ACM Symposium on Operating Systems Principles, 1995
19. Ripeanu, M., Foster, I., Iamnitchi, A.: Mapping the Gnutella Network: Properties of Large-Scale Peer-to-Peer Systems and Implications for System Design. IEEE Internet Comput. J. **1**(Jan-Feb), Springer-Verlag (2002)
20. ROOT: An Object-Oriented Data Analysis Framework http://root.cern.ch

21. The babar collaboration home page. http://www.slac.stanford.edu/BFROOT
22. The IEEE Computer Society's Storage System Standards Working Group. http://ssswg.org/
23. The Scalla/xrootd Software Suite. http://savannah.cern.ch/projects/xrootd and http://xrootd.slac.stanford.edu/
24. Waterhouse, S.R., Doolin, D.M., Kan, G., Faybishenko, Y.: Distributed Search in P2P Networks. IEEE Internet Comput. J. **6** (2002)
25. Whitney, V.: Comparing Different File Systems' NFS Performance. A cluster File System and a couple of NAS Servers thrown in. The sixth SCICOMP Meeting, SCICOMP 6 (Univ. of Berkeley), IBM System Scientific Computing User Group, http://pdsf.nersc.gov/talks/talks.html, 2002
26. XRootd explained. Computing seminar at CERN, http://indico.cern.ch/conferenceDisplay.py?confId=38458

Chapter 11
The Planetary Brain

Giuliana Galli Carminati

Coming at the end of this book dedicated to Computing in High Energy Physics, this last chapter may seem disconnected from the main subject. Indeed it cannot be denied that there is a difference in style and focus from the preceding material, which may arise from my professional experience, which is different from that of the other authors.

The project for this book did, in fact, start with this chapter, and this is why we have found it more appropriate to place it at the end. While this subject is not more important than the others, the wish, or need, to write about the development of computing at CERN started from a reflection on what we called "the planetary brain". As often happens, this reflection was born of discussions that took place in the CERN Cafeteria and from some evenings spent in restaurants in the nearby village of Meyrin on the outskirts of Geneva, where CERN is officially located.

The evolution from a physicist programming solely for himself, to a group of physicists (and computer scientists) working for a group of physicists (and where the two groups overlapping), has seen a "jump" in complexity with the Internet and the links between the Web and the Grid.

The simulation, reconstruction and analysis tools of HEP may now be used, by design, from any point on Earth where there is (as a minimum) an Internet connection. This is the essence of the Grid project.

And it should not be forgotten, of course, that it was at CERN, with its large population of physicists, that the Web was invented.

> High Energy Physics computing has gone through a quantum jump in complexity, gaining a planetary dimension.

G.G. Carminati (✉)
UPDM-HUG, Geneva, Switzerland
e-mail: Giuliana.GalliCarminati@hcuge.ch

Over a drink in the CERN Cafeteria, we reflected that the emergence in the early eighties of planetary thinking with an associated language, based on a material layer (the Internet) and software (HTML, middleware) may be compared with the development of the human mind with its "physical" brain and specific learning systems.

Just as a child cannot develop alone, because its development is "nourished" by the links with the external and internal world (the so-called "real" world and the world of collective social archetypes) and by its interaction with them, in the same way the Web and Grid develop via their interaction with the reality in which they are immersed and to which they connect.

What seems to emerge clearly, even if we are not aware of any specific study on the subject, in the Web and Grid world, is one more example of the superiority of the "interactive" or "evolutionary" development model relative to "planned" development.

The creation of clear and rational projects for computing structures, promoted by classical software engineering, which would seem, a priori, the best way to produce predictable results, lacks the necessary flexibility to adapt to the requirements for the evolution of an experiment. The very concept of a clear and rational project seems to go against the needs for development. This is similar to the evolution of the "constraint neurosis" that gets "encysted" into obsession.

This remark drives proponents of the "cathedral" projects to despair. It would have been necessary to take into account the fact that the consciousness (of which rationality is one part), remains a small part of the archetypal and social "unconscious magma", which has had a large part in driving the evolution of humanity.

> The Web and the Grid will develop under the influence of the "good" or "bad" education provided to them by reality and the archetypal "magma".

The Web and the Grid will therefore develop under the influence of the "good" or "bad" education provided to them by reality and the archetypal "magma". We firmly believe that this is true for the "purely computational" Grid, despite the fact that in dealing with its scientific language it also draws us into concepts of rational consciousness.

11.1 Introduction

We believe that the concept of language brings with it, irrespective of the kind of language itself, a degree, large or small, of ambiguity. The point here is that language, be it natural or computational, is the link through which reality, and also our archetypal magma, enters into contact with the "other", a developing human being or a planetary brain.

The development of the Web (and of the Grid) depends upon the level of ambiguity of the languages used, which are the transport layers of the real but also of the unconscious environment. Too much rigidity or too much "uncertainty" in these languages could cause "pathologies" in the developing "planetary brain".

What is certain is that we are in uncharted territory, and importantly, as is in raising a child, we need a bit of luck to guide us.

- *Can we describe our brain as a "thinking machine"?*
- *Can a theory of the brain as a thinking machine explain the evolution of global connectivity?*
- *Are thoughts physical objects? Are thoughts matter?*
- *Are the elements of the WWW and the Grid and their interconnections evolving toward a planetary brain?*

Here we would like to take you on an excursion into a field of knowledge that is still to be fully discovered and which may be excessively vast. If you are lost (as indeed the authors may be, at some points) we wish to reassure you that we are definitely in the domain of hypothesis and you can cast a polite but sceptical glance on the entire chapter without offending us.

In any case, the Ariadne's thread to be delicately unravelled during this discussion may be described as follows:

The brain is a thinking machine but also a physical object, producing thoughts that have both a material and spiritual nature.

The galaxies and the brain both consist of the same type of matter. The human brain, a last minute detail in the huge design of Nature, produces reasoning that can observe, understand and explain (more or less) the movements of galaxies.

The brain uses language to develop itself (babies need speech to learn to speak and thinking to learn to think). Languages possess a double nature, similar to the brain, oscillating between clarity and ambiguity just as the brain's functions oscillate between cognition and emotion.

The Web is a synthesis of "clear" language (mathematics) and "ambiguous" language (slang), and here again we find the ambiguity of Nature (as in language and in the brain). In fact the binary structure of the computing programs, constituting and working on the Web is exquisitely mathematical, and the blogs are of an exquisite slang-like construction, often disobeying the grammatical rules of any language.

We may find this duplicity, or rather, this spectrum from cognitive to affective functions, in cognitive-affective psychological functions.

If we accept the idea that thoughts are physical objects, we can imagine using a model to describe them. Physics offers a rich set of models, and quantum mechanics in particular seems an appropriate starting point for a model describing our thoughts. The interest in using quantum mechanics is that we can exploit the state superposition typical of quantum mechanics, and also that we can describe the phenomena of synchronicity that are at the core of the interaction between matter and spirit.

> If we accept the idea that thoughts are physical objects, we can imagine using a model to describe them.

Another interesting point when using a physical model is that group situations may be expressed simply enough as not just individual interactions if we consider the group as a specific entity. The Web may be considered as a group entity, reproducing the functions, dreams, wishes, "evils, devils and other demons" of our "thinking machine" and, at the same time, sharing the dual nature of the brain (hardware and software being likened to neurons and thoughts).

11.2 Can We Describe Our Brain As a "Thinking Machine"?

Functionally, our brain is a connection machine, moving information from one neuron to another, and from one group of neurons to another group. In animals, and also in human babies, the brain develops from software to hardware, building itself through inner and outer stimuli, generated by internal and external environments [1–5]. The human baby evolves as a social animal, needing to communicate with other human, and even non-human, beings [6].

The development of language is the development of a communication tool, to communicate with other humans (and we suspect, but we have no proof of this at the moment, also to communicate with the "physical world") with different degrees of ambiguity, depending on the needs and emotions of the one who communicates [7, 8].

In general, human beings think of their own nature as completely different from the matter that surrounds them: believing in a fundamental difference between body and soul. The fact that the Universe and humans are made of the same atoms – even if in different proportions – is striking in its simplicity. If this similarity does not seem that striking, it is because of our incredible human *nonchalance*, or because of our incredible boldness.

In the fourth and third centuries B.C. a never-ending argument began between Plato and Aristotle, the champions of two different conceptions of the relationship between ideals and reality. Plato said that Ideas were the generators of our thinking and the Aristotle believed that our thoughts came from the observation of Reality. Plato was a deductive thinker and Aristotle an inductive one: ideas generated knowledge for Plato, reality generated knowledge for Aristotle. Knowledge was of a distinct nature from reality for Plato. For Aristotle knowledge had the same nature as reality and the two were indistinguishable.

In fact, the journey from Reality or Ideas to our mind is far from being clear because, being human "inside" ourselves, Reality and Ideas (using the word in the platonic sense, and endowing "Ideas" with a capital letter) are both remote from us and from our ideas (our internal "ideas", differentiated with a lower-case initial letter) and our reality (again, our everyday "reality"). In other words, it is difficult for us to be at the same time the observer and the observed: the reality observed by the Ideas and the ideas observed by Reality.

To further complicate the situation, there is the fact that our ideas (in the sense of our thoughts) originate from a soft grey-and-white brain (and the functions of the grey and white regions are different), full of cells and other material elements. In other words, the ideas come from a material object.

So, in going from Ideas through ideas, passing through the brain to reality and finally arriving at Reality, we are probably surfing around the Great Unification in physics with an endless movement back and forth between philosophical and physiological concepts.

The situation becomes more complicated if we consider that the brain of the human baby[1] is formed by the action of both the inner and outer environments.

To answer the question as to whether the brain is a thinking machine, the central point is probably more dependent on our concept of "machine" than the concept of "thinking". In fact we have a lower conception of machines because we created them, in comparison to our almost mystical concept of the brain, created by Nature, and to an even more mythical conception of soul, created by God (so it is said). It seems inappropriate to diminish the dignity of our brain by giving it the same nature as a machine.

> To answer the question as to whether the brain is a thinking machine, the central point is probably more dependent on our concept of "machine" than the concept of "thinking".

Here we begin a short digression to underline once again that the brain indeed "is" a physical object, made up of cells that consist of molecules, which consist of atoms, and so on. This is simple to understand but, as often happens, difficult to accept. Humanity often considers what is "spiritual" to be beyond that which is material, instead of thinking of spiritual and material as partaking of the same ontology.

There is, between the Ideas of Plato and the Reality of Aristotle, a difference in the ontology of the conceptions or ideas: they were external to the sensible world for Plato, and internal to the sensible world and inseparable from it for Aristotle. Dividing spirit and matter is felt to be a necessity: it might be a good exercise to

---

[1] But then, why not also, of (say) Fanny the beagle puppy? Probably less so in the case of one of Babylone's four kittens, because of the different socialisation of cats who, unlike dogs, hunt alone.

analyse different or similar functions, but it is necessary, then, to re-synthesise the two concepts with their difference and similarities.

Indeed, during a long period of our human history, we have seen some very crude machines that were useful, but capable only of accomplishing a few tasks, and very far from the complexity of our brain.

In the second half of the 1900s we were fascinated by space travel, and the emotional thrust of science was exemplified by the conquest of space. The fact that the conquest of space ceased when its funds were cut, may have been not only a question of money, but also primarily a question of changing international politics, driven by a change in public opinion (and public emotions).

> Around 1960, information technology was an important field, but, emotionally and socially, not as central. Often real innovation comes from the "periphery of the (knowledge) galaxy".

At that time, around 1960, information technology was an important field, but, emotionally and socially, not as central. It is quite interesting to note that, often, real innovation comes from the "periphery of the (knowledge) galaxy".

But, and this was unexpected (because it was really a "bolt out of the blue"), along came the computer science revolution. In fact, most of the activity on the Web was (and still is) something nearer to a form of entertainment than to serious science, and the revolution it brought about was more the result of leisure and creativity than of any kind of master plan.

Strictly speaking, the beginning of computing with Turing, Zuse and Von Neumann was a very serious business: the practical purpose of the computer was to compute, in other words, to create a powerful tool for calculation. Sadly, in fact it was mainly used to build bombs.

The primary goal for the computer was therefore a very practical application. We may say that, analogously, the brain in animals has a "practical" goal: to organise life processes, such as breathing, thermoregulation etc., etc.

In comparison, what happened at CERN in the 1980s with the Web seemed initially more like a multidimensional game of chess for researchers in identity crisis and for hackers in self-indulgent experimentation. The young Web was quick enough to avoid an early death from becoming a committee-driven activity (still the best way of shelving a project), and also because it was far enough from the mainstream of officially supported programs.

A lot of the early work on the Web was a borderline, "just-for-fun activity", a pastime, rather than a serious technical and philosophical construct. We may note here that the first picture posted on the Web was the one of the "Cernettes" a sexy group portrait taken with a film camera [9].

11.3 Can a Theory of the Brain As a Thinking Machine Explain the Evolution of Global Connectivity?

It seems almost certain that language operates on thoughts in the sense that the thinking functions are developed by the exercise of language, and language is not only the expression of words but also the expression of numbers, or more generally, of symbols [10, 11].

We are not discussing here the difference between spoken and written symbols (we will come back to this difference later) but the difference between different kinds of languages adapted to different subjects.

In poetry, and even more in theatre, the beauty of the text arises from the ambiguity of meaning (polysemy): we can hear and understand (which is not the same) a text in several ways, because of the superposition of different meanings of words, verbs and parts of phrases; and because of the composition of the different phrases, all influenced by the interpretation of the reader or the actor.

We need to introduce here another short aside, to explain why the Grammelot,[2] slang or argot may be considered more ambiguous than language used in written work. The answer is: "because they are not used in written work". Dialect and vernacular language – even if in the great majority of cases they are spoken – may also be used in poems or prose as in the works of Carlo Goldoni and Trilussa (Carlo Alberto Salustri) or François Villon. To follow the path from language to written communication, we might watch Fellini's Satyricon. Just listening to the sound of Latin turning into the Roman dialect and then into Italian would be an easy and pleasant way to learn philology!

In novels, words are used more precisely, because the differences in meaning come from the narrative situations, from the description of the landscape, or from nuances in the various characters.

In a scientific paper a less ambiguous communication is in order, because the goal is to express a point of view in a precise manner. We may again digress here for the benefit of readers who work in the domain of sciences and publish this kind of papers (following the need to "publish or perish"). These readers are now almost surely smiling, thinking of their last paper in which they tried to write something that was simultaneously statistically accurate, politically correct, that avoid conflicts of interest, respect the privacy of patients in case of medical papers, obeyed the laws

---

[2]Grammelot is a style of language used in satirical theatre. A gibberish with onomatopoeic elements, it has been widely popularised by the Nobel-winning Italian playwright Dario Fo, who used it in his satirical monologue play Mistero Buffo ("Comic Mystery") [12]. Fo, in his Nobel lecture, traces the origin of this language back to the sixteenth century Italian Commedia dell'Arte play-writer and actor Ruzzante, and he says it was a sort of *Lingua Franca* aimed at protecting the actors when playing satirical texts. French scholar C. Duneton [13] says that the world Grammelot itself has been introduced by the Commedia dell'arte-derived French theatre of the early part of the twentieth century.

on the ethical treatment of lab animals, and increased their chances of seeing their grants renewed.

> Often the level of perceived (and accepted) ambiguity in a field is directly proportional to the competence in the field.

By way of bringing this digression to a close, we may note that often the level of perceived (and accepted) ambiguity in a field is directly proportional to the competence in the field. For example biologists think that physics is an exact science and physicists think that biologists and chemists can exactly understand complex phenomena: the Truth is assumed to be elsewhere, in either case.

Today, physics and mathematics are probably the heirs of Aristotle and Plato, and are not really arguing about the primacy of Reality versus Ideas, nor its opposite, but oscillating between reality, explained by adequate formalism, and formalism searching for an experimental basis to be confirmed, or falsified, whichever the case may be.

Galaxies and quarks are real objects, understandable with theoretical tools such as the Cosmic Gaulois Group or Non-Commutative Geometry. Should we admit that Matter speaks the same language as us? Might we assume that the Universe and ourselves consist of the same matter and (who knows?) the same thoughts? [14, 15].

Taking the language of physics into account, we have to admit that it can present some ambiguity: for example in re-normalisation theories we use mathematical tools even if there is not always a solid mathematical consistency or proof. We may say that the back-and-forth process occurring between mathematics and physics may also be found between biology and physics: both reciprocally confident that truth is in the "other" field. The perfect certainty of mathematical language is a pious wish to use a language with the greatest possible universality. We do not want to imply in any way that mathematical language is inefficient but, on the contrary, to try to explain the unique nature of different languages and their common root.

The Web offers a synthesis of mathematical language (a programming language is a mathematical binary code interpreted by a precise algorithm), with other more ambiguous and intimate languages, such as literature and – at the other end of the spectrum – argot or a blog. To clarify the term "intimate" we emphasise that the intimacy of a language is linked to the expression of personal feeling without (or at least with no compulsion to involve) any literary filter. The "Last letters of Iacopo Ortis" by Ugo Foscolo describe the deeply intimate nature of the central character (and probably of the author), but there is a technical, literary filter that adapts the way in which this is expressed for a literary public.

If you use a blog on the Web – which, in my opinion, is a parallel to using slang or argot – you can express your intimate opinions without any intention of offering them to a literary public (or at least there is no compulsion to do so).

An interesting phenomenon is the written language of SMS, which is not simply a symptom of a lack of culture, as a purist would say, but is an extremely economically

11 The Planetary Brain

written expression of the spoken language. We could start a debate about the difference between phonetic and non-phonetic language (and culture) but we would be digressing too far from the main thrust of our argument.

> The Web offers a synthesis of mathematical language with poetry, theatre, news and images, as well as several other forms of expression.

The Web offers a synthesis of mathematical language with poetry, theatre, news and images, as well as several other forms of expressions, including pornography. of course. But the expression of "porneia" depends on what is considered "porneia" in a given context. Several paintings by Titian, for example the "Venus of Urbino" hiding her genitalia with her hand, Danaë receiving divine sperm in the form of celestial rain, and the Majas of Goya, together with a number of other artistic creations were considered pornographic when they appeared and were thus confined to bed chambers to induce (aristocratic) arousal and (noble) progeny. These paintings are now on display for even school children to see in museums and galleries. Looking at the frescoes in Pompei can offer a pleasant and educational experience about the relative view of the concept of pornography over the centuries.

It seems curious that mathematical language could be so predictive of the movement of celestial bodies in the Universe (for example). This is probably because it is the physical model, written in mathematical language, that is predictive, and not the mathematical language that is predictive in itself. The fact that mathematical language provides the tools useful to describe physics, and human mathematical language has the capacity to describe reality is, in our opinion, a hint of the link between matter and thought.

In other words. it would seem incredible that, originating in our brain, mathematical language can describe far distant galaxies, if we do not take into account the point previously noted that the Universe and ourselves are made of the same matter, and probably of the same thoughts? As Einstein said, "the most incomprehensible fact about the universe is that it is comprehensible". Might we imagine that Nature found in Humanity (a last-minute detail) a way to express its thinking? In any case, we can at least envisage such a concept.

With regard to the apparent conflict between ambiguity and clarity (precision, accuracy, specificity, and "interjudges" accord[3]) in languages, we may consider this in the same context as other eternal conflicts: rational versus irrational, classic versus romantic, cognitive versus affective.

The fact that the two components, namely cognitive and affective, are different, is true, but they are not completely separate and probably we need both, even in activities that are apparently solely affective or solely cognitive. For example, in

---

[3]This term is used mainly in psychology to express the consistency of the results obtained by different examiners using the same test independently on the same subject.

learning, without the affective competences the cognitive function does not work, and without cognitive competences, affective aspects have no application in real life: we need a sufficiently good relationship with the teacher to understand the concept, and we need a social understanding to begin any affective relationship.

> We need a sufficiently good relationship with the teacher to understand the concept and we need a social understanding to begin any affectionate relationship.

The connections between different areas within the brain itself, and its connections with other portions of the nervous system that cover different functions in other parts of the body, such as the bowels, muscles and the synapses of the motor neurons, create a network of nervous channels, connecting numerous kinds of cells (the different functions of which remain incompletely known), using several receptors and several transmitters.

Remarkably we can say that the Web system is similar to our brain, in connecting several different units together. Human beings are reproducing our brain on a planetary dimension: several major hardware systems (equivalent to cells in the brain regions and loci), connected by a wired (nerves) or a wireless channel (which in the nervous system may be an electromagnetic field), communicate with other hardware systems at a greater or lesser distance on the planet (mimicking the ganglia in the nervous system). Nevertheless hardware is mute without the language of software with its "binary base" (which may be equated with the electric signals passed by chemical transmitters in the nervous system) and multiple, very differentiated functions (thermoregulation, autonomic regulation, emotions, affects and cognition) that interact with each other.

We may ask ourselves if (as with pathological psychiatric conditions), the planetary brain could develop any pathological conditions such as delusions, hallucinations, obsessions or compulsions, or suffer from persecution, depression, etc . . .

11.4 Are Thoughts Physical Objects? Are Thoughts Matter?

It might be interesting to consider some psychological phenomena (correlations between distant minds, synchronicity effects) in the light of some phenomena observed in quantum mechanics that are at odds with classical causality, such as the Einstein-Podolsky-Rosen paradox [16].

We refer here to a recent paper by Galli Carminati and Martin [17]. This paper gives the example of twins who buy, simultaneously, two identical ties without having consulted each other beforehand. The entanglement (the correlation) appears in the "classical" world only when a human consciousness (one of the twins or a third party) becomes aware of the fact. (See also [18–20]).

The paper cites also the following example/case: *C. G. feels bad, she makes a phone call to her twin sister, a psychotherapist who tells her that she is presently treating a difficult case. C. G. believes that her feeling of sickness is the result of her quantum entanglement with her twin sister. However she needs to telephone her sister, to transmit the information by a "classical" channel, in order to confirm that her feeling of sickness is really the demonstration of her correlation with her twin sister... There is still the difference between that which is quantum entangled at the unconscious level and that which reaches insight and consciousness and appears in the "classical" world.*

The two situations are linked by there being the possibility of action, or the perception, of two distant and simultaneous events.

In the "normal" world these situations are rare and associated with "peculiar observers", such as twins, or, in any event, persons with a strong mutual bond.

We may refer here almost verbatim to part of the intervention by F. Martin at the "Conference on Wolfgang Pauli's Philosophical Ideas and Contemporary Science", in 2007 at Ascona [21].

...Jung describes the possibility of significant coincidence between a mental state and a matter state. In this case the matter state is symbolically correlated to the mental state. Both states are correlated by a common meaning. They appear not necessarily simultaneously but in a short interval of time such that the coincidence is significant. Jung thought that those phenomena appear very rarely in daily life. For him synchronicity effects emanate uniquely from an activated archetype and not from a latent one. "such phenomena happen above all in emotional situations such as death, illness or accidents..." [22–27].

However Jung also said: "I must again stress the possibility that the relation between body and soul may yet be understood as a synchronicity one. Should this conjecture ever be proved, my present view that synchronicity is a relatively rare phenomenon would have to be corrected" [28, 29].

Synchronous events between mind and matter seem difficult to explain in terms of correlations between minds (between unconscious). For Jung, synchronicity events are remnants of a holistic reality – the unus mundus – the "One World" from the alchemist Gerhard Dorn (1600). This unus mundus could be related to the World of Ideas of Plato. It underlies both mind and matter. "Jung's notion of a synchronicity of pairwise arranged events in the mental and the material domains, correlated by a common meaning, is tightly related to the idea of a broken symmetry of the unus mundus. The synchronous correlation between the events can be regarded as a retrospective indication, a remnant as it were, of the unity of the archetypal reality of the unus mundus from which they emerge" [30].

Moreover one can possibly see synchronous events between the mental and the material domains as a consequence of a quantum entanglement between mind and matter [31].

As Jung and Pauli we adopt here a dualistic view of mind and matter. Mental and material domains of reality will be considered as aspects, or manifestations, of one underlying reality in which mind and matter are un-separated [30].

Synchronicity phenomena, especially those involving a correlation at a distance between several psyches, lead us to postulate a non-localisation of unconscious mental states in space and time. Mental states are not exclusively localised in the human brain. They are correlated to physical states of the brain (possibly via quantum entanglement) but they are not reducible to them. Since we are going to study the analogy between synchronous events and quantum entanglement, we treat mental states (conscious and unconscious), as quantum states (as matter states are), i.e. as vectors of a Hilbert space [32].

Von Neumann [33], *Wigner* [34] *and Stapp* [35] *set the boundary between the observed system and the observing system in the observer's brain.*

In quantum mechanics there are two levels of reality. First there is the quantum level of reality, in which there exists superposition of quantum states, evolving in time in a deterministic way. The second level of reality is called classical reality. It is the level of the unique reality that we observe with our consciousness. It is also the level that in physics is given by the (single) result of a measure. The border between the quantum and the classical reality occurs through an operation "the reduction of the wave packet".

An event that reaches our consciousness might be similar to the photon that is registered by a detector but, in the case of our consciousness, it could be a different particle (a boson of information?). The point is that all the conscious events (choices, thoughts, and decisions) could affect (change) the entropy of our system and, through the environment, of the whole Universe, and this is a priori a serious problem, because we have to face an unbalanced diminution of entropy at every step of ordered thought, cognition and understanding.

> An event that reaches our consciousness could be similar to the photon that is registered by a detector, but in the case of our consciousness it could be a different particle (a boson of information?)

We can imagine that something similar happens for psychological processes, and that enables us to explain some behaviour in consciousness within the context of quantum theory. To do that we can try to build a quantum model of the correlations at a distance that shows links between several minds, for example between two people or within a group of people (group correlations). We can also try to model the awareness of unconscious components through the present theories of quantum measurement. We shall see that the model of Cerf and Adami [36], in which there is no collapse of the wave function and entropy is brought into equilibrium. This seems to fit the phenomenon of awareness better, because it does not alter the state of the unconscious to so great an extent.

We should briefly remember here that the problem of measurement in quantum physics is still hotly debated. To try to find way out of the *impasse*, in 1932 von Neumann suggested [33] dividing the evolution of the wave function, as a function of time during a measurement, into two processes.

The first process is the unitary and deterministic evolution of this wave function. The second process is the collapse of this wave function on one of the eigenstates of the measured observable. If the first process is continuous and deterministic, the second one is discontinuous and non-deterministic (it is probabilistic). The theory of quantum decoherence explains how, because of interaction with the environment, the quantum system composed of the observed object and of the detector goes from a coherent superposition of quantum states to a statistical mixture of states relative to a given basis (the reduced density operator). Some theories (such as the "Relative State" theory of H. Everett, and the quantum information theory of N. Cerf and C. Adami) try to avoid the collapse of the wave function [36–40].

In a recent work, F. Martin and G. Galli Carminati proposed a model in which they assumed that minds could be entangled like quantum particle states. They proposed that, in the case of two quantum entangled minds (e.g. Alice and Bob's), if (at a distance) Alice becomes aware of some information concerning Bob, Alice knows the quantum state of some part of Bob's psyche (the one that is quantum entangled with her own psyche). However, according to the authors she does not know the "classical state" of Bob's psyche, i.e. what Bob becomes aware of. " ... It could be that what Bob becomes aware of is related to that part of his psyche which is quantum entangled with the one of Alice. In this situation there would be correlation between the two consciousnesses (the one of Alice and the one of Bob). But it could also be that what Bob becomes aware of does not at all concern that part of his psyche, which is quantum, entangled with the one of Alice. In this situation the appearance of quantum entanglement (the correlation) of which Alice becomes aware remains unconscious for Bob" [17].

The situation between two persons could be extended to a group, this group being of greater or lesser extent. Let us consider a state built from the components of the subconscious of the persons in the group: it is the quantum state that is created starting from the various individual subconscious of the group members. This group state may be compared to a Bose-Einstein condensate, insofar as the group situation could cause a large majority of the group's subconscious to be in the same quantum state, which tends to "orientate" in a homogeneous manner.

The observation of the conscious behaviour of the group influences ideas in an indirect manner, via the functioning of the group. Say we want to build an experiment to understand the unconscious mechanisms, or at least the unconscious orientation, of a group. The difficulty with this is the performance of a methodologically rigorous study on a very large number of persons. Moreover, a study of the unconscious mechanism would be quite unrealistic on the Web, but we can at least consider the Web as a very large group situation with exchanges between persons, subgroups and different entities all over the planet.

The main point is that we can model the transfer of unconscious and/or conscious information in the terms of "bosons" of information, and thus build a formal model of interaction from unconscious to conscious and vice versa, in a person, between two persons, or in a group. We have to remember that the mechanisms of a group are not only the sum of the interactions between members of a group but also "group" interactions.

According to Foulkes [41] the psychological medium of the exchanges between several minds, the net of the individual mental processes, communications and interaction is the "matrix" of the group.

According to Bion [42, 43] the explanation of several group phenomena may be found in the "matrix" of the group and not only in the persons forming the group.

If we hypothesise a model for the interaction of information in a similar manner to that for physical particles (photons in a detector in a physics experiment, for instance) we can also hypothesise that thoughts, consisting of bosons of information, are physical objects, and are thus "matter" in the wider sense of the word.

> If we hypothesise a model for the interaction of information in a similar manner to that for physical particles we can also hypothesise that thoughts are physical objects in the wider sense of the word.

We may also ask ourselves about the nature of the Web, which we may consider as a physical object with an energy and an entropy, or even with a temperature and an equation of state – if we prefer the thermodynamics approach. Or as a quantum object, if we consider it as the superposition of different states, on the basis of the actual parallel elements of information in the Web.

An interesting concept is that the Web, like the brain, is a material object, with its own hardware, producing and exchanging information: its software. In the same way as we did for the psyche, we can hypothesise that the rules for the transmission of this information may be formally described in terms of Quantum Information Theory.

Could we suppose that matter and information are governed by similar mechanisms?

11.5 Are the Elements of the Web and the Grid and Their Interconnections Evolving Toward a Planetary Brain?

This question seems somehow to have a trivially positive answer: Yes, the Web and the Grid are evolving, and we may make the conjecture that they may evolve toward a planetary brain. The less trivial question (with a less trivial answer) is: should we assume, because self-consciousness exists in the brain (above and beyond the system of connections), the existence of self-consciousness within the Web?

One possible way to tackle this question is to suppose that the planetary brain, being a brain created by men and women in the image of the human one, is possibly reproducing the human being's development where, at a given moment, self-consciousness appears.

> Can we postulate that the Web, being created by man, follows a development similar to the one of the human being from infancy to adulthood and old age?

We have to clarify our view of consciousness. As Penrose says [44] "It is not without doubts that we are the only living creature with the gift of the consciousness. How the lizard or the codfish could have their own consciousness we don't know". Just as with the lizard or the codfish, it is difficult for us even to imagine *how* the Web could have its own consciousness. So, most probably, we are not even in the position to be able to decide *whether* this consciousness already exists [44,45].

In other words, we have no evidence about the development of the consciousness of the Web, but we could try to postulate that if humanity in its entirety follows a development similar to the one followed by the individual human beings, with "infancy", "adolescence" and "youth", the Web too, being created by man, might follow similar steps in its development. Of course, we are not forgetting the existence of adulthood and old age, but for the sake of our argument, we consider here only the first stages of human development. It is true that the development of the human beings differs according to environment and culture, but nevertheless overall development largely follows similar lines.

To provide support for our argument, let's turn for a moment to human development as it is described by psychology and let us consider the neurological and psychological aspects of human development. During our psychological development we acquire neurological, anatomical and psychological functions.

Piaget, for instance, describes the development of a child as occurring in different stages. The sensorimotor stage (from birth to the age of two) is subdivided into six substages: (1) simple reflexes; (2) first habits and primary circular reactions; (3) secondary circular reactions; (4) coordination of secondary circular reactions; (5) tertiary circular reactions, novelty, and curiosity; (6) internalisation of schemes. In the preoperational stage (from ages 2 to 7) magical thinking predominates, with acquisition of motor skills. In the concrete operational stage (from ages 7 to 12) children begin to think logically, but are very concrete in their thinking. At the formal operational stage (from age 12 onward) there is development of abstract reasoning in that children develop abstract thought and they can think logically [46].

Melanie Klein, from the psycho-developmental point of view, distinguishes two fundamental attitudes in development: the schizo-paranoid (or persecutory) attitude and the depressive attitude. These two psychological attitudes reappear and have different forms at different periods in one's life. During childhood at the anal stage for example, the schizo-paranoid attitude is expressed by refusal and opposition.

Here the depressive attitude allows a transition from refusal and opposition to acceptance, thanks to the children's love for their parents. For the adolescent, the persecutory attitude appears in the refusal to accept parental prohibitions and in transgressions, while the depressive attitude corresponds to mourning the loss of the basic parental imagoes and to their idealisation. In the adult, the schizo-paranoid attitude appears as the refusal of maturity, and the depressive attitude helps the individual to accept that youth is gone and to cope with the death of the previous generation [47, 48].

Might the Web behave in a similar fashion, starting with the earliest stages of its existence – taking into account the fact that we actually have models of only a child's evolution – and be oscillating between a persecutory/transgressive attitude and a depressive/de-idealisation attitude?

Observing the evolution of the Web, we see that the beginning of this project was the need to link different computers to exchange what today would be called "multimedia" information. As explained in detail in Chap. 2, thanks to a fertile environment, the idea of developing a protocol to exchange images, texts and messages was able to be implemented. In fact, at CERN there were "spare" facilities that could be used. These facilities were not "centrally managed" by "a clever and efficient" administration. Surplus and luxury often are the mulch for creativity, and that was apparently the case.

Although at the beginning the Web was developed on a small machine and by a very small number of persons, this development could happen because there was also some margin of manoeuvre. We could relate this phase to Piaget's sensorimotor stage, thus defining the Web as being at the beginning of its development.

The next step was to develop the Web infrastructure, on the one hand based on the existing Internet one, and, on the other, on the newly developed exchange protocol, and thus creating an increasingly sophisticated system based on the use of algorithms primarily adapted to the HEP environment. We may draw a parallel here with Piaget's preoperational stage: a sort of acquisition of "motor skills" with "basic" interactive tools.

These interactive tools were used to produce search engines looking for information at the nearest possible available location, following the bazaar's philosophy to "pick up and carry away just what you need" rather than the classical, software engineering hierarchical model, with established formal rules. This we may equate with Piaget's concrete operational stage of development that is accompanied by a beginning of logical thought.

Out of the Web came the Grid. The Grid is aimed at the distributed mining and processing of data and information. This structure needs nodes of exchanges: centres of computing that are dynamically connected. In a similar way, during its development, the brain, interacting with its environment (both the external and internal ones), develops circuits for different functions from autonomic to cognitive ones, from learning to walk, to eat, to speak, to, eventually, studying philosophy.

With the Grid, we have arrived at Piaget's formal operational stage, with the development of abstract reasoning. And we note also that, the digital and binary nature of the computers is increasingly hidden because the operations are increasingly "analogical" in nature.

> The introduction of parallelism at all levels in the processing units and in the Grid itself narrows the differences from the human brain.

The introduction of parallelism at all levels in the processing units and in the Grid itself, narrows the differences between them and the human brain. For instance, if a Grid centre is down, another can replace the loss, in a way similar to the one used by the brain to preserve its different functions in case of lesions. We cannot stop the brain and wait for a replacement unit, the activities have to continue, and on the Grid it is pretty much the same.

The Grid works by hiding the differences in hardware, for example by virtualisation (see Chap. 6). Serotoninergic and adrenergic neurons and glia are different components of the brain that work together "on the same program", with different functions, and more or less harmoniously. In the case of a "breakdown", the brain can rely on its relative plasticity by using "sleeping neurons" to compensate for the malfunction.

Considering all this, we see an intriguing parallel between the development of a child and that of the planetary brain as they both pass through the different stages that are characteristic of the human neurological development. If we extend the comparison to psychological development, we have a transition from a "free" initial phase, the transgressive (but also schizo-paranoid and persecutory) one through the refusal of "parental order" to the present need for an "ordered structure", demanding regulation and even censorship, similar to a depressive attitude.

Continuing the parallel with a human being's development, if we accept that human history continues to repeat an individual's developmental path, could we assume that the planetary brain is passing through the different phases of an individual's life and, will pass through the different phases of human history as well?

Human history may be seen as a continual oscillation between transgressive and depressive attitudes, which are in alternation and opposition: rationalism versus romanticism, science versus spirituality, poetry versus mathematics, West against East, North against South ... If we look at the evolution of the Web in ts present state, we find that we are regretfully giving up some of the initial "freedom", and giving priority to "parental controls" of different kinds for "our own protection" perhaps from several different real or imaginary dangers, "vortexes", attractions, vanities and temptations: "evils, devils and other demons", most of which seem to be in the "eye of the beholder". This is may be the true essence of education: moral, ethics and legal help is needed to avoid sin (of the computing kind) and to be safe.

It is probably too early to understand all this. Our conception of consciousness probably needs to evolve from an anthropocentric view to a more objective conception, which may be observed and even measured. But the difficulty is that the instrument here is our own consciousness of self. To increase the complexity, the best way to observe and monitor the Web, is to use the Web itself. Thus it becomes the observer and the subject of observation. With the human psyche, the only way to observe it is via another human psyche. Is this another interesting parallel or just one more anthropomorphic projection?

It would be interesting (and even necessary) to observe the evolution of the Web from a "libertarian" approach (involving Linux, open-source software, GNU, and similar programs) to an increasingly rigid design: "from bazaar to cathedrals" [49]. In other words, from extreme-computing to computer engineering, from the hacker world to political correctness. It is sometimes reassuring for large number of people to participate in large planetary events, examples being the Olympic Games, the Football World Cup or social events such as the Miss World competitions and the European Song Contest. This is a good incentive to get together peacefully on a large scale, and infinitely preferable to wars and genocide and other activities of mass destruction. Similarly in science: the attraction of a huge prestigious collaboration is irresistible. The need to assemble large groups of people working together comes with the need to organise the work as a whole.

As a result, it is probably impossible to escape passage through a form of medieval ex-cathedra philosophy, because the importance of this period of scholarly culture in human evolution is undeniable. However we need to pay attention not to spend too many years (in manpower units), nor too much energy or too many resources (in units of tax-payers' money) in religious conflicts.

We have seen that scientific language may be considered slightly less ambiguous than other "more human languages", but some ambiguity is intrinsic in human communication and probably in human nature. The point is that science, which is a special kind of human activity, especially when it becomes a group activity, might follow group mechanisms (conscious or unconscious ones) and be seduced by auto-definition and self-substantification, extinguishing the individual genius of sceptical spirits and this might probably be more dangerous than the "evil, devil and other demons" that have been mentioned earlier.

11.6 Conclusions and Lessons Learned

Every civilisation aspires to attain its desire for survival (or for becoming eternal). "The cloud-capped towers and gorgeous palaces are such stuff as dreams are made on." [50]. These are the dreams of all humanity, not of an elite minority. What our ancestors left behind: cave drawings, carvings in the rocks of mountains, and writing in books of stone, papyrus, paper and now in digital form, are their legacy and our heritage, and a way of leaving an enduring mark of our existence.

It is natural that, having such a planetary tool in our hands, we are tempted to build a gigantic construct. There is something very naïve in this attitude. The problem is that the need to build gigantic constructs becomes an article of faith and a norm, and far more than just a tool of knowledge.

Fashion might be considered as the "colour" in music or the "style" ("maniera") in painting or in literature: the trappings of an ideology, the mood of a time, or the spirit of a particular period in history. In computing, first there was FORTRAN, then C++, Java, Python, Perl, Ruby. Fashions have the merit and the disadvantage of existing in parallel with or in contradiction to one another: usually there are different fashions at any one time, for example, mini-shorts and long skirts may both simultaneously be a trend (but you can't wear both at the same time). This becomes a problem when fashion rules our way of behaving, and, even worse, of thinking. When an idea becomes a dogma, the theory unfortunately overwhelms the real needs of the time (for example the idea of centralising the management of the Web, as was the initial intention with the Grid, was necessary in theory but useless in practise).

When developments in information technology drift towards a monolithic state – and this is often the case for gigantic structures – all the energy of the group is expended in following the dream of immortality, infallibility, and omnipotence.

We have to admit that the Web (and the Grid), representing such a huge innovation in planetary life and knowledge, and developing so fast, suffers from a lack of interdisciplinary vision and self-criticism. This is probably intrinsic in the development of a relatively new science.

Computer science is in its early seventies and the Web is still in its infancy. Expansion into other fields needs a preliminary period of experimentation and consolidation. The possible interpretation of the phenomena of the Web in terms of a psychological dynamic has, by contrast, not yet been born.

The sole ambition of this chapter, with its "blue-sky thinking", has been to reflect on the unknown (and inevitable) mechanisms that exist in the planetary brain.

References

1. Abraham, I.M., Kovacs, K.J.: Postnatal handling alters the activation of stress-related neuronal circuitries. Eur. J. Neurosci. **12**(12), 3003–3014 (2000)
2. Fenoglio, K.A., Chen, Y., Baram, T.Z.: Neuroplasticity of the hypothalamic-pituitary-adrenal axis early in life requires recurrent recruitment of stress-regulating brain regions. J. Neurosci. **26**(9), 2434–2442 (2006)
3. Noriega, G.: Self-organising maps as a model of brain mechanisms potentially linked to autism. IEEE Trans. Neural Syst. Rehabil. Eng. **15**(2), 217–26 (2007)
4. Moore, D.R.: Auditory processing disorders: acquisition and treatment. J. Commun Disord. **40**(4), 295–304 (2007)
5. Card, J.P., Levitt, P., Gluhovsky, M., Rinaman, L.: Early experience modifies the postnatal assembly of autonomic emotional motor circuits in rats. J. Neurosci. **25**, 9102–9111 (2005)
6. Bauman, M.D.: Early social experience and brain development. J. Neurosci. **26**(7), 1889–1890 (2006)
7. Salvador, A.: La guerison du point de vue du psychiatre. Action et Pensée **50** (2007)

8. Popper, K.: Three Worlds, The Tanner Lecture on Human Values, Delivered at The University of Michigan. http://www.tannerlectures.utah.edu/lectures/documents/popper80.pdf, April 7 1978
9. http://musiclub.web.cern.ch/MusiClub/bands/cernettes/firstband.html
10. Piaget, J.: The Origin of Intelligence in the Child. Routledge and Paul Kegan, London (1953)
11. Wittgenstein, L., Anscombe, G.: Philosophical Investigations. Blackwell, Oxford (2001)
12. Fo, D., Rame, F. (eds.) Mistero buffo. Einaudi tascabili, teatro (2005)
13. Duneton, C.: L'art du grommelot. Le Figaro, April 20, 2006
14. Cartier, P.: A mad day's work: from Grothendieck to Connes and Kontsevich. The evolution of concepts of space and symmetry. Bull. Amer. Math. Soc. **38**(4), 389–408 (2001)
15. Connes, A., Marcolli, M.: From physics to number theory via noncommutative geometry. II: Renormalization, the Riemann – Hilbert correspondence, and motivic Galois theory. Front. Number Theor. Phys. Geom. **2**, 617–713 (2006)
16. Einstein, A., Podolsky, B., Rosen, N.: Can quantum-mechanical description of physical reality be considered complete? Phys. Rev. **47**(10), 777–780 (1935)
17. Carminati, G.G., Martin, F.: Quantum psyche. Physics of Particles and Nuclei (2007). Accepted
18. Carminati, G.G., Carminati, F.: The Mechanism of Mourning: an Anti-Entropic Mechanism. NeuroQuantology J. (Online) **4**(2), 186–197 (2006)
19. Marshall, I.N.N.: Consciousness and Bose-Einstein condensates. New Ideas Psychol. **7**, 73–83 (1989)
20. Pitkanen, M.: Quantum Mind Archives. http://listserv.arizona.edu/archives/quantum-mind.html, August 1998. n. 88
21. Martin, F., Carminati, G.G.: Synchronicity, Quantum Mechanics and Psyche. In: Springer-Verlag (ed.) to be published in Proceedings of the Conference on Wolfgang Pauli's Philosophical Ideas and Contemporary Science, Ascona Switzerland, May 20-25, 2007
22. Jung, C.G., Pauli, W.: The Interpretation of Nature and the Psyche. Pantheon, New York (1955). Translated by P. Silz
23. Meier, C.A. (ed.): Atom and Archetype: The Pauli/Jung Letters 1932-1958. Princeton University Press, Princeton (2001). Traduction franaise: Correspondance 1932–1958, Albin Michel (ed.) (2000)
24. Atmanspacher, H., Primas, H.: The hidden side of Wolfgang Pauli: an eminent physicists extraordinary encounter with depth psychology. J. Conscious. Stud. **3**(15), 112–126 (1996)
25. Atmanspacher, H.: Quantum approaches to consciousness. In: Zalta, E.N. (ed.) The Stanford Encyclopedia of Philosophy. Stanford University, Stanford, CA, Winter (2006)
26. Beck, F., Eccles, J.C.: Quantum Aspects of Brain Activity and the Role of Consciousness. In: Proceedings of the National Academy of Sciences of the United States of America, vol. 89 (23), pp. 11357–11361, Dec. 1, 1992
27. Hameroff, S.R., Penrose, R.: Conscious events as orchestrated space-time selections. J. Conscious. Stud. **3**(18), 36–53 (1996)
28. Primas, H.: Synchronizitaet und Zufall. Zeitschrift fuer Grenzgebiete der Psychologie **38**, 61–91 (1996)
29. Jung, C.G.: The Structure and the Dynamics of the Psyche. In: The Collected Works of C.G. Jung, vol. 8. Princeton University Press, Princeton (1969). Second Edition, parag. 938, footnote 70
30. Atmanspacher, H., Primas, H.: Paulis ideas on mind and matter in the context of contemporary science. J. Conscious. Stud. **13**(46), 5–50 (2006)
31. Primas, H.: Time-entanglement between mind and matter. Mind Matter **1**(1), 81–119 (2003)
32. Baaquie, B., Martin, F.: Quantum Psyche: Quantum field theory of the human psyche. NeuroQuantology [Online] **3**, 1 (2007)
33. von Neumann, J.: Mathematische Grundlagen der Quantenmechanik. Springer Verlag, Berlin (1932) English translation: Mathematical Foundations of Quantum Mechanics, Princeton: Princeton University Press (1955)
34. Wigner, E.P.: chapter Remarks on the mind-body question. In: Symmetries and Reflections, pp. 171–184. Indiana University Press, Bloomington (1967)

35. Stapp, H.P.: chapter A quantum theory of the mind–brain interface. In: Mind, Matter, and Quantum Mechanics, pp. 145–172. Springer, Berlin (1993)
36. Cerf, N.J., Adami, C.: Quantum mechanics of measurement. Preprint MAP-198, KRL (1996)
37. Zurek, W.H.: Pointer basis of quantum apparatus: into what mixture does the wave packet collapse? Phys. Rev. D **24**, 1516 (1981)
38. Zurek, W.H.: Decoherence and the transition from quantum to classical. Phys. Today **44**(10), 36 (1991)
39. Everett, H.: Relative state formulation of Quantum Mechanics. Rev. Mod. Phys. **29**(3) (1957)
40. Wheeler, J.A.: Assessment of Everett's "Relative state" formulation of Quantum Theory. Rev. Mod. Phys. **29**, 463 (1957)
41. Foulkes, S.H.: La groupe-analyse: Psychothérapie et analyse de groupe. Payot-poche (2004)
42. Bion, W.-R.: Experiences in Groups and Other Papers. Ballantine, New York (1974)
43. Bion, W.-R.: Recherches sur les petits groupes, 3rd edn. Presses universitaires de France, Paris (1976). Trad. fr.
44. Penrose, R.: The Emperor's New Mind Oxford University Press, New York (1990)
45. Penrose, R.: Shadows of the Mind Oxford University Press, New York (1994)
46. Piaget, J.: Introduction à l'épistémologie génétique. Tome I: La pensée mathématique. Tome II: La pensée physique. Tome III: La pensée biologique, la pensée psychologique et la pensée sociale. Presse Universitaire de France, Paris (1950)
47. Klein, M.: Psychanalyse d'enfants. Payot, coll. "Petite Bibliothèque Payot", 2005 (ISBN 2228899992)
48. Klein, M.: Le transfert et autres écrits Presse Universitaire de France (1995) (ISBN 2130472206)
49. Raimond, E.S.: The Cathedral & the Bazaar. O'Reilly (1999). http://www.firstmonday.org/issues/issue3_3/raymond/
50. Shakespeare, W.: The Tempest, New edn. Penguin Classics, New York (2007)

Glossary

2-D graphics representation two-dimensional objects, as opposed to 3-D
3-D graphics representation of three-dimensional objects, with perspective and lighting effects, as opposed to 2-D
3D CERN project for distributed condition Database
3270 a class of terminals made by IBM since 1972 and their communication protocol
Accelerator a device used to accelerate elementary particles or nuclei via an electromagnetic field
ADA programming languages developed under the auspices of the U.S. Department of Defence
ADAMO a system for scientific programming based on the Entity-Relationship (ER) model
ADL ATLAS Description Language
ADSL Asymmetric DSL, a DSL where the upload data connection speed (from the user to the server) is less than the download one (from the server to the user)
AEGYS Operating System developed by Apollo
AFS Andrew File System, a distributed file system providing a homogeneous, location-transparent file name space to all the clients
AIS Accounting Information System
ALEPH A Large Electron Positron experiment at the CERN LEP accelerator
ALICE A Large Ion Collider Experiment, a major experiment at the CERN LHC
AliEn ALICE Environment, a set of middleware tools developed by the ALICE experiment
ALPHA 64-bit RISC computer architecture developed by DEC
Alphanumeric attribute of an output device capable of displaying letters and numbers
AlphaServer a server computer base on the Alpha architecture produced by DEC
Amazon a U.S.-based multinational electronic commerce and software company
AMD Advanced Micro Devices is one of the leading semiconductor companies together with Intel

AMD Advanced Micro Devices, a CPU and other computer component manufacturer

Amdahl's law mathematical formula governing the expected speedup of a parallel algorithm with respect to its sequential version

AMSEND Automatic Mail SENDing, a electronic mail distribution system used by ALEPH and DELPHI in the eighties

Apache The Apache Software Foundation is a software development group who is producing a widely used HTTP server

API Application Programming Interface

Apollo a computer manufacturer

Apple a computer manufacturer

ARC Advanced Resource Connector, a middleware developed by the NorduGrid project integrating computing resources and storage facilities, making them available via a secure common Grid layer.

Archetype a universal symbol behaviour pattern, from which others are derived, often used in myths and traditional storytelling

ARDA Architectural Roadmap towards Distributed Analysis, a CERN Requirements Technical Assessment group setup in 2003 and charged to draw up a blueprint for the Grid components needed to effectively run the analysis

ARPA U.S. Advanced Research Projects Agency, renamed the Defense Advanced Research Projects Agency (DARPA) in 1972

ASA American Standard Association

ASCAP American Society of Composers, Authors and Publishers, an U.S. not-for-profit performance-rights protection and license fee collection organization

Assembler low-level programming language

AT&T the largest provider of fixed telephony, broadband and subscription television services in the U.S.

ATHENA Software framework developed by the ATLAS collaboration and based on GAUDI

ATLAS A Toroidal LHC ApparatuS, a major experiment at the CERN Large Hadron Collider

Atomic transaction a transaction where all operation involved either all succeed or none succeed

Authentication process by which a system verifies the identity of an entity (user or process) who wishes to access it; it involves the authority of a third party called Certification Authority

Authorisation verification that an authenticated entity is allowed to perform certain operations or access given resources; it has to be preeceded by authentication

B-tree a sorted data structure in form of a tree allowing searches, sequential access, insertions, and deletions in a time that grows with the logarithm of the data size

BaBar experiment at the PEP-II storage ring at SLAC

BARB, BARBascii a computerised documentation system developed in the 1970s

Batch operating mode of a computer where a request for work is executed asynchronously till completion with little or no control by the user

Baud Unit of transfer of information corresponding to one binary digit (bit) per second

BBC a point-to-point network file copy application developed at SLAC the BaBar collaboration to transfer files at approaching line speeds in the WAN

Big Bang theory according to which the universe originated from extremely hot and dense state whose expansion, which is continuing today, caused the universe to cool and resulted in its present state

Bind variable a place-holder variable in a SQL statement that must be replaced with a valid value to execute the statement

Bit Binary digit that can assume the values of 0 or 1

Bitmap display a computer screen where each pixel on the monitor corresponds directly to one or more bits in the computer's video memory

BITNET network protocol developed by IBM

Block device a computer component (device) through which the Operating System moves the data in blocks to increase transfer efficiency; examples are disks or CDs

Blog a type of website where different individuals write regular entries of commentary, descriptions of events, including graphics or video

BMI Broadcast Music Inc., an U.S. not-for-profit performance-rights protection and license fee collection organization

BNL Brookhaven National Laboratory

Boinc an Open Source software platform for using volunteered computing resources

BOS Bank Organisation System

Bose-Einstein condensate a quantum state that can be reached a class of subatomic particles where all components occupy the lowest energetic level

Boson a subatomic particle that can participate to a Bose-Einstein condensate

BSD Berkeley Software Distribution is a Unix operating system derivative

Bubble Chamber a container filled with a super-heated transparent liquid where electrically charged particle moving through it create small bubbles that allow to reconstruct their trajectory

Buildbot a system to automate the compile and test cycle required by most software projects to validate code changes

Bus a device for data transfer between the components inside a computer

Byte Sequence of binary digits (bits), usually of size 8

C programming language

C++ computing language

Cache a component that transparently holds data which has been requested, so that data can be served faster in case of further requests

cache miss request for data which is not in the cache, and which therefore it will take longer to retrieve

Calcomp a brand of pen plotters

CAP Computing for Analysis Project, a FNAL project with the goal of enabling rapid HEP analysis and event selection from multi-TeraByte datasets

CASE Computer Aided Software Engineering, a set of programs, methods and automated tools meant to improve the quality of existing software or to improve the efficiency and correctness of the software development process

CASTOR he CERN Advanced STORage manager is a hierarchical storage management system developed at CERN for physics data files

CBM Compressed Barionic Matter experiment at the FAIR facility, GSI, Darmstadt

CCIN2P3 Centre de Calcul de l'Institut National Pour la Physique Nucléaire et des Particules, a computing centre in Lyon, France

CCRC Common Computing Readiness Challenge, exercise simulating the operating conditions of WLCG to test its readiness for real operation

CDC Control Data Corporation, a computer manufacturer

CDF The Collider Detector at Fermilab Tevatron proton accelerator

CE Computing Element is a gateway operated at each Grid site that provides services for the submission and control of jobs

CERN The Conseil Européen pour la Recherche Nucléaire, or European Council for Nuclear Research

CERNET CERN developed network system, comprising both hardware and protocol

CERNLIB CERN Program Library

CERNVM public interactive service on the CERN IBM mainframe running VM/CMS, active from 1984 to 1992

CernVM CERN Virtual Machine, a virtualisation environment developed at CERN

Certificate Authority an entity that issues digital certificates (thrusted third party) certifying the ownership of a public key by the named subject of the certificate

CHEETAH a data management system produced at SLAC

CHEOPS a batch data dissemination system developed at CERN and based on the OLYMPUS satellite

CHEP Computing in High Energy Physics conference series

Chip an integrated circuit on a thin semiconduttor material

CINT an interpreter for C and C++ code part of ROOT

CIS Corporate Information System

Cisco a company producing network equipments

CLI Command Language Interface, a way to interact with a computer programme by typing commands on a keyboard to perform specific tasks

Client an application or system requesting services from another programme on the same machine or across the network ("server")

Cloud Distributed computing infrastructure based on multiple server-based computational resources accessed and used via a network. Cloud users access the servers ubiquitously from different platform, from a desktop computer to a palm-held device or smart-phone. Applications, data and resources are maintained within and managed by the Cloud servers

Glossary

Cluster computing system composed by several similar units, usually workstations, communicating via a common network to support a large number of users and applications

CMT code maintenance system

CMZ Code Management with ZEBRA

CN Computing and Networks, a CERN division, formerly known as Data Division (DD) and later as Information Technology (IT) division

CNUCE Centro Nazionale Universitario per il Calcolo Elettronico, an Italian computing centre located in Pisa, Italy

COMIS FORTRAN interpreter, part of CERNLIB

COMMON BLOCK fixed size memory are shared between different routines in a FORTRAN program

COMPASS COmmon Muon and Proton Apparatus for Structure and Spectroscopy, a multi-purpose experiment taking place at CERN's Super Proton Synchrotron accelerator

Condition Database Database holding the status of the experiment, including its alignment and calibration constants

Condor a project to provide tools and policies to support High Throughput Computing on large collections of distributively owned computing resources

Consciousness objective awareness of oneself and of the world

COOL a CERN developed condition Database for LHC

CORAL COmmon Relational Abstraction layer, a CERN project to provide functionality for accessing data in relational databases using a C++ and SQL-free API

CORBA Common Object Request Broker Architecture, a standard to enable software components written in multiple computer languages and running on multiple computers to interoperate

CORE Centrally Operated Risc Environment, a CERN project to provide computing resources based on RISC workstations that followed the SHIFT project

CORE graphics system developed in the 1970s by A. van Dam, Brown University

Core an independent processor part of a multi-core or many-core system

COS Cray Operating System

CPP C Pre-Processor

CPU Central Processing Unit, set of electronics circuits executing the arithmetic and logical instructions contained in the computing programmes

Cray a computer manufacturer

Crossbar interconnect an interconnect connecting multiple input and output nodes in a matrix manner, with one switch at each crosspoint

CUDA a parallel computing architecture and programming environment for Nvidia GPUs

Cut Selection of a portion of the events based on their characteristics

CVS Code Versioning System, a code management programme

CWN Column Wise N-tuple

Cyclotron a device that accelerates elementary particles or heavy nuclei along a curved path via a high-frequency alternating electric field, while the particle trajectories are bent by a magnetic field

D0 D0 Detector at Fermilab Tevatron proton accelerator

daemon a computer programme performing a service and running in "background" and not under the direct control of a user

DAG Direct Acyclic Graph, a representation of a set where some ordered pairs of the objects are connected by links and there is no directed cycle

DAQ Data AcQuisition System, the hardware and software used to collect and record in digital format the signals coming from the detectors

Data caching transparent and temporary storage of data so that future requests for that data can be served faster

Database an organised collection of digital data

DataSynapse a provider of Grid and distributed software products

DATE Data Acquisition Test Environment, data acquisition system developed by the ALICE collaboration

DB2 proprietary RDBMS software developed by IBM

DBA Data Base Administrator

DBL3 DataBase for the L3 experiment at CERN

DBMS DataBase Management System, a software program that manages the information contained in a Database

dCache a system for storing and retrieving huge amounts of data, from a large number of heterogeneous servers, under a single virtual filesystem tree

DD Data Division, a CERN division, then named Computing and Networks (CN) and finally Information Technology (IT) division

DDL Data Definition Language, it is used to create and modify the structure of database objects in Database

DEC Digital Equipment Corporation, a computer manufacturer

DECnet DEC proprietary network protocol

DeCSS a program running on Unix systems used to decrypt and display commercially produced DVD video disc

DELPHI DEtector with Lepton, Photon and Hadron Identification at the CERN LEP accelerator

Desktop computer fixed personal computer

DESY Deutsches Elektronen-SYnchrotron, HEP laboratory in Hamburg, Germany

DIANE DIstributed ANalysis Environment, the result of R&D in CERN IT Division focused on interfacing semi-interactive parallel applications with distributed GRID technology.

Dirac a set of middleware tools developed by the LHCb experiment

Distributed lock manager a software system that provides distributed software applications with a means to synchronise their accesses to shared resources, such as a Database

DML Data Manipulation Language, it is used to manipulate data in a Database

DNA DeoxyriboNucleic Acid, an acid containing the genetic information for the development and functioning of most known living organisms

DOC IBM documentation system

DOD U.S. Department of Defence

Domain proprietary networking system developed by Apollo

DoXygen automatic documentation generator from the code comments

DPM Disk Pool Manager is a lightweight solution for disk storage management developed at CERN

DRDC CERN Detector Research and Development Committee

driver an Operating System component allowing applications to interact with hardware devices such as disks, screens, keyboards or printers

DSL Digital Subscriber Line, a data communications technology enabling data transmission over copper telephone lines by using frequencies that are not used by voice telephony

DZDOC A ZEBRA Bank Documentation and Display System, a CERNLIB package

EARN European Academic and Research Network launched by IBM in 1983

EASI European Academic Supercomputer Initiative launched by IBM in 1987 to establish partnerships with leading European academic and research organisations to foster supercomputing education and research

EASINET IBM sponsorship for European Networks part of the EASI initiative

EBU European Broadcasting Union, the largest association of national broadcasters in the world, promoting cooperation between broadcasters and exchange of audiovisual content.

EC2 Elastic Compute Cloud, a web service provided by Amazon that offers resizable compute capacity in the Cloud in order to make Web-scale computing easier for developers

ECFA European Committee for Future Accelerators

ECP Electronics and Computing for Physics, a CERN division

EDG European Data Grid, (2001-2004)is a project funded by European Union with the objective is to build the next generation computing infrastructure providing intensive computation and analysis of shared large-scale databases, from hundreds of TeraBytes to PetaBytes, across widely distributed scientific communities

EGEE Enabling Grid for E-sciencE, a Grid project funded by the European Union that aims to build on recent advances in Grid technology to develop a service Grid infrastructure for science

EGI European Grid Initiative, a foundation aiming at creating and maintaining a pan-European Grid infrastructure to enable access to computing resources for European researchers

EGS Electron Gamma Shower, radiation transport programme for electrons and photons

Eiffel programming language

electron-Volt one electron-Volt (eV) is the energy acquired by one electron falling through a potential energy gap of 1 Volt

Eloisatron Eurasiatic Long Intersecting Storage Accelerator, a particle accelerator project proposed by A. Zichichi

EMDIR Electronic Mail DIRectory, an Oracle-based mail address database setup by the CERN Computing Division in the eighties
Emulator a computer that is able to reproduce via software the behaviour of the hardware of another computer
Entanglement when a quantum system in a given state becomes divided in two or more subsystems these are entangled: each one of them cannot be described separately from the others, no matter how far they are in space-time
Entropy a thermodynamic quantity that measures the disorder of a system, but also the amount of information and of useful work that can be extracted from it
ER Entity Relationship Model, where relations between data are modelled in tabular form
ESA European Space Agency
ESD Event Summary Data
Ethernet a family of computer networking technologies for local area networks based on the exchange of data packets
Ethernet networking technology for local area networks defining wiring, signalling standards and a common addressing format for computer networks
ETICS Einfrastructure for Testing, Integration and Configuration of Software, a E.U. funded CERN project aimed at developing and maintaining a distributed infrastructure and tools to manage Grid software configuration, builds and tests
EUnet a company providing network services
Exa prefix meaning 10^{18}
ExaByte one ExaByte corresponds to 10^{18} bytes
exploit a software or procedure exploiting a defect in a computer system in order to cause unexpected, and often harmful, behaviour, such as gaining unauthorised access to protected information
Expresso a CERN project to develop an in-house ODBMS
F77 common abbreviation for the 1977 FORTRAN standard revision
Facebook a social networking service and website
FAIR Facility for Antiproton and Ion Research at GSI
FATMEN CERNLIB package for file cataloguing
FDDI Fiber Distributed Data Interface, an optical standard protocol for local area network based on optical fibers providing 100 Megabit/s to a distance of up to 200 km
FDT Fast Data Transfer, an application developed by the MonALISA project capable of reading and writing at disk speed over wide area networks
FERMI FERMI experiment at SLAC
Fermilab FNAL laboratory
File catalogue a logical list of the files on the Grid where each entry can correspond to one or more physica replicas on different Grid Storage Elements
File replica on a Grid the same file can be present in several site for improved availability and access performance; each of these copies is called a replica
File system a (usually hierarchical) representation of a collections of data in a form that is human-readable
FIND documentation system developed by IBM

Floating point a system used on computer for representing numbers with a wide range of values using the form $significant\ digits \times base^{exponent}$

FLUKA FLUctuating KAscade, a fully integrated particle physics MonteCarlo simulation package

FNAL Fermi National Accelerator Laboratory near Chicago, Illinois, U.S.

FORTRAN FORmula TRANslation, a computing language invented in the sixties

FPACK a tape and file cataloguing system developed at FNAL

Framework Programme European Union program to fund collaborative Research and Development project across the European Union

FroNTier a CMS-developed server that distributes data from central databases to be read by many client systems around the world

FTD File Transfer Daemon, part of the AliEn middleware that manages file transfers between sites

ftp File Transfer Protocol, an Internet protocol used to transfer files between computers over the network

FTP Software a software company established in 1986

FTS File Transfer Service is a Grid service that transfer reliably files between two Grid locations

FZ ZEBRA package for sequential I/O

G4ROOT implementation of the GEANT 4 navigator system using a ROOT based detector geometry

Ganga The Gaudi/Athena and Grid Alliance is a front-end for the managing output collection, and reporting of computing jobs on a local batch system or on the Grid used by the LHCb and ATLAS experiments at LHC

GAUDI Software framework used by ATLAS and LHCb experiments at LHC

GD3 Graphics library part of CERNLIB

GEANT CERN developed detector simulation package part of CERNLIB

GEM CERN project for a data manager program that was never completed

Geoplex a collection of computing clusters geographically distributed and continuously mirrored for augmented availability and reliability

GEP Graphics system developed at DESY

GGF Global Grid Forum, a conference series dedicated to the Grid

GHEISHA General Hadronic Electromagnetic Interaction SHower, simulation programme to transport hadrons in matter

Giga prefix meaning 10^9

GigaByte one GigaByte corresponds to 10^9 bytes

GKS Graphical Kernel System, first ISO standard for low-level computer graphics introduced in 1977

gLite Middleware suite produced by the EGEE project

Globus Globus Toolkit is an open source software toolkit used for building Grids

Gnome graphics environment for Unix workstations

GNU an Open Source mass collaboration project started in 1983 by R. Stallman at the Massachusetts Institute of Technology aimed at developing "a sufficient body of free software [...] to get along without any software that is not free."

Google a company invested in Internet search, Cloud computing, and advertising technologies

GPGPU General Purpose GPU, the use of GPU for solving computational problems not necessarily linked with graphics

GPL GNU General Public License, widely used Open Source license

GPS Global Positioning System, a space-based satellite system that provides location and time information. It is maintained by the United States government and is freely accessible by anyone with a GPS receiver.

GPU Graphics Processing Unit, a specialised circuit designed to build and manipulate images intended for output to a display

Great Unification a putative physics theory that explains and encompass in the same model all known physical phenomena, and can predict the result of any experiment that can be actually done

Grid A platform that allows direct, seamless and transparent access to high-end computational resources (computers, network, software and data), as is required by a range of collaborative problem solving and resource-brokering strategies emerging in industry, science, and engineering

Grid service a service specific to Grid computing

GRIDFTP a Globus project that produces high-performance, secure, reliable data transfer technologies optimised for high-bandwidth wide-area networks

GriPhyN Grid Physics Network, a Grid project funded by the U.S. National Science Foundation aimed at developing the concept of Virtual Data

Group analysis psychoanalytic doctrine that combines psychoanalytic interpretation with an understanding of interpersonal functioning according to a number of *basic assumptions*

GSI Gesellschaft für SchwerIonenforschung, centre for research on heavy ions at Darmstadt, Germany

GSI Globus Security Infrastructure, part of the Globus middleware providing secure access to the Grid resources via public key cryptography

GTSGRAL commercial implementation of the GKS standard

GUI Graphics User Interface, application allowing to interact with a computer via images and graphic elements rather than written text

HADES High Acceptance Di-Electron Spectrometer at GSI, Darmstadt

HARP HAdRon Production experiment at the CERN Proton Synchrotron

HBOOK Histogram BOOKing, histogramming and statistical analysis package part of CERNLIB

HCP High Ceremony Processes, the software development methodologies elaborated by classical Software Engineering

HDD Hard Disk Drive, a storage device to store persistent data in form of files using a spinning magnetic disk read by a head that moves on its surface

Heap internal memory pool set-up at the beginning of an application and used to allocate memory as needed

HEP High Energy Physics

HEPCAL HEP Common Application Layer, a specification of the common Grid functionality needed by the four major LHC experiments

HEPDB CERNLIB package for bookkeeping and Database management

HEPDDL HEP Data Description Language, an abandoned proposal made at the 14th Eloisatron Project workshop held in Erice, Sicily in 1990 for a language do describe HEP data structures

HepODBMS a class library developed for internal use by the CERN RD45 project to facilitate the development of HEP applications that use a commercial ODBMS

HIGZ High Level Interface to Graphics and Zebra programme part of CERNLIB

Hilbert space vector space with an infinite number of dimensions

HiPPI High Performance Parallel Interface, a computer bus for high speed communication between computers and storage devices

HP a computer manufacturer

HPLOT graphics interface to the HBOOK programme part of CERNLIB

HPSS High Performance Storage System, an IBM developed software to manage PetaByte of data on disk and robotic tape libraries via hierarchical storage management that keeps recently used data on disk and less recently used data on tape

HTML HyperText Markup Language, system of textual annotations used to format, present and control the behaviour of web pages

HTTP HyperText Transfer Protocol, a networking protocol for information systems used as the foundation of data communication for the World Wide Web

HTV graphic programme part of CERNLIB

HTVGUI adaptation of the HTV graphics programme to bitmap displays, part of CERNLIB

Hudson a tool for continuous integration of software written in Java

HYDRA memory and data structure manager part of CERNLIB

Hypertext text on a computer or other device containing references (hyperlinks) to other text that can be readily accessed, by a mouse click or key sequence

Hypervisor a software that allows a computer to run one or ore Virtual Machines (VM)

I-WAY Information Wide Area Year was a project conceived in 1995 as a large-scale test-bed for innovative high-performance and geographically distributed applications at the Argonne National Laboratory and at the University of Illinois at Chicago

I/O Input Output

I/O bound an application whose performance is limited by the data Input/Output

IBM Industrial Business Machines, a computer manufacturer

icc a C++ compiler from Intel

ICL a computer manufacturer

IEEE Institute of Electrical and Electronics Engineers, a non-profit professional association dedicated to technological innovation related to electricity and electronics

IETF Internet Engineering Task Force

IN2P3 Institut National Pour la Physique Nucléaire et des Particules

INDEX a serial twisted pair network connecting terminals on the CERN site to central Gandalf system

INESC Instituto de Engenharia de Sistemas e Computadores, Lisbon, Portugal

INFN Istituto Nazionale per la Fisica Nucleare, Italian national agency for nuclear physics

Ingres a company producing Relational DataBase Management Systems

Intel CPU and other computer component manufacturer

Interactive operating mode of a computer where a request for work is executed synchronously and under full control by the users

INTERCOM Control Data Corporation operating system

Internet a global open system of interconnected computer networks Internet Protocol (TCP/IP) and linking billions of systems worldwide

Interpreter in computing jargon it is a programme that can read a source code and execute the instruction contained in it

IP Intellectual Property

ISO International Standard Organisation

ISR Intersecting Storage Ring, a CERN apparatus where protons are circulated in opposite directions and brought to collide, in operation from 1971 to 1984

IT Information Technology

Itanium a family of 64 bit Intel microprocessors that implement the Intel Itanium architecture

ITU International Telecommunication Union, United Nations specialised agency for information and communication technologies

JADE A compact magnetic detector at PETRA, experiment at the PETRA facility at DESY

Java programming language

Job In the computing jargon, a self-contained portion of work for a computing system

Journal File a log file of computer I/O operation that allows to recovery from a failure replaying the operations recorded in it

KAPACK a CERNLIB package providing random access to data records using keywords

KDE graphics environment for Unix workstations

kernel the main component of a computer Operating System that bridges between the applications and the actual operations done at the hardware level

kilo prefix meaning 10^3

kiloByte one kiloByte corresponds to 10^3 bytes

KUIP Kernel User Interface Package CERNLIB package to implement an interactive language to interact with a programme

L3 detector at the CERN LEP accelerator

LAN Local Area Network

laptop portable personal computer

Latency a measure of time delay experienced in a system; in networks is the time elapsed from the emission of a signal to its reception by the intended receiver

LCB LHC Computing Board, CERN committee charged with the preparation of the software for the LHC era experiments
LCG LHC Computing Grid, synonymous for WLCG
LCP Low Ceremony Processes, the software development methodologies elaborated by modern Software Engineering
LEAR Low Energy Antiproton Ring, a CERN experiment
LEP The CERN Large Electron Positron accelerator, in operation from 1989 to 2000, colliding electrons and positron at a centre-of-mass energy of up to 209 GeV
LEP DB LEP Database Service
LFC LCG File Catalogue, a Grid file catalogue developed by CERN and EGEE
LHC Large Hadron Collider
LHC++ toolkit written in C++ that should have replaced CERNLIB
LHCb LHC beauty experiment, a major experiment at the CERN Large Hadron Collider
Linux Open Source version of the Unix Operating System
LIP Labóratorio de Instrumentação e física experimental de Particulas, Lisbon, Portugal
LISP LISt Processing, a programming language
LLVM project for the development of a collection of modular and reusable compiler and toolchain technologies
LUSTRE an Open Source parallel distributed file system, intended for large scale cluster computing
Machine learning a scientific disciple that studies algorithms allowing computers to evolve their behaviour based on data coming from sensor or databases
MacOSX Operating System developed by Apple
mainframe large high-performance computing system supporting numerous users and applications simultaneously
Makefile configuration file used by the Unix make build system
Many-core a multi-core system with a large number of cores
Mass storage persistent storage of large amounts of computer-readable data
Master in parallel computing, the process that starts, distributes the work to, possibly collect results from and stops the *worker* processes
MCNP MonteCarlo for Neutrons and Photons, transport programme for neutrons and Photons
MEDDLE MEeting with Data Division and LEP, a committee set-up to discuss the LEP experiments computing needs
Mega Prefix meaning 10^6
Mega prefix meaning 10^6
MegaByte one MegaByte corresponds to 10^6 bytes
Megatek company producing high-end graphics displays in the 1980s
Message passing a form of communication where information is contained in messages passed from a sender to one or more receivers

Meta-computing this expression used to indicate in the 1980s and 1990s experiments with high-performance computationally intensive computing applications over wide are networks

Metadata data describing a set of data, usually in a Database

metafile a file containing graphical elements in textual format from which a picture can be produced on a graphic device

MeV one MeV (Mega electron-Volt) corresponds to 10^6 electron-Volts

MFC Microsoft Foundation Classes, C++ API for part of the Microsoft Windows system functionality and services

Micro-processor single integrated circuit that contains all the functions of a central processing unit of a computer

Microsoft software company producing the Windows Operating System and other applications for personal computers

Middleware software used to connect other programs or people and their applications across networks.

MIMD Multiple Instructions Multiple Data, parallel execution where different instructions are executed simultaneously on different data

Mini-computer a multi-user, multi-application middle-size computer system, between the mainframes and the single-user systems (microcomputers or PCs)

MINOS Main Injector Neutrino Oscillation Search, a long baseline experiment sending neutrinos from Fermilab to Soudan, Minnesota

MonALISA networked distributed monitoring system developed at the California Institute of Technology

MONARC Models of Networked Analysis at Regional Centres for LHC Experiments, project aiming at designing a distributed computing model to process LHC data

Monte-Carlo computational algorithm based on the repeated sampling of functions or distributions via computer-generated pseudo-random numbers

Moore's law Empirical law enunciated by G. Moore stating that the number of transistors that can be placed on a single integrated circuit, and hence the performance of the CPU, doubles every 18 months

Moose nickname for the RD41 project

Mortran parametric (macro) extension to FORTRAN

MOSAIC first X11-based Web browser

MOTIF graphical user interface and toolkit for computer graphics based on X11 and introduced by the Open Software Foundation (OSF)

MoU Memorandum of Understanding

MPI Message Passing Interface, a message library to program parallel computers

MPP Massively Parallel Processor systems

Multi-core (processor) a computing component with two or more independent processors ("cores"), which read and execute program instructions independently

Multimedia media and content using a combination of different content forms

Multistreaming a data transmission technique where the data to be transmitted is divided in chunks which are transmitted in several independent streams in parallel, for example when transmitting Web page images together with the Web page text

MVS Monitoring Virtual System, an operating system introduced by IBM

Myrinet a commercial high-speed local area network, designed to interconnect machines forming computer clusters

MySQL Open Source implementation of a RDBMS system

n-tuple ordered sequence of n elements

NA3 CERN experiment at the North Area experimental hall

NA4 CERN experiment in the North Area experimental hall

NA48 CERN experiment in the North Area experimental hall

NA49 CERN experiment in the North Area experimental hall

NA57 a CERN experiment in the North Area experimental hall

NA60 a CERN experiment in the North Area experimental hall

NADIR NAme DIRectory, an Oracle-based database of the members of an experiment used by ALEPH and DELPHI in the eighties

Namespace a logical container providing a context giving a precise meaning to the identifiers it holds and preventing ambiguity with the same identifiers in another namespace

NAT Network Address Translation, the process of modifying the address of data flowing onto the Internet, usually to address multiple computers sharing a single Internet address

NCSA National Centre for Supercomputing Applications, University of Illinois at Urbana-Champaign

Network a set of computers and devices linked by communications channels that allow sharing of resources and information

Neurosis a type of mental disorders involving distress without delusions nor hallucinations, and where behaviour remains within socially acceptable standards

NexT a computer manufacturer

NexTStep operating system of NexT computers

NFS Network File System, a protocol by Sun allowing a client computer to access files on a server over a network as if they were located in its own local storage

NIC Network Interface Card (or Controller), a computer hardware element that connects a computer to a network

Non-regression procedure that verifies that the next version of a software production is "not worse" than the previous

NorduGrid the NorduGrid Collaboration coordinates the development of ARC and provides the best-effort support through community mechanisms

Norsk-Data a Norwegian computer manufacturer

Nvidia a company producing graphics processing units (GPU) and driver for computer graphics

O2 a commercial object-oriented Database Management System

Object an element of a computer programme that associate data and the algorithms that operate on them, which is the basic component of object-oriented programming and Design

Object Databases Database system where information is stored as object and not as records

Objective C programming language

Objectivity commercial object-oriented DataBase Management System

ObjectStore a commercial object-oriented Database Management System

ObjectTime a commercial CASE tool

Obsession an idea maintained so firmly that it resists any attempt to modify it

OCCI Oracle C++ Calling Interface, an API to access the Oracle database from programmes written in C++

ODBMS Object DataBase Management System, a databse system manipulating objects instead of records

ODMG a consortium aiming to put forward a set of specifications that allowed a developer to write portable applications for object database and object-relational mapping products

Offline in HEP jargon it is the software used to process HEP data after that they have been recorded onto mass storage

OGF Open Grid Forum, an open community committed to driving the rapid evolution and adoption of applied distributed computing

OGSA Open Grid Service Architecture, a grid architecture blueprint based at first on Open Grid Service Infrastructure and later on Web services

OGSI Open Grid Services Infrastructure, a set of mechanisms for creating, managing, and exchanging information among entities called Grid services

OID an unique number that identifies every object in an object-oriented DataBase application

OL Object Life, a Software Engineering technique

OLYMPUS an ESA telecommunications satellite launched on July 12, 1989

Omega Spectrometer at CERN West Area experimental hall operated as a facility for some 48 experiments from 1972 to 1996

OMG Object Management Group, a consortium, originally to set standards for distributed object-oriented systems

OMT Object Modelling Technique, a Software Engineering technique

OMTool Object Modelling Tool, a commercial CASE tool

OMW Object Management Workbench, a commercial CASE tool

On-line in HEP jargon it is the software and the components that collect the events from the detector, format and write them onto mass storage

OO object-oriented design and programming

OOADA object-oriented Analysis and Design with Applications, a Software Engineering technique

OPAL Omni-Purpose Apparatus at LEP, one of the experiment at CERN LEP accelerator

OPCAL OPAL experiment CALibration Database

Open Office suite of applications for office use

Open Source Software Open-source software indicates software available in source code form provided under a software license allowing inspecting, changing and possibly distributing the software

OpenLab a collaboration between CERN and industrial partners to develop new knowledge in Information and Communication Technologies through the evaluation of advanced solutions and joint research to be used by the worldwide community of scientists working at the Large Hadron Collider

OpenMP Open Multi-Processing, an API for programming on multi-platform shared memory multiprocessing in C, C++, and FORTRAN based on compiler directives and environmental variables controlling run-time behavior

OPN Optical Private Network, a virtual circuit over an optical network transport layer providing guaranteed capacity between two or more end-points

Opteron processor line produced by AMD with a 64 bit architecture

OQL standard query language for object-oriented databases derived from SQL

Oracle a software company producing relational database management systems

OS Operating System: software that manages the computer hardware providing services to run the user applications and to control the computer status and operations

OSF Open Software Foundation, nonprofit organisation established in 1988 to define open standard implementation for the UNIX system. It has now been absorbed by the OpenGroup

OSG Open Science Grid is a consortium of software, service and resource providers and researchers, from universities, national laboratories and computing centres across the U.S., funded by the National Science Foundation and the Department Of Energy.

OSI Open Systems Interconnection model proposed by the International Organisation for Standardisation based on a layered description of communication systems

OSM Open Storage Manager, a disk and tape management system developed at DESY

P2P peer-to-peer or point-to-point, simplest network topology consisting in direct links between all the devices on the network

PA-RISC HP implementation of the RISC computer architecture

PAF Pico Analysis Framework developed for the BaBar experiment and based on ROOT

PAM PAtchy Master file

Panda a set of middleware tools developed by the ATLAS experiment

Panda experiment at the FAIR facility, GSI, Darmstadt

PAR Proof ARchive, a single file in compressed archive format containing the files to execute some user code under PROOF

Parallel computing simultaneous execution of different parts of a programme to reduce the running time

Pascal a computer language

PASS Petabyte Access and Storage Solutions, a SSC project aimed at solving the data storage and access problem for the LHC and SSC experiments

PATCHY early CERNLIB code management system

PAW Physics Analysis Workstation programme part of CERNLIB

PC Personal Computer, computer destined to be used by a single user

PDC Swedish Centre for Parallel Computing
PDP 11 a computer model manufactured by DEC
PEP, PEP-II Positron Electron Project for a storage ring and collider at SLAC in operation since 1980
Perl programming language
PERQ a computer manufacturer
Persistency attribute of data that outlives the process that crated them, usually by means of writing them into a file
Peta prefix meaning 10^{15}
Peta prefix meaning 10^6
PetaByte one PetaByte corresponds to 10^{15} bytes
PetaByte one PetaByte corresponds to 10^6 bytes
PETRA storage ring based X-ray source operated at DESY, Hamburg
PHIGS Programmer's Hierarchical Interactive Graphics System, an standard for 3-D computer graphics developed in the 1990s
PI Physicist Interface, LCG Application Area project aimed at providing the interfaces and tools by which physicists will directly use the software
PIAF Parallel Interactive Analysis Facility developed at CERN
Pilot jobs a Grid job that acquires its workload only after a communication with some central service holding the job queue
PIONS Partial Implementation of Our New System, CERN project for a graphics package
PL1 programming language
Platform Computing a provider of Grid and distributed software products
plotter a computer output device capable of displaying vector graphics, i.e. drawing lines
Poet a commercial object-oriented Database Management System
POOL Pool Of persistent Objects for LHC, LHC Application Area project aimed at providing a neutral persistency layer for LHC experiments
Posix Portable Operating System Interface for UniX, a set of IEEE standards defining API and interfaces for software intended to run on a variant of the Unix Operating System
PostresSQL an open source RDBMS
PowerPoint application developed by Microsoft to prepare computer-based presentations
PPDG Particle Physics Data Grid, a collaboration of High Energy Physics laboratories in the United States, which was set up in 1999 to gain experience in applying this technology to the problems of particle physics
process a thread of activity on a computer
PROOF Parallel ROOT Facility, an extension of ROOT enabling interactive analysis of large sets of ROOT files in parallel on clusters of computers or many-core machines
proxy a server computer system or application program acting as an intermediary from clients requesting resources.
PSS-05 Software Engineering Standard of the European Space Agency

pthreads Posix standard for threads
PTT Postal, Telegraph, and Telephone service government agency responsible for postal mail, telegraph, and telephone services, usually working in a regime of monopoly
PVM Parallel Virtual Machine, a message library to program parallel computers
Python programming language
QCD Quantum CromoDynamics, a theory of the force describing the interactions of the quarks and gluons (strong interaction) that composed the hadrons, such as the proton, neutron or pion
QT GUI toolkit produced by TrollTech
Quantum Information Theory the theory of the information referred to the quantum world
R602 a CERN experiment
RAC Real Application Cluster, allows multiple computers to run Oracle RDBMS software simultaneously while accessing a single database, thus providing a clustered database
RAL Rutherford Appleton Laboratory in U.K.
RAM high-speed Random Access Memory
RARE Réseaux Associés pour la Recherche Européenne is a research organisation founded in 1986 by several European networking organisations to promote the OSI protocol)
Rational Rose a commercial CASE tool
RB Resource Broker
rBuilder a tool to build and maintain Virtual Machine images
RD41 CERN R&D project to study the viability of the object-oriented software approach for reconstruction code for the LHC experiments
RD44 CERN R&D project to implement in C++ a new version of the GEANT detector simulation programme (GEANT 4)
RD45 CERN R&D project to investigate object-oriented databases solution for the coming experiments, proponent of the use of Objectivity for physics data
Red Hat one of the major provider of Open Software, including a version of Linux
Refection possibility for a programme to dynamically access the description of its own data structures
REFLEX C++ reflection system, supposed to be the main deliverable of the SEAL Project
Resource Broker synonymous for Workload Management System (WMS)
RHIC Relativistic Heavy Ion Collider facility at BNL
RIOS Remote Input Output Station, a room equipped with a card puncher and a printer connected to the CERN main computer system
RIPE Réseaux IP Européens is the Regional Internet Registry (RIR) for Europe, the Middle East and parts of Central Asia
RIR Regional Internet Registry, an organisation that manages the allocation and registration of Internet addresses within a particular region
RISC Reduced Instruction Set computer architecture

RLS Replica Location Service, a Grid service that handles the information on the location and status of File replicas on the Grid

ROOT High Energy Physics data manipulation and visualisation package developed at CERN

rootcint utility programme part of the ROOT package to generate the reflection dictionary and the I/O methods for the root classes

Router a device to steer data packets over computer networks

RPC Remote Procedure Call, a network protocol allowing a programme running on a computer to execute code on another computer

RS232 name for a set of standards for serial binary signals using in computing serial communication ports

RSA Rivest, Shamir and Adleman, the inventors of an algorithm for encrypting a computer message via public-key cryptography in a way that it can only be read by the intended recipient, posessing the corresponding private-key

RSCS Remote Spooling Communications Subsystem Networking is an IBM proprietary networking protocol

RTAG Requirements Technical Assessment Group, technical subcommittees mandated by WLCG to explore a particular problem area

RWN Row Wise N-tuple

RZ ZEBRA package for direct access I/O

S3 Simple Storage Service, a service provided by Amazon to store and retrieve data from any location on Internet at any time

SaaS Software as a Service, a software access model where software is delivered on demand over the Internet to users requesting it via a thin client, usually a Web portal, typically in the context of a computing Cloud

SAN Storage Area Network, a dedicated storage network that provides access to disk

SASD Structured Analysis and Structured Design, a Software Engineering technique

SCANBOOK a bookeeping system used by the ALEPH experiment in the nineties

Scheduler a computer programme supervising the concurrent execution of different tasks by a computer

Schema evolution possibility to change the model of data or objects stored in a Database preserving the capability to read the old data

SCI Scalable Coherent Interface, a standard for a shared-memory high-speed interconnect used in multi-processor systems and for message passing

SCRAM code maintenance system

SE Storage Element provides a view of the data stored at each Grid site, maps Grid names to the local namespace, and supports services for data management and transfers

SEAL LCG Application Area project aimed at providing a C++ reflection system

Search engine a system designed to search and retrieve information on the Web and FTP servers

SEFT Helsinki Reserach Institute in High Energy Physics

Self-scheduling a mechanism where each worker gets a decreasing amount of work, based on its performance in executing it, to ensure that all thread terminate as closely as possible

Serialisation conversion of a set of data in memory into a format where the objects are placed one after the other, so that it is possible to store them into a file or transmit them across a network, together with the information that allows the original structure to be "expanded" in its original form when it is read or received

Server a programme running to serve requests of other programmes ("clients") on the same computer or across a network. By extension, a server also denotes the computer that runs such programmes

Service in computer science a service is a mechanism that allows one electronic device to discover, request and obtain a service from another over a network

Service machine a computer that act as a server

SGI a computer manufacturer

SGML Standard Generalised Markup Language, an ISO standard for generalised markup languages for documents

SHIFT Scalable Heterogeneous Integrated Facility, a CERN project aimed at providing computer resources to the experiments in the form of networked computers and storage

SIGMA array manipulation and visualisation package part of CERNLIB

SIMD Single Instruction Multiple Data, parallel execution where the same instruction is executed synchronously on multiple data.

SLAC The Stanford Linear Accelerator Centre at Stanford, California, U.S.

SmallTalk programming language

SMP Symmetric Multi Processor, a parallel computer where every processor has access to the shared memory via a single shared bus

SMS Short Message System, text message communication on mobile communication system, mostly phones

SMTP Simple Mail Transfer Protocol an Internet standard for e-mail transmission over IP networks

socket logical endpoint of a bidirectional inter-process communication across a network

Software Engineering a "systematic approach to the analysis, design, assessment, implementation, test, maintenance and re-engineering of software, that is, the application of engineering to software."

Solaris proprietary operating Unix based operating system developed by Sun

Sony a Japan-based leading manufacturer of electronics products for consumers and professionals

SPIRES Stanford Physics Information Retrieval System, a physics database containing information on HEP pre-prints

SPS CERN Super Proton Synchrotron particle accelerator

SQL Structured Query Language, used to manage data in relational databases

SQL/DS a company producing Database systems

SQLite software library that implements a self-contained, serverless, zero-configuration, transactional SQL database

SQS Simple Queue Service, a service provided by Amazon that offers a reliable, highly scalable, hosted queue for storing messages as they travel between computers

Squid Web Cache a caching proxy supporting a range of Web protocols aimed at reducing bandwidth and improving response times by caching and reusing frequently-requested web pages

SRM Storage Resource Manager is a protocol for Grid access to mass storage systems resulting from an collaboration of HEP laboratories

SSC Superconducting Super Collider Laboratory, a planned particle accelerator complex to be built in the vicinity of Waxahachie, Texas set to be world's largest and most energetic, surpassing the current record held by LHC

SSD Solid State Drive, a storage device for persistent data in form of files and using solid-state memory rather than spinning disks

STELLA communication satellite launched at the beginning of the eighties

STK Storage Technology Corporation, a data storage technology company focussing on tape backup devices and the associate software

STL Standard Template Library, a C++ library of container classes, algorithms, and iterators

StorageTek alternative name for STK

STOREGATE ATLAS data access framework

StP Software through Pictures, a commercial CASE tool

Subconscious according to Sigmund Freud's theory, it is the repository for memories, thoughts and emotions excluded from the conscious mind by the psychological mechanism of repression

SUMX statistical analysis package developed at Berkeley

SUN a computer manufacturer

SURL Storage URL, a string that identifies a file on the Grid with its physical location and access protocol

SVN SubVersnioN system, a code management system which is an evolution of CVS

Switched interconnect a connection between multiple elements based on switches

Synchronicity meaningful coincidence events, that are apparently causally unrelated or unlikely to occur together by chance

Synchrotron a cyclic particle accelerator in which the magnetic and electric fields used to guide the particles are synchronised with the circulating beam

syslog Unix monitoring service

Tag collection of important features of each event in a given set, which are used to select the interesting events for further processing without reading the full event collection

Task queue a list of workloads to be performed by a computing system, possibly on a Grid

TATINA programme to simulate the interaction of hadrons with nuclei

TCP/IP Transmission Control Protocol/Internet Protocol, a protocol to manage the connection of computer systems to the Internet

TCP/IP window scaling the amount of data transferred in a TPC/IP connection is limited by the receive window set by the receiver between 2 and 65,535 Bytes; if a receiver is ready to receive more, it can indicate a scaling factor to go beyond the maximum specifiable size

Tektronix a U.S. company producing electronic test measurement equipment which used to produce also computer graphic displays

telnet a protocol used on Internet or local area networks to provide a text-oriented virtual terminal connection

Tera prefix meaning 10^{12}

TeraByte one TeraByte corresponds to 10^{12} bytes

TeV one Tera electron-Volt corresponds to 10^{15} electron Volts

thread a unit of processing within a computer programme, which can possible be executed in parallel with other threads

Throughput amount of computing work performed in a given amount of time

Tier 0 in the MONARC model the Tier 0 (T0) is the laboratory where data originate (in case of LHC this is CERN) that performs initial processing of the data and maintains master copies of the raw and other key data-sets to be exported to Tier 1's and Tier 2's.

Tier 1 in the MONARC model a Tier 1 (T1) is a large computing centre providing long term data warehousing, holding synchronised copies of the master catalogues, used for the data-intensive analysis tasks, and act as data servers for smaller centres

Tier 2 in the MONARC model a Tier 2 (T2) is a centre used for end-user analysis and simulation.

TIFR Tata Institute for Fundamental Research, Mumbai, India

Tk/TCl Tool Command language, a scripting language for rapid prototyping of user interfaces

TMS Tape Management System, a system to catalogue and manage computer magnetic tapes

Token ring network protocol using a three-byte frame called token travelling in a ring and providing permission to transmit to the medium.

Trigger in HEP jargon a trigger is a very fast signal generated by the detector that activate the recording of the current event used in order to reduce the data volume and only record interesting events

Tru64 a 64 bit Unix operating system for the Alpha architecture

TSO Time-Sharing Option, an IBM Operating System component

UA1 CERN detector aimed at studying the proton-proton collisions and installed in the Underground Area 1 experimental hall of the CERN Super Proton Synchrotron. In 1982 with this apparatus, C. Rubbia discovered the Z and W bosons, discovery for which he was awarded the 1984 Nobel Prize for Physics

Ultrix Unix based Operating System developed by Digital Equipment Corporation

UML Unified Modelling Language, a modelling language for object-oriented software engineering

Unix multitasking, multi-user operating system developed in 1969 at AT&T Bell Labs

UNOSAT United Nations institute for training and research Operational SATellite applications programme, to deliver integrated satellite-based solutions for human security, peace and socio-economic development, in keeping with the mandate given by the UN general assembly since 1963

Unus Mundus One World, the concept that everything belongs to a unified reality from which everything comes and returns to

UPDATE code management system on CDC computers

URI Uniform Resource Identifier set of characters identifying a resource on Internet

URL Uniform Resource Locator, a set of characters indicating how to find a resource on Internet

Usenet a worldwide Internet discussion system precursor of nowadays blogs, built on the general purpose UUCP architecture of the same name

Usenix the Advanced Computing Systems Association

UUCP Unix-to-Unix Copy, a collection of programmes and protocols for remote execution of commands and transfer of files, email and netnews

VAX Virtual Address eXtension, an instruction set computer architecture developed by Digital Equipment Corporation

VAXcluster a cluster of DEC computers running the VMS operating system

VCR Video Cassette Recording, a domestic video format designed by Philips and based on magnetic tapes

VDT The Virtual Data Toolkit is an ensemble of Grid middleware software produced by the OSG project as its software distribution

Vector graphics graphic composed by lines of different colours and thickness, as opposed to pixel or bitmap graphic, where the graphical representation is obtained by assigning a colour to each point (pixel) on the surface

Vectorisation Optimisation of the code where the same arithmetic operation is performed simultaneously on several values. It allows to obtain performance improvements on vector machines such as the Cray

Versant a commercial object-oriented Database Management System

Virtualisation procedure by which a real hardware components are replaced by programmes reproducing their functions via software emulation

VM Virtual Machine, a software programme that emulates the behaviour and components of a complete computer

VM/CMS Virtual Machine / Conversational Monitoring System, proprietary name of an IBM Operating System

VMC Virtual MonteCarlo, a virtual interface to MonteCarlo particle transport programme developed by the ALICE collaboration

VME VME or VMEbus a standard for computer component connections developed for the Motorola 68000 CPUs, widely used for many applications and later standardised by the American National Standard Institute

VMS Virtual Memory System, an Operating System introduced by DEC

VMSS Virtual Mass Storage System, the XROOTD global distributed namespace that is seen by a site like a virtual mass storage system

VMware application to run virtual machines

VO structure and associate software tools that define and manage a community of Grid users

VO box Virtual Organisation box, a machine running Virtual Organisation specific Grid services

VOMS Virtual Organisation Membership Service, a Globus component for managing authorisation data within multi-institutional collaborations

VT100 video terminal produced by DEC, whose attributes became the de facto standard for terminal emulators

W3C World Wide Web Consortium, principal standards organisation for the World Wide Web

WAN Wide Area Network

Web service a service using the Web communication protocols

Website a collection of related Web pages hosted on one or more servers and accessible from a top-level page usually called "home"

Widget a standard configurable and reusable element in computer graphics such as a drop-down menu or a dialog box

Windows Microsoft proprietary operating system

WIPO World Intellectual Property Organisation, a U.N. agency created in 1967 "to encourage creative activity, to promote the protection of intellectual property throughout the world."

Wireless network a computer network where the computers communicate via radio signals

WISCNET early implementation of the TCP/IP protocol by Wisconsin University

WLCG Worldwide LHC Computing Grid

WMS Workload Management System

WN Worker Node, one physical or virtual machine that participates to the execution of the Grid workload

Worker in parallel computing one of the several processes that are controlled by the *master* and that execute the workload in parallel

Workload management distribution of computing tasks amongst different computing elements to optimise application performance

Workstation medium size computing system supporting a limited number of users and application simultaneously

X-Windows synonym for X11

X.25 an international standard protocol for packet-switched wide area networks

X11 X Windows System is a software that provides the basic elements for a graphics interface for networked computers

Xerox a company producing multifunction printers and photo-copiers

XFIND extension of the FIND system developed at CERN

XNS Xerox Network Services, a network protocol developed by Xerox

XP eXtreme Programming, a software development methodology aiming at improving software quality via lightweight procedures and rapid response to changing customer requirements

XRDCP high performance file copy utility part of the XROOTD software

xrootd software package developed at SLAC and aiming at giving high performance, scalable fault tolerant access to data repositories of many kinds

XWHO electronic phone-book developed at CERN

Y2K Year 2000, shorthand notation for the problem which resulted in the practice used during the twentieth century of abbreviating four-digit years to two digits, both for computer programs and non-digital documentation

YouTube a video-sharing website on which users can upload, share and view videos

ZBOOK memory and data manager programme part of CERNLIB

ZCEDEX ZEbra based Command EDition and EXecution programme

ZEBRA CERNLIB package for dynamic memory management

ZOO proposal for Zebra object-oriented presented in 1994

Index

3D, 257

Absolute zero, 2
Accelerators, *see* Particle accelerators
Access
    random, 21
ADA, 11
ADAMO, 43, 123, 230–232, 237
Advanced Micro Devices, *see* AMD
AFS, 272
Airbus, 134
Alignment, *see* Detector, alignment
AliRoot, 138, 149
Amazon
    Elastic Compute Cloud, 168, 172
    Scalable Storage Service, 172
    Simple Queue Service, 172
AMD, 9
    Opteron, 138
Amdahl's law, *see* Parallel computing, Amdahl's law
AMSEND, 232
Analysis, 3
Antimatter, vi
Apache, 94
API, 240
Apollo, 7, 25, 27, 56, 63, 108
    AEGYS, 8
    DN 10000, 8, 71
    DN 400, 8
    DN 600, 8
    Domain, 8, 28
    joint project with CERN, 189
Apple, 10
    Macintosh, 29
    MacOSX, 10, 169

Archetype, 290
Arena, 60
Aristotle, 292, 293
ARPA, *see* DRPA61
ASCAP, 213
Aspirin, 219
Assembler, 10, 17
Astronomy, 156
AT&T, 66
ATHENA, 39
Athens, 236
ATLAS Description Language, 39
Atom, vi
Atomic transaction, 275
Authenticatio, *see* Grid, authentication
Authorisation, *see* Grid, authorisation

B-tree, 274
BaBar, 248, 251, 258
BARB, 17
Batch processing, 6, 19, 26, 27, 32, 33, 45, 73, 167, 172, 179, 180, 226, 319
    virtual, 168
BBCP, 279
Beaujolais, 25
Bern Convention, 204
Big Bang, viii, 72
Biology, 296
Bitmap display, 8, 28
BITNET, *see* IBM
BMI, 213
BNL, 38, 192, 236
Boinc, 172
BOS, 21

Brain, 291, 293
   as a physical object, 293
   development, 291, 303
   development (M. Klein), 303
   development (Piaget), 303
   neurons and thoughts, 292
   parallelism, 305
   versus galaxies, 291
   virtualisation, 305
British Telecom, 191
Buildbot, 167

C, 14, 188
C++, 12, 14, 16, 19, 22, 26, 35, 38, 43, 45, 115, 148, 192, 241
   abstract class, 149
   interpreter, 192
   templates, 16
   transition from FORTRAN, 148
Cadillac, 219
Calcomp, 32
Calibration, *see* Detector, calibration
CAP project, 247
CASTOR, 73, 80, 234, 236, 250, 259
CCIN2P3, 234
CDC, 27, 69
   CDC 3100, 5
   CDC 6400, 6
   CDC 6500, 6
   CDC 6600, 4
   CDC 7600, 6, 17
   CERNScope, 5
   FOCUS, 5, 61
   INTERCOM, 6
   Scope, 5
CERN, vii
   Cafeteria, 289
   Program Library, *see* CERNLIB
CERN experiments
   ALEPH, 21, 43, 230, 233, 235, 238
   ALICE, 26, 27, 98, 99, 133, 148, 170, 249, 253, 257, 268, 271
      Computing Model, 282
   ATLAS, 2, 19, 35, 38, 94, 98, 99, 107, 111, 170, 244, 247, 249, 257, 258
   Bubble Chamber, 18, 20, 31, 33
   CMS, 19, 38, 98, 244, 249, 257
   CNS, 99
   COMPASS, 151, 244, 250, 252, 258
   DELPHI, 22, 230, 235, 238
   HARP, 98, 151, 250
   L3, 21, 25, 230, 235, 238, 241
   LEAR, 233
   LHC, 240
   LHCb, 19, 39, 98, 99, 233, 249, 257
   NA3, 24
   NA4, 6, 24
   NA48, 239
   NA49, 9, 35
   NA57, 151
   NA60, 151
   Omega, 24
   OPAL, 7, 24, 25, 71, 230, 235, 238
   R602, 31
   RD45, 244
   UA1, 7, 24, 28, 33
CERNET, 8, 61
Cernettes, 294
CERNLIB, 6, 11, 16–19, 123, 129, 135, 146, 150, 234
   COMIS, 34
   FATMEN, 15, 233, 236, 239–241
   GD3, 27, 32
   GEM, 129
   GENEVE, 24
   HBOOK, 22, 27, 32
   HEPDB, 15, 235, 239–241, 245
   HIGZ, 25, 29, 34
   HPLOT, 27, 32
   HTV, 8, 27
   HTVGUI, 28
   HYDRA, 20, 21, 28, 135, 231
   KAPACK, 228, 230
   KUIP, 12, 26, 33
   PATCHY, 18
   PAW, 8, 12, 22, 25, 28, 33, 44, 123, 135, 140, 146, 148, 188, 232
   PAW++, 29
   PIAF, 34, 186, 188–192
      architecture, 188
      performance, 191
   SIGMA, 27
   ZBOOK, 21, 231
   ZCEDEX, 25, 33
   ZEBRA, 13, 21, 135, 231, 237, 259
      FZ, 230, 239, 241
      RZ, 230, 236, 241, 244
   ZOO proposal, 22
CERNVM, 229, 232
CernVM, 170
Certificate Authorities, 222
Chanel, 220
CHEETAH, 237
CHEOPS, 236
CHEP, 225
   CHEP 1991, 238
   CHEP 1992, 14, 18, 240

Index

CHEP 2000, 248
CHEP 2006, 263
CHEP 2010, 263
Cisco, 64
CLI, 240
Cloud computing, 67, 91, 93, 100, 155, 163–165, 284
    definition, 167
    field test, 109
    for HEP applications, 104–112
    private, 168
    public, 168
    storage cloud, 173
    virtualisation, 167, 170
Cluster, computing, 9
CMT, 19
CMZ, 19
CNUCE, 61
Code development
    distributed, 49
Code management systems, 18–19
Code optimisation, 186
CodeMe, *see* CMZ
Coding conventions, 50
Commedia dell'Arte, 295
Computer
    desktop, *see* Desktop computer
    mainframe, *see* Mainframe computer
    many-core, *see* Many-core
    mini, *see* Mini-computer
    multi-core, *see* Multi-core
    PC, *see* PC
    personal, *see* PC
    supercomputer, *see* Supercomputer
    workstation, *see* Workstation
Computer science, 120
Computing, viii, 294
Computing in High Energy Physics, *see* CHEP
Consciousness, 290, 300, 305
Control Data Corporation, *see* CDC
COOL, 257
Copyleft, *see* Copyright, Copyleft
Copyright, 202, 204, 209–216
    "works-made-for-hire", 210
    Copyleft, 212
    Database, 202, 216–218
    Database protection law, 202, 204
    derivative work, 212
    Digital Millennium Copyright act, 216
    duration, 214
    Fair use, 214
    infringement, 214
    Internet, 213
    Internet search engines, 216
    moral rights, 213
    Preprint servers, 222
    Software
        Creative Commons, 212
    software, 211, 212, 215
        reverse engineering, 215
    Statute of Anne, 210
    Web hyperlinks, 213
CORAL, 258
CORBA, 192
CORE, 28, 190
Cosmic rays, vi
CPP, 19
Cray, 7, 9, 69, 108
    COS, 64
    T3D, 184
    X-MP, 183
    XMP, 7, 64
    Y-MP, 183
cvs, 19
Cyclotron, vii

daemon, 197
DAG, 275
DARPA, 61
Data
    access, distributed, 48
    acquisition system, *see* Data, DAQ
    analysis, 180–182, 268, 277
    analysis, interactive, 38
    caching, 71
    curation, 106, 109
    DAQ, 7, 150
    dictionary, *see* Reflection
    ESD, 3, 105, 247
    File catalogue, *see* Grid, File catalogue
    long term preservation, 170, 250, 261
    management, 79
        distributed, 71, 271
    mass storage, 71, 173
    metadata catalogue, *see* Grid, metadata cagalogue
    placement, 106, 268
    raw, 3, 72, 105, 181
    replication, 105, 268
    serialisation, 281
    storage, 71, 79, 106
        virtualisation, 269
    store, transient, 23
    structure, 11, 13, 20
    structure managers, 20, 229, 236
    transfer, 106
    visualisation, 181

Data
    storage, 314
Data Definition Language, see DDL
Data management
    distributed, 267–285
Data Manipulation Language, see DML
Database, 111, 236
    bind variables, 261
    commercial vs home-grown, 240
    commercial, in HEP, 228
    DataBase Management System, see
        Database, DBMS
    DBMS, 182, 241
        in HEP, 228
    definition, 227
    distributed lock manager, 254
    ECFA working group, 227, 261
    EMDIR, 232
    for LHC, 256
    in HEP, 225–264
    licensing, 260
    management limit, 227
    metadata, 73, 80, 226, 242, 262
    NADIR, 232
    Object DataBase Management System, see
        Database, ODBMS
    ODBMS, 15, 23, 44, 242–248
    PetaByte scale, 240
    protection law, see Copyright, Database
        protection law
    SCANBOOK, 232
    schema evolution, 23, 37, 45, 244
        automatic, 40
    security, 250
    services for the Grid, 262
    SQL, 34, 228
DataSynapse, 164
DATE, 151, 249
DBL3, 230, 235, 239, 240
dCache, 47, 73, 80, 259
DDL, 243
DEC
    ALPHA, 30
    AlphaServer, 138
    DECnet, 62, 63
    PDP 11, 6
    Tru64, 138
    Ultrix, 8
    VAX, 28, 179, 231
    VAXCluster, 254
    VAXcluster, 108
    VAX 11/780, 226
    VAX 750, 28
    VAX 780, 7
    VMS, 7, 56, 62, 226, 234, 235, 239, 240
    workstations, 8
DECnet, see DEC, DECnet
DeCSS encryption, 216
Desktop computer, 60, 86, 169, 277
Desktop computers, 173
DESY, 229, 235, 236
Detector, viii
    alignment, 3
    calibration, 3, 228, 257
    geometry, 21
        package, 24
    simulation, 2, 24–27, 74, 109, 169, 180
Development
    evolutionary versus planned, 290
DIANE, 95, 96
Digital Equipment Corporation, see DEC
Direct Acyclic Graph, see DAG
DML, 243
DOC format, 17
DoXygen, 136
Dsektop computer, 10
Dual-core, 9
Duality, vi
    Body and Soul, 292

EARN, see IBM, EARN
EBU, 96
EC2, see Amazon, Elastic Compute Cloud
ECFA, 227, 238
EGS, 24
EGS, EGS 4, 25
Eiffel, 14
Einstein, Albert, 297
Eloisatron, 236
Emulator
    168E, 7
    3081E, 7
Entanglement, 298
Entity Relationship Model, see Software
        Engineering, ER
Entropy, 300
Erice, 236
ERNLIB
    DZDOC, 237
Erwise, 60
ESA, 123
ESD, see Data, ESD
Espresso, 252
Espresso project, 242, 246–247
Ethernet, 8, 59, 63
    fast, 179
    Gigabit, 195

ETICS, 167
EUnet, 62, 64
EuroGraphics, 28
European Broadcasting Union, see EBU
European Committee for Future Accelerators, see ECFA
European Space Agency, see ESA
Event Summary Data, see Data, ESD
Event tags, see Tags
Experiments
    at CERN, see CERN experiments
    BaBar, 23, 38, 44, 98, 272
    CBM, 27
    CDF, 22, 235
    D0, 38, 235, 241
    FAIR, 27
    FERMI, 19
    HADES, 27
    JADE, 261
    MINOS, 27
    Panda, 27
    PETRA, 25
Experiments, High Energy Physics, vii

Facebook, 67
FATMEN, 235
FDDI, 179, 189
FDT, 279
Fermilab, see FNAL
Ferranti
    Mercury, 4, 69
File replica, see Grid, File replica
File system, 275
    distributed, 198, 273
Filesystem, 42
FIND, 17
Firefox, 60
FLUKA, 25–27, 135, 140
FNAL, 22, 235, 257
Fondazione Bruno Kessler, 136
FORTRAN, 5, 10, 12, 15, 26, 148, 188, 192, 226, 228, 237, 241
    COMMON BLOCK, 21
    F77, 11
    FORTRAN 77, 44
    FORTRAN 90, 12, 22, 43
    interpreter, 34
FPACK, 235
Framework, 40
Framework Programme, 78
FroNTier, 257
ftp, 64
FTP Software, 63

Gandalf, 61
Ganga, 80, 95, 96
Garbage collection, 22
GAUDI, 39, 41
GEANT, 6, 21, 24, 142, 146
    GEANT 1, 6
    GEANT 2, 6, 24
    GEANT 3, 7, 8, 21, 25–26, 32, 123, 135, 140
    GEANT 4, 15, 19, 26, 32, 44, 94, 98, 129, 135, 242
General Motors, 219
Geometry, detector, see Detector geometry
GEP, 33
GHEISHA, 25
GKS, 28, 33
Global Grid Forum, 158, 166
Global Positioning System, see GPS
Globus, see Grid middleware, Globus
GNU, 306
Gnu Public License, see GPL
Google, 66
    Trends, 164, 166
Gopher, 17, 61
GPGPU, see GPU
GPL, 61, 212
GPS, 97
GPU, xiii, 49, 198
Grammelot, 295
Gran Sasso National Laboratory, 27
Graphical Kernel System, see GKS
Graphics, 27
Graphics Processing Units, see GPU
Graphics User Interface, see GUI
Gravitational constant, 218
Great Unification, 293
Grid, ix, xii, 23, 39, 44, 45, 49, 67, 69–88, 91–113, 155–162, 226, 234, 251, 289
    application porting, 93
    authentication, 75, 110, 221, 273
    authorisation, 75, 273
    baseline services, 82
    CE, see Grid, Computing Element
    Computing Element, 75, 166
    Data Grid, 159
    database services, 77, 104, 106
    definition, 163
    File catalogue, 77, 166, 269, 270
    File replica, 77, 268, 275
    funding, 112, 156
    HEPCAL, 81
    information system, 76
    Job agents, see Grid, pilot jobs

metadata catalogue, 269
operational cost, 107
overlay, 171
pilot jobs, 77, 162, 169, 172
proposal for LHC, 78
Replica Location Service, *see* RLS
Resource Broker, *see* Grid, Workload management system
RLS, 235, 255
SE, *see* Grid, Storage Element
security framework, 75
services, 159
SRM, 80
standards, 166
Storage Element, 76, 166, 283
Storage management, 76
SURL, 236
Task queue, 162
Tier 0, 74, 105, 111, 240
Tier 1, 74, 76, 79, 80, 83, 105, 111, 240
Tier 2, 74, 79, 83, 85, 105, 240
Virtual Organisation, 76, 93, 111, 167
virtualisation, 162, 171
VO, *see* Grid, Virtual Organisation
VO box, 168
WMS, *see* Grid, Workload management system
WN, *see* Grid, Worker Node
Worker Node, 166
Workload management system, 77, 159, 166, 269
Grid
  authorisation, 335
Grid middleware, 80, 82–83, 163, 172, 259
  Advanced Resource Connector, *see* Grid middleware, ARC
  AliEn, 138, 149, 160, 271, 282
  ARC, 82, 160
  Condor, 75, 81, 172
  Dirac, 160
  Disk Pool Manager, *see* Grid middleware, DPM
  DPM, 80, 259
  File Transfer Daemon, *see* Grid middleware, FTD
  File Transfer Service, 236
  File transfer service, 76, 227
  FTD, 282
  FTS, *see* Grid middleware, File transfer service
  gLite, 83, 160, 162
  Globus, 75, 78, 80, 82, 156
  Globus Alliance, 159
  Globus Security Infrastructure, *see* Grid middleware, GSI
  GSI, 157
  in Cloud computing, 172
  LFC, 236, 259
  OGSA, 159, 165
  Open Grid Service Architecture, *see* Grid middleware, OGSA
  Open Grid Service Infrastructure, *see* Grid middleware, OGSI
  OSG, 160
  Panda, 160
  Resource Broker, *see* Grid, Workload management system
  resource broker, 159
  SRM, *see* Grid middleware, Storage Resource Manager
  Storage Element, 227
  Storage Resource Manager, 77
  versus virtualisation, 165
  virtualisation, 160, 167
  VOMS, 260
Grid Middleware, FTS, 259
Grid projects, 157, 158
  EDG, 78, 137, 159, 235, 255
  EGEE, 81, 96, 107, 137, 159
  EGI, 81, 96, 99, 107, 108
  European Data Grid, *see* Grid projects, EDG
  GriPhyN, 78
  NorduGrid, 82, 160
  OSG, 81, 99, 108
  PPDG, 75
  VDT, 81, 82
GRIDFTP, 279
Group analysis, 302
GSI, 27, 283
GTSGRAL, 28
GUI, 30, 40, 48

Hard Disk Drive, *see* HDD
HDD, 198
Heap
  multiple, 22
HEPDDL, 237
HepODBMS, 244
Hewlett-Packard, *see* HP
Hilbert space, 300
HiPPI, 34, 179
HP, 9, 34, 71, 108
  Bristol laboratory, 189
  Convex, 183, 185
  HP 720, 9

Index 343

HP 750, 9
Itanium, 138
joint project with CERN, 189
PA-RISC, 9, 179, 189
HPSS, 251
HTML, 55, 59, 290
HTTP, 55
Hudson, 167
HyperText Markup Language, see HTML
HyperText Transfer Protocol, see HTTP

I-WAY, 75
IBM, 6, 17, 69, 108, 159, 231
   370/168, 6
   3081, 6
   3270, 63
   BITNET, 8, 62, 65
   Blue Gene, 186
   DB2, 259
   EARN, 62, 65
   EASINET, 65
   HPSS, 236
   IBM 370
      assembler, 234
   IBM 704, 10
   MVS, 6, 235
   RISC 6000, 9
   RSCS, 65
   SP-2, 185
   Token Ring network, 63
   TSO, 6, 28
   VM/CMS, 6, 56, 62, 63, 164, 226, 234, 235, 239
ICL, 7
Ideas, in Plato, 292
IEEE, 207, 235
   1244, 272
IETF, 58
IN2P3, 236
Industrial Business Machines, see IBM
Industrial rights, 204
INESC, 236
INFN, 26
   INFN-GRID, 75
Information Technology, 122, 294
Ingres, 230
Intel, 9, 227
   icc, 186
   Paragon, 184
   Pentium, 43
   Pentium Pro, 9, 73
   x86, 9

Intellectual Property, 201, 203, 204
   rationale, 204
   software, 201–222
      HEP, 205
Interactive processing, 226
International Communication Union, see ITU
Internet, 49, 58, 61–66, 110, 289
   identity verification, 221
   law of the, 220–222
   search engines, 67, 216, 220, 304
Internet Protocol, see IP
Interpreted code, 48
Intersecting Storage Rings, see ISR
IP, 58
   address, 171
ISR, 31
ITU, 94, 96

Java, 16, 40, 43, 115, 232, 241

KDE, 30

LAN, see Network, local area
Language, 290, 291, 306
   of the Web, 296
Language, ambiguity, 295
Laptop computer, 10, 23, 127
Large Electron Positron accelerator, see LEP
LCB, 36, 44, 248
LCG, see WLCG
Learning, 298
LEP, 7, 9, 60, 63, 70, 71, 108, 127, 188, 190, 192, 225
   "Green Book", 256
   "green book", 239
   Computing, 231
   Computing "Green Book", 231
   Database service, 226, see LEP, DB
   DB, 231, 254
   File and Tape management, 233
   MEDDLE, 233
LHC, 69, 70, 72, 170, 192, 226
   Computing requirements, 72, 159
   Computing, funding of, 74
   Data size, 72, 79
LHC Computing Board, see LCB
LHC Computing Grid, see WLCG
LHC++, 36, 129, 242, 246, 261
License, 202

Linux, 9, 30, 73, 121, 137, 169, 227, 248, 253, 306
    DVD player, 216
    Red Hat, 138
    RedHat, 30
LIP, 236
LISP, 12
LLVM, 48
Logitech, 8
Luis Vuitton, 220
LUSTRE, 272
Lynx, 60

Machine learning, 181
Mainframe computer, 9, 58, 60, 108, 126, 178, 234
Many-core, 197
Mass storage, *see* Data, mass storage
Matter
    anti, *see* Antimatter
    structure of, vi
MCNP, 24
mcvax, 64
Megatek, 28
Memory
    bus interconnect, 184
    crossbar interconnect, 184
    distributed architecture, 184
    shared architecture, 183
    switched interconnect, 184
Memory management, 13, 20
Message passing, 184
Meta-computing, 155
Meyrin, 289
Microsoft, 9, 159
    MFC, 30
    PowerPoint, 10
    Windows, 10, 60, 73, 137, 169, 211, 220
middleware, 290
Mini-computer, 179
MonALISA, 16, 95, 197
MONARC, 74, 83, 240
Monte-Carlo methods, 2, 169
Moore's law, 9
Moose project, 14, 242
Mortran, 11
MOSAIC, 18
Mosaic, 60, 94
MOTIF, 18, 29
Motorola 68000, 7, 8
Mozilla, 60
Multi-core, 9, 46, 197
Multi-process, 197

Multi-Resolution Pyramid Image, 97
Multi-threaded, 197
Music files, 269
Myrinet, 179
MySQL, 258, 259

n-tuple, 33, 34
    column wise, 188
    column-wise, 34, 40, 182
    row-wise, 34, 182
NAT, 171
National Centre for Supercomputing Applications, 18
Nervous system, 298
Netscape, 60, 94
Network, 7, 62, 83–84, 179, 182, 230
    address, 171
    Address Translation, *see* NAT
    ADSL, 276
    bandwidth, 23, 47, 67, 83, 106, 276
    CERN, 8, 61, 62, 64, 65, 86
    computing, 58, 71, 94
    evolution, 74
    file sharing, 273
    in Europe, 65
    Internet, *see* Internet
    latency, 23, 267, 277, 281
    load, 76
    local area, 61, 267
    Network Interface Card, *see* Network, NIC
    NIC, 189
    OPN, 84
    PC, 63
    problems, 76
    protocols, 8, 43, 58, 61–64
    research, 155
    standards, 8, 43, 59, 62
        open, 56
        proprietary, 56
    switch, 84, 262
    TCP/IP, *see* TCP/IP
    token ring, 8, 63
    Unix, 64
    virtualisation, 164, 171
    wide area, 48, 267, 276, 285
    wireless, 10
    X.25, 58
Network
    protocols, 318
Network Address Translation, *see* NAT
Network Information Centre, *see* NIC
Network, protocols, 159
Neurosis
    constraining, 290

Index

NexT, 14, 18, 59
   NexTStep, 59
NFS, 190, 241, 273
NIC, 64
Nokia, 30
Norsk-Data, 6
   Nord10, 6
   Norsk50, 6
Nuclear Power Reactors, 10
Nvidia
   CUDA, 198

O2, 243
Object Data Management Group, see ODMG
Object Databases, see Database, ODBMS
Object IDentifier, see OID
Object Management Group, see OMG
Object Query Language, see OQL
Object-oriented, see Software, object-oriented
Objective C, 14, 59
Objectivity, 23, 37, 44, 243, 249, 251, 252, 258
ObjectStore, 243
Obsession, 290
ODBMS, 245
ODMG, 93, 243, 245
OGF, 163
OID, 37, 244
OLYMPUS satellite, 236
OMG, 243
OPCAL, 230, 235, 239
Open Grid Forum, see OF163
Open Office, 10
Open Source Software, 134
Open Source software, see Software, Open Source
OpenLab, 253
Optical Private Network, see Network, OPN
OQL, 243
Oracle, 111, 226, 227, 234, 246, 249, 252, 253
   CERN central service, 229, 231
   geoplex, 255
   Parallel Server, 254
   RAC, 254
   Real Application Cluster, 235
   Streams, 253, 257
OSI, 8, 59
OSM, 236
OSS, see Software, Open Source

P2P, 171, 273
PAF, 38
Parallel computing, 9, 46, 70, 100, 177–199
   algorithm scalability, 181
   Amdahl's law, 177, 191
   architecture, 182
   data access, 182
   data layout, 187
   data locality, 182
   data parallelism, 186
   distributed shared memory, 186
   hybrid architectures, 185
   impact of I/O, 187
   inter-processor communication, 187
   load balancing, 187
   Massively Parallel Processors, see Paralell computing, MPP
   master-worker paradigm, 188
   Message Passing Interface, see Parallel computing, MPI
   MIMD, 183
   MPI, 185, 186
   MPP, 9, 184, 187
   Multiple Instruction Multiple Data, see Parallel computing, MIMD
   OpenMP, 184, 186
   Parallel Virtual Machine, see Parallel computing, PVM
   pull architecture, 193
   push architecture, 188, 193
   PVM, 185
   self-scheduling, 187
   SIMD, 183, 198
   Single Instruction Multiple Data, see Parallel computing, SIMD
   SMP, 183, 187
   super-linear speedup, 191
   Symmetric Multi Processors, see Parallel computing, SMP
   task parallelism, 186
Particle, vi
   accelerators, vi
   collisions, vii
Pascal, 7, 11
PASS project, 242
Patent, 202, 204–209
   "prior art", 205
   business model, 208
   claim, 205
   Doctrine of Equivalents, 209, 215
   human genome, 208
   infringement, 206, 214
   invention vs discovery, 207
   software, 208, 209
PAW, see CERNLIB
PC, 9, 73, 86, 108, 151, 226, 253, 277
Perl, 16, 115

PERQ, 7
Persistency, 26
    data structures, 22
    object, 37, 38
Personal Computer, see PC
PETRA, 21, 261
PHIGS, 29
Photoelectric effect, vi
PI Project, 39
Pico Analysis Framework, see PAF
PIONS, 29
PL1, 12, 33
Platform Computing, 164
Plato, 292, 293
Poet, 243
Pompei, 297
POOL project, 39, 242, 249
Posix, 184, 272
PostgreSQL, 259
Programming errors, 50
Programming language, 120
Property, 202–204
Providenciales, 245
proxy, 98, 171
pthreads, 184
Python, 16, 48

QCD, 94, 180, 182
QT, 30, 40
Quantum ChromoDynamics, see QCD
Quantum decoherence, 301
Quantum Information Theory, 302
Quantum Mechanics, vi

Radioactive decay, vi
RAL, 234
RARE, 65
Rays, cosmic, see Cosmic rays
rBuilder, 170
RD41, 14, 242
RD44, 15, 35, 242
RD45, 15, 23, 35, 37, 242–248, 261
Reconstruction, 3, 180
    track, 3
Reflection, 14, 39, 40, 48
Remote Procedure Call, see RPC
RHIC, 192
RIPE, 65
ROOT, x, 12, 19, 26, 34–40, 123, 129, 135, 138, 140, 142, 146, 148, 150, 192, 242, 249, 261
    CINT, 37
    G4ROOT, 27

PROOF, 47, 172, 186, 192–198
    architecture, 193
    data access strategy, 194
    design goals, 193
    monitoring, 196
    PAR, 196
    performance, 197
    resource scheduling, 195
    user interface, 196
    rootcint, 37
    TFolder, 41
    THhtml, 136
    Tree, 40
    TSelector, 195
    TTask, 41
    TXNetFile, 280
Round-trip, 283
RPC, 59
RS232, 61, 63
RTAG, 39
    ARDA, 160
    Blueprint, 39
    RTAG 1, 249

S3, see Amazon, Scalable Storage Service
SAN, 253
Scalla, see xrootd
SCANBOOK, 235
Schema evolution, see Database, schema evolution
SCI, 179
SCRAM, 19
SEAL Project, 39
    REFLEX, 39
SEFT, 236
Service machine, 226
SGI, 9, 35, 108
    Origin, 185
    Power Challenge, 183
    PowerSeries, 9
SGML, 59
SHIFT, 9, 58, 71, 190, 231, 234
SIGGRAPH, 28
Silicon Graphics Industry, see SGI
Simulation, viii, see Dtector, simulation2
SLAC, 15, 57, 235, 236, 248
SmallTalk, 12
SMS, 296
SMTP, 64
Software
    Copyright, see Copyright, software
    crisis, 116, 119
    development cycle in HEP, 128, 143

Index 347

development factors, 132
development, Darwinist, 130
documentation, 134, 135
HEP developers, 141
HEP requirements, 124
intellectual property, see Intellectual
     Property, software
license, see License, software
maintenance, 139
management, 40
   in HEP, 142
non-regression test, 134
object-oriented, 12, 23, 35, 241
offline, 141
On-line, 141
Open Source, ix, 19, 56, 60, 121, 145, 212,
     306
patent, see Patent, software
portability, 137, 168
project management, 140
prototyping in HEP, 128
re-factorisation, 140
release cycle, 129, 134
test procedure, 134
test procedures, 167
Software as a Service, see SaaS
Software Engineering, 116–151, 290
   Agile, ix, 145
   Cathedrals versus Bazaars, 290
   crisis, 120
   ER, 43, 123, 231
   extreme programming, see Software
       Engineering, XP
   HCP, see Software Engineering, High
       Ceremony Processes
   High Ceremony Processes, 118, 129
   in HEP, 126–144
   LCP, see Software Engineering, Low
       Ceremony Processes
   Low Ceremony Processes, 146
   Object Time, 123
   OL, 123
   OMT, 123
   OMTool, 123
   OMW, 123
   OOADA, 123
   PSS-05, 123
   Rational Rose, 123
   reverse, 136
   roles, 118
   SASD, 123
   spagetti code, 136
   spiral model, 118
   StP, 123

UML, 123, 136
waterfall model, 117
XP, 146
Solid State Drive
   seeSSD, 198
Sony, 215
SPIRES, 57
Spirituality, 293
SPS, 63, 233
SQL/DS, 228–230, 234
SQLite, 258
SQS, see Amazon, Simple Queue Service
Squid Web cache, 257
SSC, 238, 240, 242
SSD, 198
Standard Template Library, see STL
STELLA Satellite, 61
STK, 235
STL, 26, 35, 40
Storage
   mass, 71
Store, data, 21
   dynamic, 21
   multiple, 22
STOREGATE, 39
Subconscious, 290
SUMX, 31
SUN, 8
Sun, 108
   Solaris, 227, 229, 253
Supercomputer, 64
Superconductivity, 2
svn, 19, 149
Swan, Black, 67
Swedish Centre for Parallel Computers, 159
Synchronicity, 291, 299, 300

Tags, 258
Tape, magnetic, 5, 73, 76, 80, 108
TATINA, 24
TCP/IP, 8, 62, 188, 239
   multistreaming, 273
   socket, 197
   streams, 276
   window scaling, 279
Tektronix, 27
   4002, 6
   4010, 5
telnet, 64
Thomas Jefferson Laboratory, 236
Tier 0, see Grid, Tier 0
Tier 1, see Grid, Tier 1
Tier 2, see Grid, Tier 2

TIFR, 241
Track, 3
  reconstruction, *see* Reconstruction, track
Tracking, *see* Reconstruction, track
Trade Secret, 204
Trade Secrets, 202
Trademark, 202, 204, 218–220
  expiration, 219
  software, 220
  Web pages, 220
Trigger, 180
TrollTech, 30

UML, *see* Software Engineering, UML, 149
Uniform Resource Identifier, *see* URI
Uniform Resource Locator, *see* URL
United Nations Institute for Training and Research Operational Satellite Applications Programme, *see* UNOSAT
University of Edinburgh, 159
Unix, 8, 59, 62, 63, 94, 108, 226, 234, 235, 239, 240
  BSD, 63
    license, 212
  syslog, 191
UNOSAT, 94, 96
Unus Mundus, 299
UPDATE, 19
URI, 55
URL, 55, 58, 236
Usenet, 60, 62, 64
Usenix, 64
UUCP, 62, 64

VCR, 215
Vector
  arithmetic, *see* Vectorisation
Vector graphics, 27
Vectorisation, 25, 70
Venice, 205, 218
Versant, 243
Viola, 60
Virtual MonteCarlo, *see* VMC
Virtualisation, 10, 164
  benefits, 165, 170–173
  hypervisor, 164, 171
  image generation, 171
  in Grid computing, 155–173
  legacy system support, 165
  para-virtualised drivers, 171
  performance, 170

security, 171
server, 155, 164, 173
use cases, 167–170
Virtual Machine, *see* Virtualisation, VM
VM, 164, 169, 171
VMC, 26–27
VME, 7
VMware, 10
Volunteer computing, 169
VT100, 29, 33

W3C, 56, 58
WAN, *see* Network, wide area
Wave function
  collapse, 300
Wave particle duality, *see* Duality
Web, 14, 18, 55–67, 94, 110, 289, 294
  blog, 296
  first picture, 294
  self-consciousness, 302
  services, 159
Web evolution
  analogy with M. Klein, 305
  analogy with Piaget, 304–305
Windows, *see* Microsoft, Windows
WIPO, 222
WISCNET, 63
WLCG, 39, 79, 84, 92, 96, 99, 107, 155, 163, 236, 240
  alarm system, 102
  CCRC, 109
  Memorandum of Understanding, 84
  operation, 81, 99
  service challenges, 109
  service interruptions, 104
  service states, 104
  service targets, 102
  site dependency, 104
Workload management, 71
Workstation, 7, 58, 179
World Intellectual Property Organisation, *see* WIPO
World Wide Web, *see* Web
World Wide Web Consortium, *see* W3C
Worldwide LHC Computing Grid, *see* WLCG
WWW, *see* Web
Wylbur, 6

X-Windows, *see* X11
X.25, 64
X11, 18, 29, 30, 34
Xerox, 63, 219

Index

XFIND, 17, 56
XROOTD, 80, 194, 268, 273, 276
   cmsd, 273
   Fault tolerance, 275
   VMSS, 283
   XrdClient, 280
   XRDCP, 279

xrootd, 38, 47
XWHO, 17, 56

Y2K, 236
YBOS, 22
YouTube, 67

Name Index

Adami, Chris, 300
Addis, Louise, 57
Amdahl, Gene, 177
Anderson, Carl David, vi
Andreessen, Marc, 18
Aristotle, 292

Böck, Rudy, 12, 28, 33
Backus, John W., 10
Baden, Drew, 238
Baessler, Erik, 33
Banerjee, Sunanda, 241
Baroncelli, Toni, 24
Barone, Luciano, 238
Bassler, Erik, 12
Beck, Ken, 132
Beolco, Angelo
    seeRuzzante, 295
Bequerel, Henri, vi
Berezhnoi, Vladimir, 34
Berners-Lee, Tim, 14, 17, 55
Bettels, Jurgen, 28
Bion, Wilfred, 302
Blöbel, Volker, 21
Booch, Grady, 123
Bosack, Len, 64
Bose, Satyendra Nath, 301
Brooks, Gary, 132, 138
Brun, René, xiv, 192
Bruyant, Francis, 21, 25
Buncic, Predrag, xiv
Butterworth, Ian, 21

Cailliau, Robert, 57
Carminati, Federico, xiv, xvii, 25

Carpenter, Brian, 63
Cerf, Nicholas, 300
Cockroft, John D., vi
Connor, Michelle, xvii

de Broglie, Louis-Victor-Pierre-Raymond, 7th Duc, vi
Dekeyser, Jean-Luc, 25
Delfino, Manuel, 78
Dirac, Paul, vi
Dorn, Gerhard, 299
Duneton, Claude, 295

Easterbrook, Frank, 220
Einstein, Albert, vi
Eisenhower, Dwight, 121
Everett, Hugh, 301

Fellini, Federico, 295
Ferrari, Alfredo, 26
Fesefeldt, Harm, 25
Fo, Dario, 295
Foa, Lorenzo, 36, 44
Foscolo, Ugo, 296
Foster, Ian, 75, 113, 155, 158, 163
Foulkes, Siegfried Heinrich, 302
Furano, Fabrizio, xv

Gagliardi, Fabrizio, 77
Galli Carminati, Giuliana, xv, 289, 298, 301
Galois, Évariste, 296
Gamble, John, 65

Gaponenko, Igor, 258
Gheata, Andrei, 27
Gheata, Mihaela, 27
Glenn, Patrick, 25
Goldoni, Carlo, 295
Goto, Masaharu, 37
Goya, Francisco, 297
Grant, Alan, 24
Gray, Jim, 227, 238, 252, 255, 263
Groff, Jean-François, 56
Grossman, Bob, 238
Grote, Hans, 234

Hagedorn, Rolf, 27
Hansom, Andy, 12
Hanushevsky, Andy, xv, 38
Hertz, Rudolf, vi
Hess, Victor, vi
Hilbert, David, 300
Hoffmann, Hans, 38
Holl, Burhardt, 241
Howie, Mike, 27

Johansen, Jan, 216
Johnson, Tony, 57
Johnstadt, Harald, 29
Jung, Carl Gustav, 299

Karrenberg, Daniel, 64
Kesselman, Carl, 75, 155
Klein, Melanie, 303
Klein, Wim, 4
Kowalski, Andy, 12
Kunz, Paul, 15, 57
Kunze, Marcel, 38
Kyte, Tom, 259

Lassalle, Jean-Claude, 24
Lawrence, Ernest O., vii
Lessig, Lawrence, 212

Méndez Lorenzo, Patricia, xv
MacPherson, Andy, 25
Maire, Michel, 25
Malon, David, 247
Martin, François, 298, 301
May, Ed, 247
Mazzucato, Mirco, 36
Mellor, Stephen, 123

Metcalf, Mike, 13
Moore, Gordon, 9
Myers, David, 12, 28

Nassim, Taleb, 66
Nathaniel, Alfred, 33
Newman, Harvey, 21, 74
Newton, Isaac, 218

Pagiola, Emilio, 129
Palazzi, Paolo, 31, 43, 57, 237
Pape, Luc, 33
Pauli, Wolfgang, 299
Pellow, Nicola, 18, 56
Penrose, Roger, 303
Piaget, Jean, 303
Pinsky, Lawrence, xvi
Plato, 292, 299
Pohl, Martin, 21
Pollerman, Bernd, 17

Rademakers, Fons, xvi, 36
Rambaugh, James, 123
Renshall, Harry, 264
Revol, Jean-Pierre, 28, 33
Robertson, Les, xvi, 62
Rubbia, Carlo, 24
Rutherford, Ernest, vi
Ruzzante, 295

Sala, Paola, 26
Salustri, Carlo Alberto, see Trilussa
Sandoval, Andres, 35
Schaile, Otto, 237
Segal, Ben, xvi, 71
Sendall, Mike, 56
Shiers, Jamie, xvii
Shlaer, Sally, 123
Sphicas, Paris, 248
Stapp, Henry, 300
Stepanov, Igor, 35

Ting, Samuel, 115
Titian, 297
Trilussa, 295
Turing, Alan, 294

Urban, Laszlo, 25

Name Index

van Dam, Andy, 28
Van de Graaf, Robert, vi
Vandoni, Carlo, 27, 33
Vaquero, Luis, 167
Vecellio, Tiziano, *see* Titian
Vialle, Jean-Pierre, 28
Villon, François, 295
Von Neumann, John, 294, 300

Walton, Ernest, vi
Watkins, Howard, 32
Watson, Thomas J., 113
Wenaus, Torre, 39

Wendell Holmes, Oliver, 220
White, Bebo, 57
Wigner, Eugene, 300
Williams, David, 35

Yourdon, Ed, 123

Zanarini, Pietro, 25, 32, 33
Zichichi, Antonino, 317
Zoll, Julius, 18, 21
Zuse, Konrad, 294

Titles in this Series

Quantum Mechanics and Gravity
By Mendel Sachs

Quantum-Classical Correspondence
Dynamical Quantization and the Classical Limit
By Josef Bolitschek

Knowledge and the World: Challenges Beyond the Science Wars
Ed. by M. Carrier, J. Roggenhofer, G. Küppers and P. Blanchard

Quantum-Classical Analogies
By Daniela Dragoman and Mircea Dragoman

Life - As a Matter of Fat
The Emerging Science of Lipidomics
By Ole G. Mouritsen

Quo Vadis Quantum Mechanics?
Ed. by Avshalom C. Elitzur, Shahar Dolev and Nancy Kolenda

Information and Its Role in Nature
By Juan G. Roederer

Extreme Events in Nature and Society
Ed. by Sergio Albeverio, Volker Jentsch and Holger Kantz

The Thermodynamic Machinery of Life
By Michal Kurzynski

Weak Links
The Universal Key to the Stability of Networks and Complex Systems
By Csermely Peter

The Emerging Physics of Consciousness
Ed. by Jack A. Tuszynski

Quantum Mechanics at the Crossroads
New Perspectives from History, Philosophy and Physics
Ed. by James Evans and Alan S. Thorndike

Mind, Matter and the Implicate Order
By Paavo T. I. Pylkkänen

Particle Metaphysics
A Critical Account of Subatomic Reality
By Brigitte Falkenburg

The Physical Basis of The Direction of Time
By H. Dieter Zeh

Asymmetry: The Foundation of Information
By Scott J. Muller

Decoherence and the Quantum-To-Classical Transition
By Maximilian A. Schlosshauer

The Nonlinear Universe
Chaos, Emergence, Life
By Alwyn C. Scott

Quantum Superposition
Counterintuitive Consequences of Coherence, Entanglement, and Interference
By Mark P. Silverman

Symmetry Rules
How Science and Nature Are Founded on Symmetry
By Joseph Rosen

Mind, Matter and Quantum Mechanics
By Henry P. Stapp

Entanglement, Information, and the Interpretation of Quantum Mechanics
By Gregg Jaeger

Relativity and the Nature of Spacetime
By Vesselin Petkov

The Biological Evolution of Religious Mind and Behavior
Ed. by Eckart Voland and Wulf Schiefenhövel

Homo Novus - A Human Without Illusions
Ed. by Ulrich J. Frey, Charlotte Störmer and Kai P. Willführ

Brain-Computer Interfaces
Revolutionizing Human-Computer Interaction
Ed. by Bernhard Graimann, Brendan Allison and Gert Pfurtscheller

Extreme States of Matter
on Earth and in the Cosmos
By Vladimir E. Fortov

Searching for Extraterrestrial Intelligence
SETI Past, Present, and Future
Ed. by H. Paul Shuch

Essential Building Blocks of Human Nature
Ed. by Ulrich J. Frey, Charlotte Störmer and Kai P. Willführ

Mindful Universe
Quantum Mechanics and the Participating Observer
By Henry P. Stapp

Principles of Evolution
From the Planck Epoch to Complex Multicellular Life
Ed. by Hildegard Meyer-Ortmanns and Stefan Thurner

The Second Law of Economics
Energy, Entropy, and the Origins of Wealth
By Reiner Kümmel

States of Consciousness
Experimental Insights into Meditation, Waking, Sleep and Dreams
Ed. by Dean Cvetkovic and Irena Cosic

Elegance and Enigma
The Quantum Interviews
Ed. by Maximilian Schlosshauer

Humans on Earth
From Origins to Possible Futures
By Filipe Duarte Santos

Evolution 2.0
Implications of Darwinism in Philosophy and the Social and Natural Sciences
Ed. by Martin Brinkworth and Friedel Weinert

Probability in Physics
Ed. by Yemima Ben-Menahem and Meir Hemmo

Chips 2020
A Guide to the Future of Nanoelectronics
Ed. by Bernd Hoefflinger

From the Web to the Grid and Beyond
Computing Paradigms Driven by High-Energy Physics
Ed. by René Brun, Federico Carminati and Giuliana Galli Carminati